Geographic Information Systems For Resource Management: A Compendium

William J. Ripple, Editor

ISBN 0-937294-89-6

Published by
American Society for Photogrammetry and Remote Sensing
and
American Congress on Surveying and Mapping
210 Little Falls St.
Falls Church, VA 22046

Table of Contents

Part 6
Vegetation Resource Management

Part 7
Global Studies

Bibliography

PREFACE

Historically, the spatial relationships of land, its characteristics, and all the cultural structures and boundries imposed on it, have been depicted through the medium of printed maps. Maps have been, and will continue to be, important tools for storing and conveying spatial information. Maps have also presented the user with severe limitations, particularly those resulting from the need to associate information displayed on separately prepared maps. Additionally, the value of the printed map for representing current land status is greatly diminished as time, land cover, and land use changes proceed. Rates of these changes have been accelerating with population and technological growth. The acceleration in the rates of change place a greater and greater value on the timeliness and quality of data and a similar increase in the cost of having outdated information. Today, it is possible to address these problems with the capabilities of digital computers and software for handling geographic data. Geographic Information Systems (GIS) are increasingly being used to input, store, manipulate, and display geographically referenced data to support the decision making process. With the current explosion of activities in the use of geographic information systems, there is a need to assemble various published works into a readily available form to open lines of communication and identify key participants.

The objectives of this book are to 1) introduce the reader to the techniques and functional capabilities of geographic information systems, and 2) bring together recent information on the use of geographic information systems for a variety of resource management applications. It is intended that this book will be of use to ecologists, foresters, geologists, geographers, hydrologists, landscape architects, soil scientists, and urban and regional planners, as well as individuals working in the mapping sciences. It can also be used as a reader for introductory university courses on geographic information systems.

This volume is divided into seven parts. In Part 1, the first article provides an overview of the nature of geographic information systems. A detailed description of the techniques required to create a computerized spatial database is included in the second article. The third article provides a review, on a function by function basis, of the capabilities of computer-based geographic information systems. Parts 2 through 7 include articles regarding the applications of geographic information systems for managing various types of resources, with articles on land suitability studies, urban studies, water resources, soil resources, vegetation resources, and global studies. A selected bibliography on the applications of geographic information systems can be found at the back of the book.

This book is the first in a series of volumes planned by the Geographic Information Management Systems (GIMS) committee. The GIMS committee is composed of members from both the American Society for Photogrammetry and Remote Sensing (ASPRS) and the American Congress on Surveying and Mapping (ACSM). This committee is responsible for providing the membership with a forum for discussing and disseminating information relating to techniques associated with the design, development, application and maintenance of spatial databases. The next volume planned by the committee will involve the topic of multipurpose land records information systems.

Acknowledgements are in order to the members of the GIMS committee and to the past ASPRS President, Tamsin G. Barnes, for conceiving the idea for this volume. Appreciation is expressed to the following organizations for providing assistance and contributing materials: Environmental Systems Research Institute (ESRI), Earth Resources Data Analysis Systems (ERDAS), and the Environmental Remote Sensing Applications Laboratory (ERSAL)-Oregon State University. Special thanks are due to Joseph A. Bernert for assistance in the literature search, to Donald F. Hemenway, Jr., Director of Communications at the Society Headquarters, for offering his technical expertise in producing this book, and to Gerald Churchill for contacting authors and publishers for copyright permission.

William J. Ripple, Editor
Oregon State University
November, 1986

PART 1

INTRODUCTION TO GIS TECHNIQUES
AND
FUNCTIONAL CAPABILITIES

GEOGRAPHIC INFORMATION SYSTEMS: AN OVERVIEW

Dr. Duane F. Marble

Depts. of Geography & Computer Science
State University of New York at Buffalo

ABSTRACT

This paper provides a general overview of the development of computer-based systems for spatial data handling. It examines the nature of these geographic information systems (GIS) and explores certain links in their development to developments in associated fields such as computer graphics, software engineering, and photogrammetry and remote sensing. The paper concludes with a brief enumeration of certain critical research areas and institutional developments which are felt to be necessary to insure continued, effective development of the field.

INTRODUCTION

Researchers and practitioners in geography as well as in other disciplines have dealt for many years with problems relating to the analysis and manipulation of entities which exist within a specific space-time framework. The most common medium for storing and displaying such coordinate-based information has traditionally been the analog map document (most commonly seen in the form of a paper reproduction). The first map was apparently created before the first alphabet, so it is apparent that we have been working with these analog, storage and display devices for spatial data over an extended period of time. During this time, these devices have evolved to a high level of sophistication and today's map combines high density data storage with complex, color-based displays.

Spatial data elements, recorded on maps as points, lines and areas, are commonly recorded on the basis of a standard coordinate system (latitude, longitude and elevation with respect to sea level). The identification of spatial data elements, the determination of their locations in a standard coordinate system, the measurement of their aspatial attributes, and the subsequent storage and portrayal of these data elements on maps is a common function in nearly all societies. Retrieval and analysis of these map data normally involves visual inspection of the map document coupled with intuitive analysis, which is occasionally aided by simple measurement tools (e.g., scales and planimeters). The information stored on maps is often of critical importance, but experience has demonstrated that while it is easy to retrieve small amounts of data, the retrieval of larger numbers of map elements or attempts to determine in a quantitative fashion the complex relationships which exist between map elements is a very slow process.

For example, in attempting to assess the recreational potential of a specific study area, a planner may desire to know which water bodies are not more than thirty-five minutes driving time from a specific urban area (and are of a size greater than five acres), and which lie within one-half mile or less of a developed road and have at least 50% of their shoreline in public ownership. Determining the answer to this simple query requires the comparision of three spatial data sets, one showing the size and distribution of water bodies, another the spatial pattern and nature of the road network, and a third displaying the pattern of land ownership. Yet another coverage, showing the distibution of urban areas, is necessary to determine the spatial base point for the analysis. It is not uncommon in situations such as this for the area of interest to cover several map sheets, thus requiring that the coverages be joined at the sheet edges; an operation which may be impeded by differences in the scale of the maps and in the time at which the data were compiled.

Traditionally, integration of spatial data sets is carried out by transforming the two or more spatial data sets to a common map scale, creating a transparent or translucent overlay for each data set, registering these overlays so that the coordinate systems are alined, and then manually creating a composite overlay sheet that shows those locations where the various phenomena classes being studied occur in spatial juxtaposition. The time involved in this process has generally been so great that it is utilized far less than one might expect. The complexity of the map display represents a significant factor in slowing down the analysis process and in contributing to error generation.

Analog map documents also display another major problem as a data storage device; they are expensive and time consuming to change when updates need to be made to the spatial database.

Updating the analog spatial data base requires that manual changes (restripping, cutouts, etc.) be made to the film master of the map sheet. No mechanism exists to insure that changes in one spatial data element are reflected in other elements which are logically linked to it. For example, a single spatial entity (e.g., a portion of a highway) may also serve as both a political and statistical boundary. If the highway right-of-way is moved, this change may be reflected in updates to the transportation overlay, but there is nothing in the analog data handling system to indicate that there may be possible changes in other spatial entities (e.g., a census tract boundary) as well.

The technology involved in the creation of these analog devices for the joint storage and display of spatial data has reached a high level of development, but it has never succeeded in overcoming these, and other, basic handicaps. Therefore, the advent of the digital computer as a data handling device soon raised the question of its applicability to the storage and manipulation of spatial data.

The initial attempts to apply computer technology to the reduction of the substantial data handling problems encountered with spatial data were, as one might expect, associated with military problems and produced useful results only after the massive application of computing resources. This was a function not only of the state-of-the-art in computer technology in the late '50's and early '60's, but also an early demonstration of the special problems encountered in digital spatial data handling.

About two decades ago the first serious attempt to handle substantial amounts of spatial data in the computer was instituted by what has become the Canada Geographic Information System (CGIS).[1] Today CGIS is still in operation and it remains one of the most cost effective examples of large scale spatial data handling. Many other systems failed (e.g., the Land Use and Natural Resources (LUNR) system of the State of New York), largely through poor design and a failure to anticipate the special technical problems encountered in the computer handling of spatial data; problems which are significantly enhanced by the large of volumes of data which must be manipulated.[2]

During the first half of this period, all spatial data handling systems were custom built. During the mid-1970's, and increasingly of late, general purpose, turnkey systems began appearing and the use of such systems has become the rule rather than the exception.

THE DEFINITION OF A
GEOGRAPHIC INFORMATION SYSTEM

The substantial improvement in computer systems during the last two decades has made it much easier to apply computer technology to the problem of storing, manipulating and analyzing large volumes of spatial data.[3] Today many organizations make routine use of what are called 'geographic information systems' to undertake tasks such as forecasting potential market areas, analyzing factors contributing to seismic hazard levels in the Eastern United States and finding what oceanographic data is available off the coast of Newfoundland. These geographic information systems comprise some quite sophisticated computer software, but they all contain the following major components:

1. A data input subsystem which collects and/or processes spatial data derived from existing maps, remote sensors, etc.

2. A data storage and retrieval subsystem which organizes the spatial data in a form which permits it to be quickly retrieved by the user for subsequent analysis, as well as permitting rapid and accurate updates and corrections to be made to the spatial database.

3. A data manipulation and analysis subsystem which performs a variety of tasks such as changing the form of the data through user-defined aggregation rules or producing estimates of parameters and constraints for various space-time optimization or simulation models.

4. A data reporting subsystem which is capable of displaying all or part of the original database as well as manipulated data and the output from spatial models in tabular or map form. The creation of these map displays involves what is called digital or computer cartography. This is an area which represents a considerable conceptual extension of traditional cartographic approaches as well as a substantial change in the tools utilized in creating the cartographic displays.

This definition of a GIS excludes a number of software systems which meet only part of the stated criteria. For example, digitizing systems which concentrate upon the problem of data capture from map documents and which provide minimal data storage/retrieval capabilities and only 'quick-look' graphics are clearly not geographic information systems. Neither, for that matter, are most remote sensing and image processing systems. Similarly, thematic mapping packages which concentrate upon the production of complex computer maps do not qualify.

Today, to be considered a GIS, the software system must include all four of the stated functions; and, furthermore, it must perform efficiently in all four areas. The mere addition of a few, inefficient spatial data handling functions to a package which is oriented toward only one of the basic functions does not turn it into a true geographic information system. Many systems today also do not include an explicit interface to spatial modelling activities; within a short period of time this too will be included as a mandatory function of any true GIS.

THE RELATION OF GIS DEVELOPMENT TO OTHER TECHNOLOGIES

The development of the computer-based technology for spatial data handling has both drawn upon and contributed to a number of other technical areas. The main conceptual development of these systems has come, of course, from geography and cartography, but their present effective status would not have been possible without a number of significant, interdisciplinary interactions.

Computer Graphics/Image Processing.

The output, or reporting, stage of the geographic information system is heavily dependent upon the availability of rapid, high resolution. graphics displays. We have been fortunate in that these displays have also been of significant utility in a substantial number of other fields. This high demand level has led to the rapid development of low-cost, sophisticated, computer graphics systems which are now capable of reproducing any desired spatial data display. Indeed, the development of this technology (especially in the area of color rendition and dynamic displays) has now outstripped our effective ability to make use of it.[4]

Computer graphics, and especially image processing, has contributed far more than cost-effective and sophisticated hardware. Many of the algorithms used in computer graphics and the data structures used in image processing have proven quite useful in spatial data handling.[5] Conversely, a number of developments pertaining to algorithms and data structures for spatial data handling have proven to be of considerable utility in the computer graphics area.[6]

Computational Geometry.

A specialized, and rather small, area of computer science deals with the analysis of algorithms for handling geometric entities.[7] The

work that has been undertaken here has led to significant improvements in geographic information systems (e.g., the recent development of the ARC/INFO system by ESRI, Inc.) and has stimulated a growing interest in the explicit analysis of the efficiency of algorithms used in spatial data handling systems.

Although the number of persons involved in this area of computer science is small, their work has had a disproportate imact on GIS development.

Database Management Systems.

In contrast to computational geometry, theoretical and practical work on systems for managing large volumes of data has occupied the attention of a substantial number of academic and commercial researchers in computer science.[8] Although a number of these systems have been applied to simple forms of spatial data (e.g., point data), their developmental emphasis on one-dimensional data has limited their utility for general spatial data handling. Current approaches tend to make use of a general database management system (DBMS) for handling the aspatial attribute information and specialized software for storage, retrieval and manipulation of the spatial data. ARC/INFO is a good case in point since it consists of INFO (a commercial, semi-relational DBMS) and ARC (a specialized spatial data handling system).

The inability of existing DBMS to efficiently handle large volumes of spatial data represents a real obstacle in the development of global databases. Similiar impacts are found in the image processing field where picture data management is also of serious concern.

Software Engineering.

Within the last decade increasing attention has been given within computer science to the problems of efficient design of large software systems. This work has become known as software engineering and, through the concept of the system life-cycle, has led to the development of conceptual models and tools for effective system design.[9] This work was badly needed because of the large number of system disasters that occured in the late 1960's and 1970's. Many systems failed and the reason for most of these failures was determined to be bad design: systems were over-budget, over-schedule and failed to work as desired.

The same problems had, of course, plagued the area of geographic information systems. Many of the early systems were held to be failures due to poor performance and, in some cases, the offending systems vanished from the scene. Other early systems managed to survive, often through a combination of good luck as well as sometimes good design. Attention was given in the GIS area to problems of system design and selection at an early date and it is interesting to note that many of the notions contained in these early design models parallel concepts found in modern software engineering practice (e.g., structured functional requirements analysis).[10]

Structured design approaches are becoming more common in the spatial data handling area and initial attempts are being made to construct the types of engineering cost estimation functions which are found today for less specialized, large software systems. The tools of software engineering are also being applied to developing more efficient structures within individual segments of GIS operation (e.g., manual digitizing).

Remote Sensing and Photogrammetry.

In a sense, the great majority of the data contained in digital, spatial data bases is derived from remote sensing. The derivation is indirect since most data are captured by digitization (either manual or automatic) from map documents which are, in turn, frequently derived from photogrammetric processing of aerial photography. However the direct utilization of remote sensing inputs (especially those based on orbital sensors) is found in only a limited number of cases at present.[11]

The reasons for this limited interaction appear to lie in misperceptions by both groups (remote sensors and GIS managers) as to the nature of the data created by remote sensing systems and used by geographic information systems. GIS managers, who are used to dealing with map data which normally carries no information pertaining to the accuracy and precision of individual elements (and hence is interpreted as being highly precise!), view remote sensing data as relatively inaccurate and hence of limited utility. Those organizations generating remote sensor data have neither understood this view, nor have they devoted much attention to the comparative economics of the two data sources (maps and remote sensors).

This lack of interaction between GIS and remote sensing systems is indeed unfortunate since significantly higher levels of interaction would improve the effectiveness of the GIS through the availability of more current data, and would improve the quality of remote sensor data through utilization of ancillary data contained in the spatial databases of the existing geographic information systems.[12]

SOME DEVELOPMENT AREAS IN

SPATIAL DATA HANDLING

As in many new scientific and developmental areas, especially those with a strong interdisciplinary component, attacks upon problems are often uncoordinated and lacking in effectiveness. In an attempt define a number of critical research areas a small group was convened by NASA in the Spring of 1983 in Palm Springs, California. This group (Jack Dangermond, Ray Boyle, David Simonett, Roger Tomlinson and myself) identified a number of critical development areas in spatial data handling. Some of these are discussed briefly here, but I must note that there have been a few modifications and wording changes which I have made (mainly reflecting recent developments) which may not necessarily represent the views of the other participants.

Theory

Spatial Relations Theory. There is, at present, no coherent mathematical theory of spatial relations. This seriously impedes both the quality of existing research and the speed at which developments can take place. The impact of this lack is especially felt in the attempts to develop true spatial database management systems and in the creation of efficient algorithms for spatial data handling. Similiar impacts are felt in the areas of image processing and computer vision.

Applied Science

Artificial Intelligence. Important developments are occuring in the field of artificial intelligence and will certainly have spatial data handling applications. The developments should be watched closely by the spatial data handling field and applied to spatial data handling technology as promptly as possible.

Expert Systems. Expert systems which include spatial data handling capabilities are likely to attract new users to the spatial data handling field. The development of expert systems helpful to spatial data handling users should be encouraged.

Data Aggregation and Generalization. The need to aggregate and generalize spatial data is a continuing technical problem in spatial data handling, and it becomes more severe as data bases grow larger and approach global coverage.

Exploratory Data Analysis. As data bases become ever larger, it is important to be able to use spatial data handling to explore them efficiently. Methods and algorithms for exploratory data analysis and for data base browsing need to be further developed. This is especially critical as large space-time databases become available.

Database Queries. It is extremely difficult to query large spatial data bases, but it is extremely important to be able to do so and do so efficiently. Research and development in this area, in addition to the basic theoretical work on spatial realations, is especially important for the creation and use of very large or global data bases.

Engineering and Technology

Global Data Bases. A series of concerns must be addressed if successful global data bases are to be created for wide public use. A variety of problems dealing with creating a global data base will need to be addressed simultaneously if progress is to be made. Pilot studies of global data bases need to be undertaken.

Improved Data Input Data input is probably the biggest bottleneck in spatial data handling systems at present and represents the greatest single cost in most projects, especially where the data base is very large. Research and development in a number of related areas may need to be done before progress in automation can be significant; these areas probably include feature recognition, cognitive science, artificial intelligence and others. Documentation of present methods, costs and throughputs needs to be obtained to provide baseline data against which potential improvements can be measured.

Data Updating. Improved methods for updating data in spatial data bases are needed. Continuing development of updating methods is of great importance to the integrity of data bases and to the maintenance of user confidence in spatial data handling's. It is here that remote sensing inputs can be of substantial value.

Economics of spatial data handling More agencies would use spatial data handling technology if they had better information about the cost/benefit ratios and the economics of spatial data handling applications. Economic and throughput analyses of spatial data handling functions are important to the development of the field, and the performance of such analyses and the publication of their results needs to be encouraged.

Benchmarking. Benchmark tests are useful in measuring the performance of a wide variety of spatial data handling functions, and should probably be more widely used in selecting systems. Persons expert in the design and conduct of benchmark tests should be urged to share this knowledge. Some publication of benchmark results

would be useful to potential system users, if accompanied with appropriate cautionary remarks.

Case Studies. Use of spatial data handling is inhibited in many cases because of a lack of reliable information about what such use entails in terms of time, cost, personnel and other resources, and a lack of information about the courses of events which application of spatial data handling's requires. Documentation of uses of spatial data handling in the form of case studies should be encouraged. If at all possible, third parties should carry out these case studies in order to ensure greater objectivity in the studies. Steps should be taken to ensure the wide distribution and availability of these case studies.

Algorithmic Analysis. Sustained and organized study of algorithms and data structures must be undertaken if efficient, large-scale, spatial data handling systems are to be constructed. Only a few analytic studies on spatial data handling algorithms have been carried out to date, and more of these need to be encouraged.

Computer Hardware. The needs of spatial data handling for hardware will probably be met through those general marketplace pressures now leading to rapid advancement in hardware capabilities and rapid declines in cost; nevertheless, some improved hardware capabilities might be valuable to the spatial data handling field. There is a clear interaction between our lack of knowledge of algorithmic efficiencies and our inability, at present, to deal with the potential of specialized hardware configurations (e.g., parallel processors).

Ergonomics. Although the theory underlying ergonomics is not mature, ergonomic studies are important if spatial data handling systems are to be made more useful and efficient. Ergonomic approaches must be put on a sound basis and then applied rigorously to spatial data handling technology. When ergonomically sound systems are designed, they need to be promptly implemented in production models.

User Friendliness. spatial data handling systemss need to be user friendly. Serious (as opposed to cosmetic) attempts at creating user friendly geographic information systems need to be continued, using the best guidance available from a wide range of fields, such as ergonomics, cognitive science, etc.

Need for Improved Efficiencies. There is a need for improved efficiencies in nearly every aspect of spatial data handling function. Efforts in software development, algorithmic analysis, data base structure design, ergonomics,

6

engineering economics and a whole range of other areas are needed in order to achieve improved efficiencies in spatial data handling function.

Institutions

NAS/NRC Definitive Study. A neutral, competent and respected body needs to examine the field of automated spatial and geographic information systems and prepare an evaluation of the problems and promise associated with it. A National Academy of Sciences/National Research Council "definitive study" of the field should be made. Federal agencies should be approached and interested in such a study and from among them a lead agency should be found. Funding for the study needs to be solicited from a number of federal agencies.

Archival Storage of Data. A problem of pressing importance is the need for archival storage of imagery and data already gathered. Immediate intense effort needs to be made to secure the archiving of the irreplaceable data and imagery we have already gathered. Longer range efforts need to be made to provide for common archiving facilities.

Improved Communication. Communication between persons within the spatial data handling field and between the field and persons and organizations outside the field needs to be greatly improved. A wide variety of communication paths needs to be adopted and concerted efforts need to be made to make communication more effective and frequent, within the field and with those outside it.

Technology Transfer and Diffusion. The transfer and diffusion of spatial data handling technology to new users, especially in the developing world, is important. Continuing attention needs to be paid to this problem and successful methods for promoting transfer and diffusion need to be found and more widely used.

Spatial Data Handling as a Discipline. The study and use of automated geographic information systems and of spatial data systems should be considered as a discipline or field of study rather than just the application of a technology. Continuing efforts are needed to foster this perception and support it with concrete actions.

A Professional Society To promote various disciplinary and professional goals and activities, a professional society dealing with the spatial data handling field and related areas is needed. Steps should be taken to found a professional society dealing with spatial data handling on an interdisciplinary basis.

Spatial Data Handling Education. Improved education and training for working with geographic information systems is needed and more students need to be prepared for the field. Additional graduate programs, especially at the Masters level, are needed. Curricula need improvement and model curricula should be devised and publicized. A variety of other educational and training opportunities are needed for refreshing and retraining those in the field and for training potential users of the technology. Equipment and software need to be provided for such programs; government and industry might cooperate to help meet these needs.

Research Institutions. Not enough research is going on in the spatial data handling field. More support needs to be provided for research in the field. Centers of excellence, devoted to research on automated geographic and automated spatial information systems, need to be created and supported on a long-term basis.

Decision Making Institutions. The use of spatial data handling technology could be usefully integrated into the decision-making processes of many institutions where it is not now employed. Continuing efforts need to be made to make decision makers aware of the usefulness of geographic information systems in decision making and related processes. Continuing studies need to be made of the best way of integrating geographic information systems into decision making, and successes need to be made widely known.

The Role of Federal Agencies. Federal agencies have important roles to play in the development of the spatial data handling field. Work should begin in a concerted way to obtain agency support for developing the spatial data handling field. Agencies which are interested should be identified, to include the names of specific persons. Sources of funds, ongoing related projects, future agency plans, and other important information should be gathered. These efforts need to be coordinated and useful information needs to be exchanged among those working in this direction.

Institutional Cooperation. No single sector of the economy has the means to provide all the support needed for the development of spatial data handling technology; cooperation among the sectors will be required. Continuing efforts should be made to foster cooperation between government. industry and the universities in support of the development of the spatial data handling field.

CONCLUDING COMMENTS

This paper has briefly reviewed the general area of geographic information systems, including some discussion of urgently needed research and institutional developments. Both geographic information systems and remote sensing systems have enjoyed significant levels of development and acceptance during the past decade. The GIS represents the most effective mechanism for making use of the data captured and reduced by remote sensing systems, and also offers the potential of increasing the effectiveness of this data capture operation through correlation of remote sensor inputs with data already held by the GIS.

Conversely, increased utilization of remote sensing inputs can significantly improve the utility of the GIS by providing more current information for use in updating the master spatial database and in moving the current, static spatial data handling systems to a true space-time basis.

REFERENCES

[1] Tomlinson, R. F., H. W. Calkins, and D. F. Marble, **Computer Handling of Geographical Data.** Natural Resources Research Report No. 13. Paris: The UNESCO Press, 1976.

[2] H. W. Calkins and R. F. Tomlinson, **Geographic Information Systems, Methods and Equipment for Land Use Planning, 1977.** Reston, VA: U. S. Geological Survey, 1977.

[3] D. F. Marble, H. W. Calkins, and D. J. Peuquet, **Basic Readings in Geographic Information Systems.** Williamsville, NY: SPAD Systems, Ltd., 1984.

[4] M. W. Dobson, "Effective Color Display for Map Task Performance in a Computer Environment," in **Proceedings**, International Symposium on Spatial Data Handling, 1984.

[5] H. Samet, "The Quadtree and Related Hierarchical Data Structures," forthcoming in **ACM Computing Reviews.**

[6] D. J. Peuquet, "A Conceptual Framework and Comparison of Spatial Data Models," forthcoming in **Cartographica.**

[7] M. I. Shamos, "Computational Geometry," unpublished Ph. D. dissertation, Yale University, 1978.

[8] C. J. Date, **An Introduction to Database Systems.** (Third edition) Reading, Mass.: Addison-Wesley Publishing Co., 1983.

[9] R. S. Pressman, **Software Engineering: A Practitioner's Approach.** New York: McGraw-Hill Book Co., 1982.

[10] H. W. Calkins, "A Pragmatic Approach to Geographic Information Systems Design," in D. J. Peuquet and J. O'Callaghan (ed.s) **Design and Implementation of Computer-based Geographic Information Systems.** Amherst, NY: IGU Commission on Geographical Data Sensing and Processing, 1983.

[11] D. F. Marble and D. J. Peuquet (eds.), "Geographic Information Systems and Remote Sensing," in **Manual of Remote Sensing**, Vol. I (second edition). Falls Church, VA: American Society of Photogrammetry, 1983.

[12] D. F. Marble, "Some Problems in the Integration of Remote Sensing and Geographic Information Systems," in **LANDSAT '81 Proceedings.** Canberra, Australia.

DESCRIPTION OF TECHNIQUES
FOR AUTOMATION
OF
REGIONAL NATURAL RESOURCE INVENTORIES

Jack Dangermond
Bill Derrenbacher
Eric Harnden

Environmental Systems Research Institute, Inc.
380 New York Street
Redlands, California

September 1982

This paper has been reviewed and edited by members of the
Geographic Information Management Systems (GIMS) Committee
of the American Society of Photogrammetry and Remote Sensing
and the American Congress on Surveying and Mapping.

INTRODUCTION

The purpose of this paper is to provide a general description of the techniques regularly used by the authors in their efforts to create large automated inventories of natural resources.

These techniques are presented in a systematic sequence illustrating a comprehensive approach for the development of an integrated geographic data base.

The Problem

There is an increasing awareness that a comprehensive survey of natural resources organized in the form of an automated data base is an important and productive investment for planning development and environmental management of regions, states, and even entire countries. Its creation is particularly important in areas where little systematic and synthesized data are available, complex and pervasive environmental constraints exist, and where substantial land use changes are desired or projected.

Background

For the past thirteen years, the authors have been conducting systematic resource inventories, automating these inventories and setting up geographic based information systems for users in planning and resource management disciplines. This has involved the inventory and data base development of almost 50 regions of the globe with over 400 man-years of effort. These inventories have been done in Venezuela, the Dominican Republic, Libya, Iran, Nigeria, Tanzania, Japan, Thailand, Canada, and selected areas within the United States; including a generalized data base for the entire country and approximately 75,000,000 acres of Alaska, the states of Maryland and Kentucky, the Sacramento and San Joaquin Valleys of Central California, Eastern Pennsylvania, several national forests, the Mojave Desert of California, the Coastal Zone of the state of Washington, several selected river basins, all of the urbanized areas of Southern California, and numerous other counties, cities and focused project sites.

These surveys have been conducted in a variety of environs including the arctic, alpine landscapes, prairies, tropical forests, subtropical regions, and arid and semi-arid deserts. They have included highly developed and complex regions, as well as simple landscapes which have remained unaltered by man's influence.

In early attempts by the authors to develop and apply the tools of automated geographic information systems to problems of planning and analysis, one of the alarming factors that surfaced was the generally poor quality of geographic data currently available for inclusion in such systems. Data problems included incomplete map coverage, inadequate map classifications with respect to decisions to be made, and inconsistent map coverages (i.e., inconsistent map unit resolution, line crenulation, scale, accuracy, dates of collection, sampling method, and classification system).

It was recognized early that, if practical use of GIS technology was to occur, it would be necessary to develop a series of methods and procedures to overcome the severe liabilities of the typical "existing data base" commonly supplied as input for an automated GIS.

Over a number of years the authors have evolved a series of practical procedures and techniques for creation of reliable, consistent and usable data bases for use with sophisticated software and hardware technologies now being made available.

THE PROCESS

This paper presents six categories of techniques which have been organized into a standardized task sequence for creation of large area natural resource inventories. These include:
1. Project Design
2. Inventory Preparation
3. Thematic Mapping
4. Map Integration
5. Editing
6. Map Automation

Project Design

In approaching the development of a large area resource inventory, it is necessary initially to develop a systematic design for the data base. This design must be sensitive to the end users; be appropriate considering the information which is already available; and be sensitive to the manpower, financial, and time resources available for creating such a data base. A three-step method for approaching a data base design is outlined below and presented graphically in Figure 1.
- Data Needs Assessment
- Collection and Evaluation of Existing Data
- Data Base Design

Conduct data needs assessment.
Data needs assessment begins by developing a clear definition of the specific uses for the inventory information. Typically, this involves meetings and interviews with the users to analyze the exact information types and outputs which are desired.

During these meetings, it is valuable to review the types of information which already have been identified; make breakdowns of the various levels of information currently used; become familiar with the methods of handling, storing, and retrieving information, user libraries and catalogs; and get an overall understanding of the geographic and thematic areas for which users have data needs.

Arrangements should be made to collect any data which the user can contribute to the inventory. These data could be in the form of maps, reports, bibliographies, air photos, satellite imagery, etc.

Interviews, although useful, are typically not enough. It

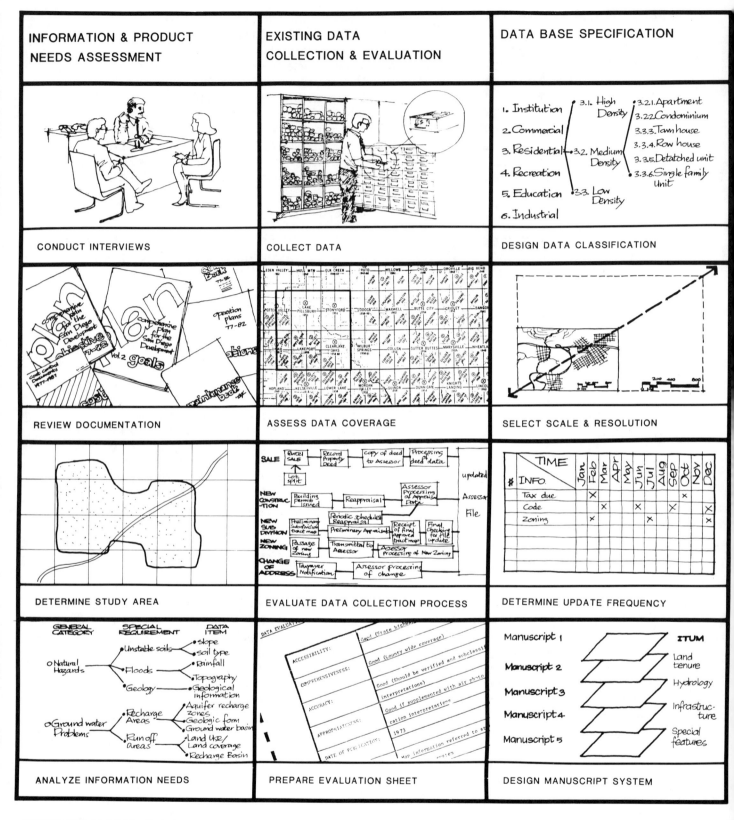

INFORMATION & PRODUCT NEEDS ASSESSMENT	EXISTING DATA COLLECTION & EVALUATION	DATA BASE SPECIFICATION
CONDUCT INTERVIEWS	COLLECT DATA	DESIGN DATA CLASSIFICATION
REVIEW DOCUMENTATION	ASSESS DATA COVERAGE	SELECT SCALE & RESOLUTION
DETERMINE STUDY AREA	EVALUATE DATA COLLECTION PROCESS	DETERMINE UPDATE FREQUENCY
ANALYZE INFORMATION NEEDS	PREPARE EVALUATION SHEET	DESIGN MANUSCRIPT SYSTEM

DESIGN DATA BASE

FIGURE 1

is necessary to review various documents outlining functions, responsibilities and mandates in order to analyze the exact data requirements.

To assist in this process, it is useful to go through an exercise of breaking down the specific analysis requirements and related data needs which are necessary to support specific functions. For example, slope and soil maps may be necessary for interpretation of soil erosion. Soil erosion evaluation may be one type of analysis necessary for water quality assessment within an organization responsible for general planning and management of the environment. A hierarchical structuring of specific data needs related to general responsibilities can be effectively represented in table, matrix or related graphic form.

The data needs assessment process is greatly assisted by clearly documenting needs and having users review, discuss and creatively participate in the final definition.

Collect and evaluate the existing data base. Data collection involves the acquisition of geographic information sources such as maps, books, reports, imagery, aerial photography, and related documents supplied directly by the user (typically during the data needs assessment interviews), or acquired from other institutions. This is accomplished by searching biblio- graphic indices, public institutions, government agencies, and private companies for information to fill the categories of data identified during the data needs assessment. Each data item collected is reviewed and evaluated to establish its usefulness and appropriateness for inclusion in the GIS.

Data are first grouped into general data need categories (i.e., natural resources administrative district information, infrastructure, etc.). Specialists in each field rate each item as important, useful, marginal, or of limited use. In addition, notes are recorded regarding the format, scale, map projection, date, media, classification categories, area of coverage and any useful considerations on how the data items match or conflict with each other.

Those data items of major value to the mapping effort for the GIS are identified for in-depth consideration during the classification and map design steps.

After a thorough review of the data has been completed and the results compared to the data needs of the user groups, data categories for which reliable coverage does not exist across the study area are identified. Separate data gathering or mapping projects are then designed to ensure that these categories are not omitted from the data base.

Data file design. The next step involves the design of the data base. It is based on information needs identified in step one and an assessment of the existing information base as developed in step two.

The data which form the basic geographic information system consist of spatial information and attribute information linked through the use of sequentially numbered identification lists, one for attributes and one for spatial components. In order to obtain maximum utility from the system, it is necessary to develop carefully both the structure of the classification systems used for attributes and the style and format of the spatial component.

In order to meet each of the data needs requirements, general categories of information and detailed classifications must be developed. The data collected are carefully reviewed and consideration is given to the level of detail needed for each category. The classification schemes are structured hierarchically to allow aggregation of classes at different levels for use at a variety of map output scales or tabular summary levels.

The classifications are typically developed toward making them as similar as possible to the classifications that the user groups are familiar with. Where new types of data are prepared, the classifications should be based upon current use within the professional writings for each given discipline.

All classifications developed for the GIS need to be expressed as numeric codes to facilitate computerized handling of the data.

Classifications represent qualitative, quantitative, or descriptive groups of individual data occurrences in a systematic order.

In some instances, classifications are developed which operate at two levels. The first level identifies the mapping unit by type; for example, a geologic formation. This is a descriptive class. The second level is an expansion upon the descriptive class intended to provide quantitative and qualitative values; for example, measurements or ratings given to the geologic formation such as age, rock type, stability, strength, etc.

The first level classifications are associated directly with the map units by use of a sequential code list. The second level values are associated with the corresponding first level classes by way of an expansion code matrix. This matrix lists the first level descriptive codes followed by their second level qualitative or quantitative values expressed as numeric codes.

This two level approach to classifying and coding the data can minimize the amount of space required to store the information in the computer and can also make the subsequent map overlay modeling efforts more efficient.

The numeric codes developed to represent each classifi-cation can be used as values by the computer to generate tabular listings, draw maps, or produce analytical models. It is important, therefore, to keep all of the classifications in a logical order; start with the smallest and end with largest; start with low and end with high; etc.

Data classes should match map units in levels of specificity (i.e., the most detailed classification levels match with the most detailed mapping unit). Unnecessary data classes which are more detailed than the ability to identify and map units should not be allowed.

All of the data codes are structured to be entered into the computer as numbers. Thought is given to the output potential of creating a variety of shades, colors, or symbols using various computer devices. Limitations imposed by those devices can be reduced by incorporating the appropriate structure into the classification.

Design of manuscript maps. What is meant herein by the term "manuscript map" is a map sheet originally prepared as an input document for automation. The manuscript map is typically related to a specific basemap size and scale. In addition to the classification structure, the design must include rules regarding minimum polygon size resolutions and line crenulations. Finally, a determination must be made regarding which variables will be placed on which manuscripts. The authors have developed an extensive set of map integration techniques associated with creation of map manuscripts.

The use of an integrated approach to mapping presents some special map design problems and distinct advantages for resource inventories. The specific integration and automation techniques used are described in subsequent sections of this paper.

Unlike manually prepared maps which rely on a variety of colors, shades, line widths and symbols to portray the information they contain; maps designed for computer data base input show data simply as points, lines, or polygons (areas). A numeric code provides the descriptive values for each map unit. Symbol recognition, color separation, clarity of shading, and line width are not factors considered when designing computer data input maps. Spacing of lines and points, line crenulation, acute angles, and minimum polygon size are important, as these are the problem areas for digital mapping programs. Since the computer has perfect logic and total recall, it is possible to pack a great deal of information onto a single map. This ability to pack information is limited only by the computer's programmed ability to discern discrete points, and from a more practical point of view, the ability of the human cartographer to prepare the maps for automation.

Inventory Preparation

Prior to beginning actual inventory mapping, it is necessary to conduct a series of basic tasks in preparation for production cartographic efforts. These tasks are graphically presented in Figure 2 and listed below:

1.   Conducting a reconnaissance field survey
2.   Collecting existing data and source mapping
3.   Selection of best data
4.   Grouping and cataloging of data
5.   Creation of basemaps

Many of these tasks are undertaken simultaneous with project design efforts.

Reconnaissance field survey. This task is intended to familiarize investigators with the actual patterning of

INVENTORY PREPARATION TECHNIQUES

FIGURE 2

environmental phenomena in the study area, thereby establishing
a sound basis for investigation, interpretation, and analysis.
This effort typically is conducted in conjunction with
acquisition of basemaps, aerial imagery and collateral data. It
involves a field visit and viewing by a team of resource
specialists. On large areas, a small airplane normally is
employed for a full overflight and overfield study area. A
series of oblique aerial photographs is taken for later use
and interpretation of various types of aerial imagery. Portions
of the study area accessible by road are subject to ground
investigations. Normally, this involves direct observation and
recording of environmental conditions, edification of photo
signatures, and select sampling of environmental phenomena.

Collection of existing data and aerial imagery. This task
involves the acquisition of existing and appropriate basemaps,
aerial imagery and collateral reports and maps. Some of the
special data materials may need to be specially ordered or
acquired, including computer information such as digital
terrain types, recent imagery such as Landsat or special flight
photography or basemaps which may require special permission
for use.

Selection of best data. This task involves an assessment of the
data collection and evaluation activities for the purpose of
selecting the best data among multiple data sets to be used for
the actual creation of maps. This effort is frequently
conducted where multiple map coverages for the same attribute
exist and must be sorted through to determine the most reliable
and consistent information to be used in the mapping.

Grouping and cataloging of existing data. Normally, an
automated card catalog is set up at the time data are collected
to ensure that data are organized properly and can be searched
and extracted rapidly.

Basemap preparation. An important base for integrated resource
mapping efforts is a well prepared set of basemaps. In the
United States the most practical system of basemaps is provided
by the U.S. Geological Survey (USGS). Topographic quadrangle
maps which are for all practical purposes equal area and equal
distant maps provide a number of locational reference points,
and can be organized into a consistent module sequence
convenient for mapping and automation. These maps range in
scale from 1:24,000 to 1:250,000, and can be transferred from
paper prints or acquired in fixed stable-based mylar film from
the USGS. The topographic contours shown on these basemaps are
extremely useful for registering map data and determining exact
locations on the earth.

Thematic Map Compilation
    Prior to creation of the final manuscript maps, a set of
Thematic Map Compilation Sheets is prepared. These sheets
consist of a map for each inventory variable or group of

compatible variables, and are drawn to the scale and map projection of the topographic basemaps. Four different methods are used to compile the manuscript maps. They are illustrated in Figure 3 and described below.

1. Direct Transfer
2. Image Interpretation
3. Photo Revision
4. Field Mapping

Direct transfer. Collateral sources which do not require reinterpretation are assembled by map module. For those items which are at the correct scale, the data are simply transferred directly to a mylar overlay of the basemap with careful rectification of the data to the new base.

If the data are at a scale other than the mapping scale, rescaling precedes the rectification and transfer steps. Rescaling may be performed manually or photographically.

Themes with simple legends or relatively gross mapping units normally are rescaled using an optical pantograph (Keuffel and Esser Kargl). This is a type of reflecting projector with a rated distortion of less than .01 percent. Work maps are placed on the Kargl platform and the basemap is projected onto the glass surface. The scale is controlled by altering the image-to-lens and lens-to-projection surface ratio. Once adjusted to the desired scale, the map categories are manually drafted onto a mylar overlay registered to the topographic basemap. The resulting mylar maps are then carefully edited to ensure that all information has been transferred correctly.

When basemaps are complex or extremely detailed, manual rescaling is too time consuming and the risk of code error through transposition or omission is increased. In those instances, it is more efficient to change scale photographically. A precision copy camera with a high resolution lens is used to produce paper positives. The proper scale is assured by using triangulation, a method in which three points of known location are marked on the work maps and then matched against the basemap on the camera's viewing glass. An attempt is made to match all three points as closely as possible. This process achieves the most satisfactory match of map scales, both horizontal and vertical.

Because the work maps are often originally prepared using a variety of cartographic techniques and mapping formats, it is necessary to rectify locations and boundaries to the topographic basemaps. This is accomplished by carefully re-registering the data map with the basemap and comparing observable points or lines common to both, such as water-bodies, roads, buildings, ridges, or stream course lines, etc. Common points, lines, and areas are drafted onto a pin-registered mylar overlay of the basemap. The resulting overlay has the scale, format, and projection of the basemap and is suitable for use in subsequent data integration steps. Common maps requiring rectification are soil and vegetation maps originally drafted from non-planimetric photos.

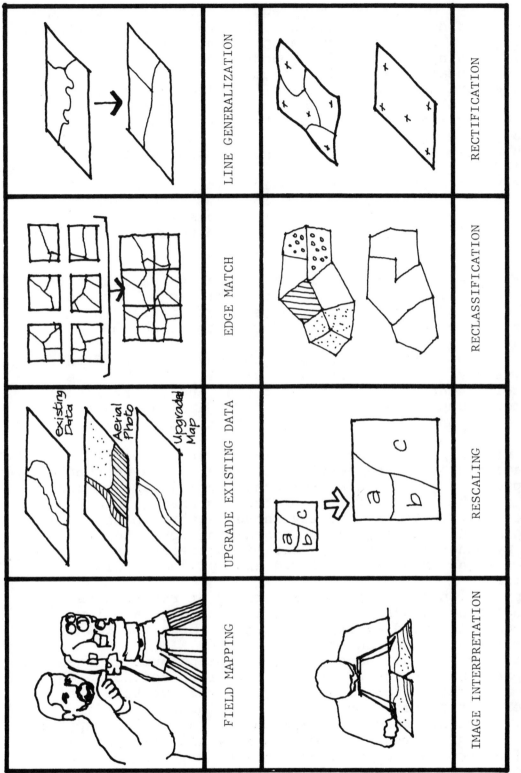

THEMATIC MAPPING TECHNIQUES

FIGURE 3

Specific uses of map data may require either 1) grouping map classes to simplify a geographic picture; or 2) splitting classes to give a more detailed perspective. The former usually is done by dropping out unneeded boundaries; the latter by adding more boundaries as derived from other data sources.

As scales and unit resolutions of mapping change, the original lines delineating boundaries and linear features may be impossible or inappropriate to retain. When this occurs, it is necessary to generalize line work. This is done using cartographic standards and keys to ensure consistency as maps are modified.

In cases where study areas contain multiple map sheets, an important thematic mapping function involves edge matching of lines crossing over adjacent map sheets. This function is performed by joining sheets and making line work and attribute codes agree, through use of source maps.

Image interpretation. Based on initial data evaluation, data deficiencies are identified. The deficiencies may be remedied through the use of a remote sensing technique called image interpretation, or more commonly referred to as photo interpretation.

Image interpretation begins with the development of an interpretive key. The key associates visually recognizable units of color, density, texture, pattern, and environmental surroundings (such as landform, elevation, drainage pattern, and relation of one area to another) with the map themes of interest.

Where stereo photography is available, stereoscopic interpretation provides three-dimensional elements to the key. This key provides guides in making consistent judgments and decisions for the interpretation of the map themes. Each interpreter contributes to the key those factors which worked best for him individually, while maintaining consistency with keys used by other interpreters. Each interpreter keeps a set of decision rules which are reviewed and updated by all interpreters.

Interpretation can rely upon a variety of remote sensing sources including Landsat MSS, radar images, high and low altitude color infrared aerial photography, high and low altitude black and white photography, etc.

Registration of the images to the basemaps must address parallax problems. Though the images are often scaled to the basemaps, a perfect match is a rarity. To resolve this, the interpreter must register one area of the image at a time, constantly re-registering the image to the basemap as mapping progresses. Accurate registration requires alignment of linear or point features on the photo-enlargement with those same features on the basemap. Such features include roads, stream courses, valleys, mountain ridges, and peaks.

Photo revision. In instances where the initial data evaluation shows a data source to be of good quality but out-of-date, subject to frequent change, or cartographically generalized, it is possible to upgrade the data through the use of image interpretation. This technique does not create new data but, rather, provides improved and more detailed thematic map sheets for use in compiling the resource inventory.

The process involves a visual comparison of the data source in question to the base imagery. A mylar overlay attached to the topographic basemap is laid atop the imagery and the observed phenomena are traced onto the overlay. As with image interpretation, registration is maintained by shifting the image to match linear or point features.

Field mapping. Where there is an absence of available thematic data, it normally is necessary to conduct in-field mapping of the data theme to be inventoried. This in-field mapping can be at a variety of intensities, and normally is done using some form of sampling method (i.e., forest plots, soil pits, borings, etc.) Subsequently, these samples are interpolated/extrapolated using the various image interpretation and related techniques described above.

Map Integration

Background. The next major technical procedure involved in a standard inventory project is the integration of multiple manuscript maps onto a single sheet. This process is most commonly used in the integration of thematic natural resource polygon maps such as soils, vegetation, slope, landform and geology.

This technique has been called integrated mapping, terrain unit mapping and integrated terrain unit mapping or ITUM. As used by the authors, it is most accurately titled Integrated Parametric Mapping (IPM) or Integrated Parametric Unit (IPU) Mapping, for it involves the systematic integration of geographic parameters mapped initially on single sheets and subsequently integrated onto a manuscript.

IPU mapping was developed to capture and coordinate the various environmental components of a natural resource data base. This type of mapping is a method of compressing a number of environmental factors from a variety of data sources onto a single map. This map displays homogeneous units which have the same general characteristics distribed throughout. This procedure overcomes a number of basic problems outlined later in this section.

Table 1 summarizes some of the basic principles which underlie the rationale for using the IPU approach to generate manuscripts for use with geographically based information systems. The table presents these principles in relation to the problems encountered from map overlay using a standard parametric approach. Also presented are the ways in which the IPU approach solves these problems. Figures 4 through 7 illustrate the problems and solutions graphically presented in Table 1.

The first problem resolved through the use of IPU mapping is a basic cartographic one and is important from the

Principle	Common Situations	Problems in Map Overlay	IPU Solutions
Classifications of Landscape are Interrelated	Parametric mapping rarely recognizes interrelationship of land attributes	Inconsistent • Map classes • Sliver errors	• reorganize classification system • Spacially derive lowest common integrated unit
Polygon boundaries reflect gradational change	Inconsistent graduational boundary delineations between parameters	• Sliver errors	Common boundary determination through multi-stage integration
Multiple scales of data can be accurate but inconsistant when integrated to common scale	Line grinulations and unit resolutions vary	• Sliver errors • Resolution inconsistancies	Remap all thematic data to common resolution and scale
Areal information changes over time	Maps are out of date and inconsistent with respect to time	• Inconsistent map classes • Inconsistent data reliability	Update information overlays to common date.

TABLE 1

IPU SOLUTIONS FOR CARTOGRAPHIC ERROR

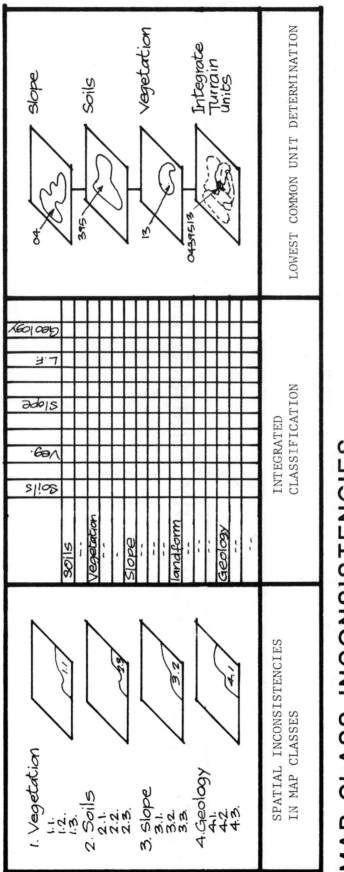

MAP CLASS INCONSISTENCIES

SPATIAL INCONSISTENCIES IN MAP CLASSES · INTEGRATED CLASSIFICATION · LOWEST COMMON UNIT DETERMINATION

FIGURE 4

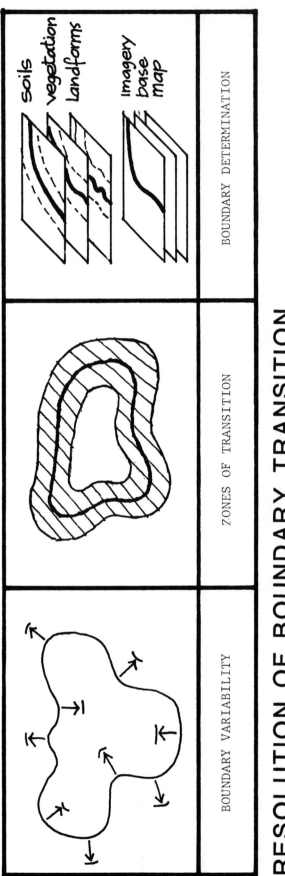

| BOUNDARY VARIABILITY | ZONES OF TRANSITION | BOUNDARY DETERMINATION |

RESOLUTION OF BOUNDARY TRANSITION

FIGURE 5

25

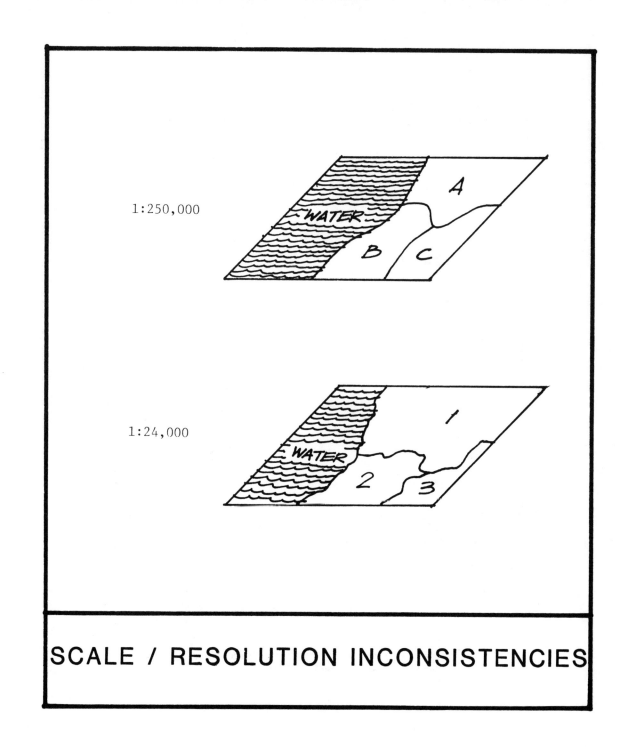

1:250,000

1:24,000

SCALE / RESOLUTION INCONSISTENCIES

FIGURE 6

25

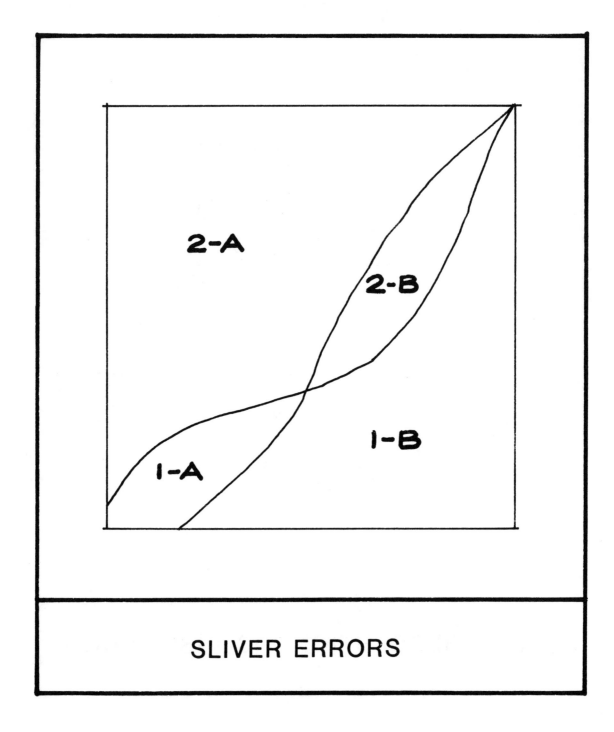

SLIVER ERRORS

FIGURE 7

26

standpoint of information, interpretation and display. Many map variables have numerous boundaries and attributes which are naturally interrelated and coincident (i.e., principle one in Table 1: Parametric classifications of landscape are truly interrelated). With parametric mapping, these variables are mapped independently by different professionals with different objectives, different map accuracies and at different times, therefore, the overlay of this information almost invariably results in classification inconsistencies and geometric sliver errors. Sliver errors are small errors created by the overlapping of coincident boundaries. A second associated principle presented in Table 1 is that polygon boundaries reflect gradational changes and are not an absolute fixed line. Therefore, the line represented on a map is actually an abstracted line representing a zone of transition. As a result, when parametric maps are overlayed, the truly coincident lines almost never exactly overlay one another. The integrated process resolves this sliver error by using a common basemap and common imagery to assist in the definition of the boundaries among and between the interrelated polygon overlays.

The second major problem resolved through integration is that all information prepared for automation is subject to a second round of verification, updating and improvement using remote sensing. By methodically applying standard principles of image interpretation to each data element, the data base variables are consistently scrutinized throughout the study area. During this process, data discrepancies resulting from multiple sources at multiple scales are often detected and corrected either through remote sensing interpretation or by referring the questionable areas to an agency or organization responsible for initial interpretation, verification and correction. Furthermore, automatic techniques are used to conduct additional error checking, including code and consistency checking between and among the variables. Completed integrated maps are coded after rigorous editing and finally submitted for automated map production and modeling.

In general, integration results in final data which are superior to the data originally input.

The third problem resolved deals with cost. The cost of data base computerization is a function of the number of maps to be automated and the complexity of the lines of these maps. Integration reduces the time required to computerize and overlay a large number of different maps. The manual overlay process has been more than paid for in the savings of computer time to automate and manipulate the data. Moreover, the storage and analysis of the data on the computer is more efficient.

Integration process. As mentioned, a composite of each of the previously prepared manuscript maps onto a single map sheet results in the creation of numerous splinter polygons which may be meaningless, confusing, and lead to substantial data management problems if automated. The mechanical process of integration involves the manual merging of data shown on the thematic map compilation sheets by comparison to remote sensor images, topographic basemaps and each other to yield a single manuscript on which no unnecessary or confusing polygons are drawn. The attribute values on the thematic map compilation sheets are associated with the polygons on the integrated maps through the use of sequential coding lists.

Integration begins by simultaneously registering the two most reliable, yet complex, thematic map compilation sheets to the basemap. A new mylar overlay is placed on top of the set which becomes an attribute code sheet, the integrated form of the compilation sheet. At the end of integration, each thematic map compilation sheet (spatial data) has a corresponding code sheet (attribute sheet). This process is illustrated in Figure 8.

Integration involves making judgments about the correlation among variables, variable reliability and registration, and deciding which lines will be used to form the final map polygons. The integrators match one compilation sheet with the others, register and compare them to the images and the basemap, and decide where to draw the polygon boundaries on the attribute code sheet. For example, a geologic unit may be identified as an old terrace deposit, and a corresponding soil type may be described as forming on old terrace deposits. Therefore, the units should have coincident boundaries. Slight adjustments to the lines drawn around the unit can be made using imagery and basemaps as guides.

This process continues until every map unit has been checked against the images, basemap and thematic map compilation sheet.

The thematic map compilation sheet is replaced by the new attribute code sheet in the set overlaying the basemap. By the same process, a new attribute code sheet is drawn for each of the other thematic compilation sheets, using those polygon boundaries already decided on from the previous thematic code sheet integration.

The process continues until all the thematic map compilation sheets are integrated and new attribute code sheets are created.

Each attribute code sheet is checked for correct transfer of information and correlation among data items. The attribute code sheets are used in the encoding and editing steps which follow.

When complete, the attribute code sheets are ready for consolidation onto the final manuscript. The manuscript is drafted by compositing all the code sheets onto a single mylar sheet, one at a time. Each polygon formed represents a terrain unit or other area of the landscape having its own unique set of characteristics separating it from the adjacent areas.

Linear and point data integration follows similar procedures but normally does not require new variable code sheets. In most cases, straight transfer of compilation sheet data onto the manuscript yields no conflicts.

Numbering and encoding. The spatial data shown on the integrated manuscript and the attribute values shown on each of

28

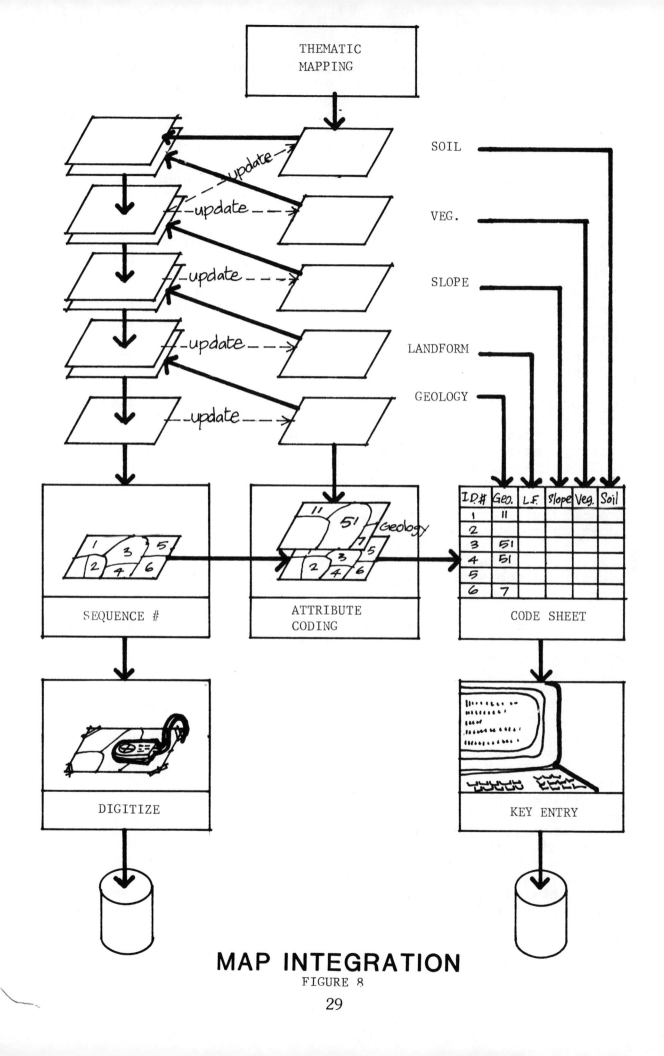

THEMATIC
MAPPING

update

update

update

update

update

SOIL

VEG.

SLOPE

LANDFORM

GEOLOGY

SEQUENCE #

ATTRIBUTE
CODING

I.D.#	Geo.	L.F.	Slope	Veg.	Soil
1	11				
2					
3	51				
4	51				
5					
6	7				

CODE SHEET

DIGITIZE

KEY ENTRY

MAP INTEGRATION
FIGURE 8

29

the integrated code sheets are prepared for machine processing in a way that allows the values of each data layer to be associated with the units shown on the integrated map. This is accomplished by sequentially numbering each integrated map unit on the manuscript and then preparing a sequentially numbered list of coded attributes which match the sequentially numbered spatial units.

After every map unit has been numbered, an encoding form is prepared. The manuscript map data are coded by associating the sequential number on the manuscript map with the sequential number on the encoding form. Data variables are assigned specific fields or sets of columns on the encoding form. Each manuscript data layer is encoded by overlaying the manuscript on top of the appropriate attribute code sheet overlay. The variables are coded by reading the manuscript's sequential number and the corresponding data code number for that polygon. The data code is then recorded in the assigned data column for that corresponding sequential number. The coding process continues until all data variables are recorded for all polygons, lines, and/or points on the manuscript.

Editing. Two basic categories of data editing are performed. These include editing of the manual manuscripts and editing of the automated data. During the manuscript preparation process, quality control checking is performed during each map step. Each overlay is checked for uncoded polygons, lines, and points, unclosed polygons, consistency of information, variables, basemap registration, photomorphic signature, and transposition errors. The maps are edge matched for line location and code consistency.

During the integration process the integrated thematic code sheets are edited for correlation between variables by overlaying the attribute code sheets and looking for areas with incompatible attribute combinations.

During the encoding process the manuscript is edited for missing lines by overlaying the manuscript on top of each attribute code sheet. The manuscript is also edited for drafting errors and unnumbered and double numbered polygons, lines, and points. The manuscripts and encoding forms are code and line edge matched by matching manuscripts to adjacent module manuscript and comparing corresponding polygon and line code values and line positions with each other.

The modular manuscripts are overlain on the map sheets to check consistency of study area boundaries and module borders. After the final review, the manuscript and expansion file are ready for digitizing through keyboard entry.

Subsequent to automation, the data base manuscripts and attributes are subjected to further editing. This editing falls into two basic categories: editing of the manuscript line work; and editing of the map attributes. The original manuscript is compared with the digitized graphic output to determine graphic errors (i.e., spike errors and incorrectly digitized locations for points, lines, and polygons). Subsequently, the computer is used to create "dropline plots" representing plotter displays of each thematic variable included within the attribute data base. These themes might be soils, vegetation, geology, landform, slope, etc. These dropline plots are compared manually with the original thematic maps created as part of the integration process. Errors in either attributes or graphic location are noted and edited.

The attribute file is edited in terms of class ranges, as well as consistency among the classes maintained within the integrated file. The class range edit consists of checking each variable against a theoretical minimum and maximum variable range which is possible for any given map element described by an attribute class. The computer simply reads the variable class within the attribute table and compares it against a theoretical class range provided in an associated table.

The second attribute verification involves a consistency evaluation comparing all of the classification variables and their subclasses against one another to determine those additions where improbable conditions exist. These are flag checked by the photo interpreter responsible for creation of the theme, and updated where necessary. Figure 9 shows a graphic summary of this automated editing procedure.

Map Automation
The following is a description of the tasks associated with data base automation.

Digitizing. All graphic data recorded on the manuscript maps (i.e., points, lines and areas) are digitized into machine readable form using an electronic device known as a digitizer. This device has a movable cursor for conversion of the location of features into x,y Cartesian coordinates. The cursor is used to trace features on each manuscript map which is mounted on the digitizer table. The cursor movements are translated and recorded as digital measurements in units of one one-thousandth of an inch. Numbered tic marks are digitized first for cartographic reference. Subsequently, digitized records are created indicating the precise location in x,y coordinates of all mapped information with respect to the tic marks. In this paper, all digitizing and subsequent data automation processes described use the ARC/INFO and GRID software developed by ESRI. This software includes a highly advanced relational data base management system for attribute data, as well as a vector and grid based data management system for spatial data.

Editing of digitized files. The first step in the edit process is to shift and scale the coordinates of each file relative to tic marks which provided geographic reference. Following this step, a series of automatic geometric and topological editing procedures take place. Also, computer plots are generated for visually checking the accuracy of the digitized and machine edited x,y coordinates against the original manuscript maps.

In parallel with the map digitizing/edit function, the numeric attribute codes are key entered into the computer and associated with their appropriate spatial feature. Each of the

31

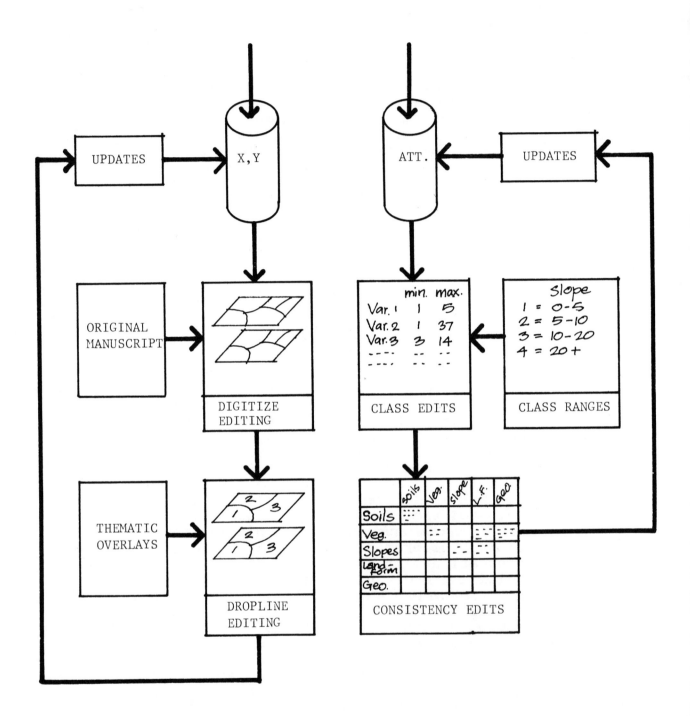

EDITING

FIGURE 9

data variables in the system is plotted out at the manuscript scale and compared against manually prepared overlays of the collateral data. These plots, termed "dropline plots", are used to ensure that each data variable is accurately delineated and coded in the computer data file. Data errors discovered in this edit process are corrected using edit software.

Final file creation. The final step in creation of an automated data base involves establishing an organized user structure for management of the attribute files and coordinate/grid spatial files. Two basic steps are required. They are: 1) merging together of the x,y coordinate file to create a continuous automated map; and 2) gridding of the coordinate data to create a continuous cell file. The complete multi-variable grid file normally contains all of the data variables on various manuscript maps and some of the interpretive data from the expansion matrices. Certain simple data items are packed into one position in the multi-variable file to save space. With extensive data bases, the files are structured into multiple multi-variable files.

SUMMARY

The techniques outlined in this paper briefly summarize the steps evaluated by the authors for creation of workable, accurate, and successful GIS data bases. Most of these methods are not new but certainly are relevant, important and sometimes essential in the conduct of an overall inventory data base methodology. It is recognized that there are many new automated technologies which are making substantial improvements in the manner in which geographic data are manipulated, displayed and communicated. However, for large area surveys of natural, as well as cultural, data a great deal of careful thought and extensive mechanical effort must still be invested to create meaningful and workable data to use with this technology.

It is recognized that the process of creating a totally integrated data base as outlined in this paper results in modification of the typical "existing data base" provided as input for such a survey. In certain cases, it can be argued that these modifications constitute the basis for disagreement regarding accuracy and may introduce the potential for mechanical error. For the overwhelming majority of cases, however, it is our experience that the opposite is true; that a well organized and integrated data base leads to consistent information agreement and focus by users on the real issues of planning, resource management, and environmental assessment, rather than inconsistencies in various mapping and data collection philosophies.

THE STATE OF DEVELOPMENT OF SYSTEMS FOR HANDLING NATURAL RESOURCES INVENTORY DATA

ROGER F TOMLINSON AND A RAYMOND BOYLE

Tomlinson Associates Ltd., Ottawa/University of Saskatchewan

ABSTRACT In early 1980, Tomlinson Associates examined nine computer-based systems to assess their ability to handle natural resources inventory data. Their client, the Saskatchewan Department of Tourism and Natural Resources, was considering the acquisition of such a system to handle their forest resource data. Cost effectiveness, reliability, ease of use, and ability to meet the volume and throughput requirements were considered, as well as technical capability. Five systems were selected for detailed standardized tests on a function-by-function basis. Except for a block adjustment procedure, the Forest Inventory Section required general-purpose spatial data handling software suitable for many users who process natural resources inventory data. The tests showed that the systems could not then meet the requirements for forest data processing in Saskatchewan. Much of the software was under development but was not yet reliable. Some packages worked well but others operated so slowly that they could put a heavy load on computing facilities. The Forest Inventory Section was advised not to purchase a system but to watch system developments with a view to acquiring a system later. The report produced allows potential users to assess the degree to which their needs for individual spatial data handling functions may be met by the current technology. It identifies the capabilities that are difficult to provide and the nature of the problems. The state of the art of systems for handling resources inventory data is shown.

Recent years have seen improvements in computer-based systems for handling spatial data that describe natural resources, geographical entities, and environmental conditions. Much work has been done using minicomputers. Some of the systems now have extensive spatial data handling capabilities, and the combination of the modest financial investment required and the prospect of bringing computer assistance to some of the more laborious manual cartographic map handling operations has made them attractive to many potential users in government and industry.

The data handling requirements of users of such geographical data are of two broad types. The first might be considered to be a "project-related" requirement, such as for a one-time analysis of a specific problem in a relatively limited area. Examples are highway location studies, town site location analyses, environmental impact assessments, power plant siting, waste disposal site studies, and similar tasks carried out by an agency responsible for carrying out the subsequent project or by a consulting organization serving such an agency. Generally, such tasks encompass relatively few map sheets but many types of data may be required to get an acceptable answer to the problem. The type of analysis needed can be quite complex, leading to an optimal use of the site in question. Systems that are suitable for such projects need not handle large volumes of data but they must be able to construct relatively complex questions, which may be addressed only once to the data concerned. The emphasis is on flexibility of operations; even if the operations themselves are awkward to set up, it must be possible to set up the appropriate queries.

The second general type of spatial data handling requirement can be thought of as an "inventory-related" need. Examples are land use surveys, topographic surveys, soil surveys, wetland surveys, and forest inventories. The number of map

ROGER F. TOMLINSON is President of Tomlinson Associates Ltd., a firm of Consulting Geographers that specializes in the design and evaluation of computer systems to handle natural resources data. Dr. Tomlinson's firm is located at 17 Kippewa Drive, Ottawa. A. RAYMOND BOYLE, a Senior Associate of Tomlinson Associates, is a Professor in the Department of Electrical Engineering, University of Saskatchewan and former President of the Canadian Cartographic Association. MS *submitted January 1981* sheets may be large, rarely less than a few hundreds and often several thousands. Most data queries are addressed to one inventory data set, and most ancillary data sets are derived from the primary data. Operations are typically simple, such as area measurement, but need to be performed on every sheet. Systems answering these needs emphasize smooth and efficient input procedures, the ability economically to store and handle large volumes of data (and often to fit the data together in one large data bank), and the ability easily to perform often-repeated data handling operations. To some extent, flexibility can be traded for efficiency.

These two broad categories of requirement represent two widely spaced points on a continuum of system capabilities. The boundary between the two types is not a hard line. Other types of system requirement also exist within the realm of spatial natural resources data, and in time, users with inventory requirements may need the flexibility now exhibited by the smaller project-oriented systems. At the moment, however, the categories represent commonly observed types of user requirement and systems that now exist, and they also serve to identify the capabilities of interest in this report, namely the requirements to handle natural resources inventory data as opposed to project-related data.

THE SASKATCHEWAN FOREST INVENTORY

In the winter and spring of 1979–80, the ability of several systems to handle natural resources inventory data was examined. Forest inventory maps were used as the basis for the investigation. These predominantly contained polygon-based data, but also some linear and point data.

The investigation was carried out on behalf of the Saskatchewan Department of Tourism and Natural Resources, who were actively considering the acquisition of a computer-based system to handle their forest resource data. The systems had to be examined from the point of view of a real-world resources data handling operation, and not as an abstract technical exercise. The computer-based system eventually chosen would have to be more cost-effective or more productive, or both, than an already well-functioning manual method of handling forest inventory. Considerations such as reliability, ease of use, and ability to meet the volume and throughput requirements had to be addressed, as well as those of technical capability. In short, the computer operations were being tested as tools for use in a working environment.

Like many natural resources inventories, the Saskatchewan forest inventory starts with the acquisition of aerial photography (6,500 prints per year). These photographs are visually examined, interpreted, and annotated by hand by photointerpreters. They are then processed by manual photogrammetric and cartographic techniques to produce 1:12,500 scale forest maps (at a rate of 160 to 200 per year, for a total of 1,800 to cover the Saskatchewan commercial forest zone). The forest maps are subsequently measured and analyzed to produce forest statistics and are used as the source of data for other cartographic products.

The type of photointerpretation and the nature of the terrain in Saskatchewan have some effect on the subsequent data handling requirements and are worth describing at this point.

The forest photointerpreter views the air photographs stereoscopically and draws lines and writes labels directly on to one photograph of each stereo pair. The interpreters have difficulty in defining polygon lines more closely than about 0.5 mm in production. Each photograph has previously been marked with the central, fiducial, and wing points, as well as bounding lines drawn to connect the outer of these. In practice, the operator uses black pen lines in two different widths to represent polygon boundaries; the first (thin line) is for actual timber stands, and the second for soil capability. It may be inferred that each of the wider lines contains a thin line. Pens of other colors are used for linear features such as roads, trails, streams, and lakeshores. The annotated photograph is thus similar to a miniature map with a mixture of polygonal and linear features, different line widths, different colors, and a variety of alphanumeric and symbolic labels.

In Saskatchewan, the similarity of the photograph to a map can be taken a step further. The terrain is unusually flat throughout the forest inventory area, so that errors on the photograph due to relief can be ignored and each photograph can be treated as a small planar map for digitizing purposes. The plane of this "map", of course, is variable, depending on the tilt of the aircraft at the time the photograph was taken, and it must later be corrected by digital calculation in a special "block adjustment and transformation" stage, but this is not a difficult process.

REQUIREMENTS FOR COMPUTER ASSISTANCE

Computer assistance was sought to reduce the time and cost associated with the manual photogrammetric and cartographic steps that *follow* photointerpretation. These can be thought of in two parts. The first concern is with converting the interpreted data to a digital form, creating the digital data base, and providing some functions for displaying and plotting the data in map form. These steps may be termed the basic system capabilities, as listed below.

Basic System Capabilities
> Digitization
> Edgematching
> Report generation and editing of lines
> Polygonization
> Labeling
> Plotting
> Data storage
> Data management
> Browsing and plotting a derived map
> Updating

In addition to these normal basic system capabilities, the Saskatchewan forest inventory requires a block adjustment and transformation function to input data directly from air photographs.

The second stage requires software to allow further handling and analysis of the digital forest resource data. Such "map data handling capability" can cover a wide variety of functions. A list of typical capabilities is given below. The list also indicates those required by the Saskatchewan Forest Inventory Section and included in the tests.

Typical Map Data Handling Capabilities
    Data manipulation*
      Reclassify – attributes*
      Generalization*
        Dissolving and merging*
        Line smoothing
        Complex generalization
      Interpolation
        Centroid allocation
        Contouring
      Scale Change*
      Distortion elimination – linear (rubber sheeting)
      Projection change
    Generation*
      Points*
      Lines*
      Polygons*
        Simple five-sided polygons*
        Irregular polygons with islands*
        Circles*
        Grid cell nets*
        Latitude and longitude lattices*
        Corridors (along linear features)*
        Other
    Data extraction*
      Search and identification*
        Attributes*
        Shapes*
      Measurement*
        Number of items*
        Distances (straight line between points, along convoluted lines)*
        Size of areas*
        Angle direction
        Volume (cubic measure)
    Comparison*
      Intersection – overlay
        Point-in-polygon*
        Polygon-on-polygon (grid cell on polygon, circle on polygon)*
        Other
      Juxtaposition (proximity)
        Shortest route

Nearest neighbor
Line of sight
Contiguity
Connectivity
Complex space-attribute-time correlation, e.g., rate of change of density of attribute*
Interpretation (may use any or all of the above functions in concert according to a prespecified set of rules)
Determination of optimum location
Determination of suitability
Determination of desirability

*Capabilities required by Saskatchewan Forest Inventory Section and included in tests.

Nine systems were considered in detail and form the basis for the general statements of overall capability. In November 1979, a list of the future data handling requirements of the Saskatchewan forest inventory was sent to potential suppliers (see Appendix A), and proposals to meet these requirements were requested.

In January 1980, proposals were received from the following agencies and companies: Autometric, Inc.; Carl Zeiss, Canada Ltd., jointly with Collins & Moon Ltd.; Comarc Design Systems; Earth Satellite Corporation; Environment Canada, Canada Land Data Systems; Environmental Systems Research Institute (ESRI); Gibbs & Hill, Inc.; M & S, Inc.; and Systemhouse Limited. The standard of proposals was high. The average length was approximately 100 pages, and some were double that. The requirements had clearly been fully understood and were addressed in specific terms.

Five of these systems were further subjected to standardized tests which tested their capabilities on a function-by-function basis. In alphabetical order, these systems were Comarc Design Systems, Earth Satellite Corporation, Environmental Systems Research Institute (ESRI), M & S, Inc., and Systemhouse Limited. A typical test took three to four days of on-site evaluation. The results are not related to individual companies in this report, but provide an analysis of the existing level of capability by function. Potential users can assess the degree to which each function meets their needs and identify the areas where capabilities are difficult to provide and the nature of the problems.

BASIC SYSTEM CAPABILITIES

Digitization
Digitizing is here regarded as the process of converting point and line data from graphic to machine-readable form. Editing and reformatting of data after digitizing are considered separately. On the small number of photographs used during the benchmark tests, different digitizing methods required times that varied from 30 to 90 minutes. This is the minimum time for actually tracing the information from the photographs on a digitizing table and does not include the time normally required for organization and set-up. This range of 30 to 90

minutes is important, because it implies that the 6,500 photographs per year used in Saskatchewan would need a minimum of 4 to 12 digitizing stations (and maybe more if one takes into account the set-up time and institutional overhead factors such as operator delays, inexperienced operators, training sessions, demonstration sessions, rest periods, etc.) and, moreover, involves similar numbers of skillful operators on a continuing basis.

It is an alleviating factor in forestry digitization that the line data need only have a precision of 0.5 mm; only the few photogeometric points are required to have a precision of 0.1 mm. An average photograph has approximately 75 polygons and labels, each polygon having about 5 line segments or arcs and each arc having about 10 coordinate points to provide proper definition.

After watching the different digitizing methods in operation and noting areas of possible improvement, one can express the goals for digitization for the Saskatchewan forest inventory as follows:

1 The tracing required for digitizing to the necessary precision from a 75-polygon photograph should eventually take 20 minutes or less for all points and lines, using optimized cursors and procedure, exclusive of set-up time and institutional overhead

2 The label entry operation should be separated from the line entry operation

3 Simplicity of operation should be provided for the operator, as well as feedback, in order to minimize errors, because these errors are usually tedious and expensive to correct

4 Over-complex and expensive hardware should not be used for the relatively simple digitization production process required in Saskatchewan.

In order to maintain a short throughput time, keep the necessary precision, and reduce errors, the digitizing operation must be simple and reliable. Errors in the output such as missed lines, lines digitized twice, and so forth, can cause considerable trouble, although it is possible to correct them later. Most systems programs tested had a good "node snap" which removed the likelihood of junction errors within the photograph area, and in all cases, the operator had to be particularly careful to digitize lines exactly to boundary nodes in order to avoid problems in later edgematching. A good reporting system during and after digitizing can aid the operator appreciably in detecting errors made, but corrections are expensive in time and use of equipment and should be avoided. Some systems provided the capability through scaling and rotating operations performed on line in the digitizing process, by use of tablets fitted with microprocessors. By this means, a photograph can be replaced and partly redigitized without difficulty.

The actual hardware, that is, the tablet size, type of cursor, and so forth, were most important. Only a small tablet is needed to accept a photograph, and is best equipped with back lighting as this improves the contrast between the black interpreted line and the less opaque black photographic background. A cross-wire precision is needed for the photogeometric points, but a pen cursor is adequate for the lines.

The times actually obtained on test with various digitizing systems were probably longer than could have been obtained with an optimized input system and a trained production operator.

The procedures of digitization were widely different in the various systems tested, mainly because the process was to be followed by polygonization. One system digitized only in full polygons (ameliorated by digitizing end points of arcs previously fully digitized), as this reduced the computer effort later when polygons were being formed. The increase in effort over the more normal method of digitizing only in arcs was not great, and, overall, had considerable merit. Another system went to the other extreme of allowing the operator to digitize lines in any disorganized manner, the program sorting out the problems later; this procedure seemed to use more CPU time than was desirable, however. Most systems used the arc method of digitization, treating each line between nodes as a separate entity. Most systems input labels separately on a low-cost alphanumeric terminal, then merged them with the polygon centroid points which were digitized at the same time as the lines. This procedure is described later under labeling. One system required the input to be done at the more expensive digitizing table, and another required entry of the polygon labels on each side for each arc. These methods were time consuming.

A number of system tests showed that potentially, input for a photograph could be digitized within 30 minutes for the lines, centroid, and reference points, and 14 minutes for the labels. With more experienced operators and optimized hardware and method, it is inferred that the input could eventually be accomplished within 20 minutes and 10 minutes respectively for an average photograph. This, in turn, implies that a minimum of four input station units and two alphanumeric terminals would accomplish the necessary throughput for Saskatchewan. On the other hand, some systems clung to methods that took at least three times longer and involved more hardware and operators as a result. In any case, allowances must be made for organizational problems, institutional overhead, and downtime. This could double the number of stations required, and gives a practical minimum of eight input stations under normal operating conditions using optimum methods with this technology.

The advantages of direct on-line display during digitizing (as opposed to display on request at an editing station) appeared to be small in the systems tested, mainly because of the relative simplicity of the photographs and the resultant short elapsed time for digitizing. Other reasons were the standardized continuous throughput possible in this typical inventory operation, and the general lack of opportunity in the Saskatchewan situation to use on-line transformation in one pass. The last situation existed because the true position of the pass points could not be known until a block of photographs had passed the block adjustment stage. The pass points, and so forth, would first have to be digitized, and then the remainder of the data after the block adjustment had been completed.

Views varied considerably regarding the best method of line digitization itself, whether in point-by-point or stream mode. Almost every alternative of this was tested. All the systems used cross-wire cursors in a point-by-point or stream mode. The stream mode appeared to be faster, but required a subsequent weeding algorithm to reduce the number of recorded points to about the same as that selected by the operator in the point-by-point mode.

For most systems it was accepted that label points should be selected and digitized by the operator. For one system, however, it was felt that this was onerous and they should best be calculated automatically at the time of plotting and display. Although this had some theoretical advantages in polygon generalization and derived maps, the CPU time required before any plotting appeared to cause considerable problems.

Block adjustment and transformation are processes that can be invoked after digitizing to correct the tilt in the photo plane. These are off-line procedures, well established and straightforward, but they were not part of the basic system capability tests. Because of the small number of true reference positions in northern Saskatchewan, a block adjustment of up to 500 photographs is frequently necessary. The process fits together all the planar photographs to create a truly positioned and scaled planar map base, removing characteristics caused by aircraft altitude and tilt. The digitization systems would supply the pass points to the batch block adjustment program to allow it to calculate the reference points for each photograph and then, in a further batch operation, transform all points of nodes and lines, etc., also digitized, to the true UTM or other positions.

Edgematching

Edgematching is a necessary process to join the polygon lines across the boundary from one photograph to the next, after corrections for altitude and tilt have been made. The block adjustment and transformation routines together should make all photographs and their edge lines fit together exactly, assuming reasonably careful interpretation and digitization. However, there is always the possibility of problems with small gaps, slight discrepancies, overshoots, and missed and double lines. The edge-match program should deal with these and report any impossibilities such as a single end line which might indicate a line missed in digitization on the adjacent photograph.

Report Generation and Editing of Lines

Some errors will always exist, and a good system should be able to report many of these to the operator, allowing corrections to be made with ease. Usually errors consist of double-digitized or missed lines. The number of errors is usually small and so a simplified method of plotting and redigitizing is not too clumsy. However, a relatively simple system on an interactive display is faster and easier to use. Because of the throughput required and the resultant need for multiple plotters to keep up with the plotting and redigitizing, the somewht high cost of an interactive display unit is probably justified. Once the editing has been done by the above means, the data can be pronounced "clean".

Polygonization

Polygonization is the process of connecting together arcs to form polygons around designated centroids. Providing the original arc data are clean and good, the process can be simple and automatic. If initial digitization creates complete polygons, little work remains at this stage. If the arcs are digitized as individual entities in any convenient order, grouping them into polygons is also fairly simple. However, use of a free-line method, without designating the position of nodes at all, requires considerable CPU time subsequently. One system required a human operator interactively to point to each arc in turn and so form the polygons; this appeared to be rather time-consuming for equipment and operators; the process of manual intervention appeared to create an appreciable number of errors at this stage, which in turn produced problems in subsequent data handling (see measurement of areas).

Labeling

Labels describing the timber, soils, and age of stand in a standardized coded form are attached to each polygon centroid by the interpreter on the original photograph. These, as well as the line and point data, must be transformed to the digital system. A few labels are attached to cartographic features and these may also be required. These labels are normally added after polygonization, although they can be entered at any time once a centroid point has been recorded.

The method that appeared to be most efficient in the systems examined was to label when a plot was first made of all digitized lines and centroid points; a reference number was then added to each centroid automatically by the system. The plot could be drawn on a fast plotter or on a display associated with a hard-copy unit. Then an operator, on a low-cost alphanumeric terminal, examined the original photograph together with the reference-numbered plot, and entered a label against each reference number. After checking, this was easily merged with the already recorded point and line data.

The term "labeling" also applies to the printing of such labels at the time of map plotting (see later).

Plotting

In general, quick-look plots are straightforward and usually consist of first producing a plot file and then using this to produce the plot on a Calcomp or similar unit. However, it would appear that multiple plotters may be necessary to meet production throughput. Because these plots are ephemeral, the process could be better replaced by a display, possibly associated with a hard-copy device.

At this stage, labels may have been added to the data but not edited, so that labels on the resultant plots can often be unreadable. In a number of systems it is possible automatically to add reference numbers to control points, arcs, or even data points to aid in correction editing; in others, designation is by pointing.

For final copy plot, forest inventory maps do not demand high precision, and a plotter of medium precision can provide adequate quality. However, because labels have been added, it seems essential to provide some interactive manipulation facility before plotting. It is perhaps possible to adjust positions of labels, and the letters in a name, by using a plotting and redigitizing approach with special label positioning routines. It is, however, more convenient and practicable, in a production operation, to make adjustments by using a display and editing system equipped with routines for symbolization and label manipulation as well as line

42

operation. Some systems were able to show this in an embryonic form; only one was able to show the full capability and flexibility. Because a major requirement is to produce a useful and readable map document, any system, to be considered viable for forest inventory, must have an interactive display capability for lines and labels.

In one system tested, it was appreciated that interactive editing times could be large. This system therefore went to some length to provide automatic labeling, and this approach was commendable. However, CPU times were very high and it could not handle some labeling situations adequately. The present state of system development seems to be on a seesaw, with slow manual interactive methods at one end and CPU-time-intensive and not altogether adequate automatic methods at the other. A truly satisfactory approach will probably be a balance of the two. Both the interactive display and the plotter itself must provide line symbolization and be able to create symbols as well as various sizes of letters at different positions. Most plotter systems have a modest capability to produce symbols.

Data Storage

The data storage would be expected to be a complete, fast-access one, for the whole province, with disk storage. Magnetic tapes would only be used for back-up. To maintain integrity, effort and thought must be given to data control and organization of data maintenance. The amount of data was in most cases appreciably larger than necessary. Data bulk is not only a problem of the number of data disk drives required but, more importantly, affects access time because times generally depend directly on the number of words or bytes used. Two examples of lack of storage reduction will be given. One is that strings of absolute coordinates were used in all cases instead of a compacted incremental form. The second was that all systems tested used polygon storage methods, in which each arc became triplicated in order to give the original arc as well as the polygon forms. Data structures for cartographic and polygon data had also been given little consideration so that the times for searches were often very slow.

Data Management

The volume of data collected in forest inventory applications accentuates the need for efficient management of the geographic data involved. Data structures and data management software suitable for such volumes of spatial data are of major importance, but no system at the time of the tests demonstrated a useful capability relative to Saskatchewan's needs.

Browsing and Plotting a Derived Map

The complete data base must be accessible to browsing and, following a request, it should be possible to show the selected spatial data on a display. There is no intention during browsing to make any modification to this data and then replace the modified data in the data base (as with the display and editing operation); browsing simply offers the ability to select and look at the data. Some browsing displays are at the standard scale, and others at a reduced scale. The problems are appreciably different.

The browsing operation may result in a desire to obtain a plot of the displayed data. This is called a "derived map" to distinguish it from the standard forestry map for a standard area. As the plotting of the derived map is similar to that for the standard maps, the browsing and initial part of producing the derived map are considered together.

The problems of producing browsing displays or derived maps are threefold. The first is that the data have to be found so that they can be displayed. If only one display or map is being made at standard scale, only one search may be necessary. If, however, it is at reduced scale, then several searches may be needed. Good data structures reduce the search times appreciably. In general, however, times of search are not very long compared with the following problem.

The second, and larger problem, is that the data when found have to be extracted. If only one area of the original standard scale data is being requested, then those data can probably be extracted in a few minutes. However, if a reduction at 10 to 1 scale with even data distribution is required for the browsing display or derived map, then 100 times the data may have to be extracted and handled, involving a time 100 times greater than the number of minutes used for one map. The importance of data compaction then becomes very obvious.

The third problem is that processing on the extracted data may be required, particularly if the display or map is at a reduced scale. Possible processing may be line generalization to reduce the number of data points in the displayed image or plotted line on a logical cartographic basis, or polygon generalization to merge and dissolve polygons and thus reduce the number of lines and labels. The importance of line generalization did not seem to be critical when demonstrated in the tests. The use of polygon generalization, which is essential for a readable reduced-scale map, must be examined in much greater detail, however. While a number of systems showed some capability for doing these, the operations were generally very slow, and much more attention has to be given to algorithms, data structures, and data compaction.

Updating

The updating of information can be thought of as two types of operation. The first, the changes made to attribute labels, usually was accomplished by a simple instruction inserted through an alphanumeric terminal. The second, changes in the line data, is rather more complex. This operation was handled in a variety of ways in the different systems, some more cumbersome than others. One method used insertion of new complete irregular polygons (see generation of polygons, under Map Data Handling Capabilities), for example, to update forest stands after a burn. This invoked the process of overlaying new polygons on the existing data base. Other systems were able to change individual arcs (line segments) found to be in error, and subsequently to reform polygons. If a polygonized arc were edited (and it could be done), then unwanted slivers would certainly be produced, as it would not be possible to change the adjacent sides of two polygons in exactly

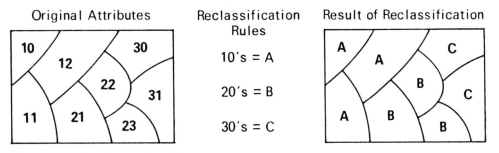

FIGURE 1. *Reclassification of attributes.*

the same manner. This is why some systems incurred triple data structures to retain the availability of the original arc lines.

General Comments on the Basic Capabilities of Systems Considered for Natural Resources Inventory in Saskatchewan

1 It seems desirable to separate input digitization units from display and editing units.

2 Digitization units can be small, simple, and very low in cost for input requirements similar to those in Saskatchewan.

3 The number of digitization stations should be kept to a minimum by optimization of input method and use of separate alphanumeric displays for labeling.

4 Display and editing units of a cathode ray tube type are preferable to multiple quick-look plotters and extra digitization.

5 Display and editing units require adequate, but not very sophisticated capabilities of editing. They must be able to handle simple digital editing, but also more complex pre-plot editing, particularly for labels and names.

6 A medium-precision plotter is adequate for output.

7 Disk drives are costly devices and the number should be kept to a minimum by data compaction processes.

8 Disk scanners are important in certain applications and were promoted as a panacea for the search problems of large bulks of data. They can be useful and attractive in situations where the main CPU can be working on calculations while a search is being made. This situation rarely arises in cartographic applications, however, and as a result the system merely exchanges being scanner-bound for being CPU-bound.

9 CPU times must be carefully examined and assessed. Batch operations should generally be done at night and daytime operations should be mainly the interactive ones of digitizing, editing, browsing, and so forth.

10 The newer color displays seem to be suitable for the browsing operation although their resolution is somewhat lower than can be obtained on the Tektronix display.

11 In the systems tested, system economy depends on minimizing the total amount of hardware, because of the amortized cost effect; and minimizing the number of interactive digitizing and editing stations in order to minimize the number of operators (an on-going cost).

45

MAP DATA HANDLING CAPABILITIES

When an acceptably error-free bank of digital spatial data is in place, it can then be, for the first time, usefully addressed by data handling software programs designed to meet the users' requirements. The types of data handling that are carried out can be described as "data manipulation", "generation", "data extraction", or "interpretation" operations.

Data Manipulation

Data manipulation refers to operations that are performed on data to make them more suitable for further processing; to improve their comparability, facilitate their retrievability, and so forth. No analysis is done on the data; at this stage they simply are massaged into another form or shape that makes them easier to handle or analyze thereafter. Sometimes this massaging results in new forms of the data which are useful in themselves, which can readily, or more readily, be visually comprehended and thus require no subsequent measurement or comparison. Typical spatial data manipulations include reclassification of the data elements on a map, generalization to produce summaries of spatial data, interpolation processes such as centroid allocation or contouring to allow one form of location identifier to be augmented or replaced by another (for better data compaction or easier visualization of the data), and scale change, map projection change, and distortion elimination (rubber sheet stretching) to allow maps to be better fitted together, compared, or scanned.

Those operations required by Saskatchewan are described below.

Reclassification of attributes. The change of an attribute or attributes found to be in error, or the addition of new data (or subtraction of old data) from part of an attribute code, are considered to be updating procedures and are described separately. Reclassification is the regrouping or change in value of a set of existing attributes based on a set of rules that specify how to regroup or change the value (Figure 1). The new file must be created without destroying the old one. Reclassification is frequently a process of simplification as well as reinterpretation. A typical reclassification required by Saskatchewan is the change of each forest stand description into a code that describes the fire fuel hazard of each stand.

Reclassification changes the attribute data set only and should not require any calculation performed on the graphic file. It is a computationally trivial operation and should be very swift. The set of rules should be easy to specify, preferably with a full range of Boolean operations. The set of data to be changed should be easy to specify. The set of rules should then operate on the chosen data without further intervention by an operator.

Some systems did not carry out this straightforward operation. In others, the entry of rules for reclassification was either laborious or error-prone. One system required the entire set of original attributes to be listed and the new codes entered alongside each unique attribute before reclassification of the data set. This is extremely laborious and useful only if few unique original attributes occur, which is not the case on a typical forest map with complex attribute descriptors. Only

46

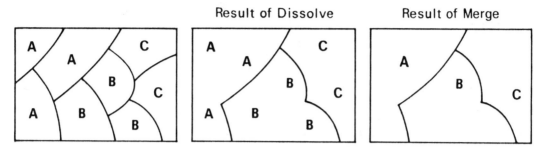

FIGURE 2. *The processes of dissolving and merging for data generalization.*

slightly less laborious are systems where the new values for each subvalue or range of values of the original attribute data have to be specified. This is time-consuming if categories have hundreds of possible subcategories or long ranges. Another system could alter only three elements of the original attribute code at a time. This required multiple set-ups and passes of the reclassification procedures. One system, which had an otherwise very efficient rule definition process utilizing a full Boolean command capability, relied on the operator mentally interpreting textual attribute names into their internal computer codes. It worked, but was extremely error-prone and could easily be modified to be automatic. Other systems proved that set-up and processing could be smooth and efficient.

Reclassification of attributes is not a technically difficult process and no significant problems inhibit the development of efficient software. At present, the general level of software development is not adapted to users, and requires modification before it is adopted.

Generalization. Dissolving and merging constitute a process used to generalize and simplify maps. It typically is used after reclassification of attributes has produced a new map with adjacent polygons now having the same codes. The boundaries between the similar polygons can be dissolved to form larger unique polygons. Similarly, similar attribute codes can be merged into a single descriptor for the larger polygons. The process is illustrated in Figure 2.

If a Saskatchewan forest inventory map was reclassified into a fire fuel hazard map, the dissolving and merging process would simplify the new map, reduce unnecessary clutter, and make the new map amenable to scale reduction from 1:12,500 to 1:50,000. Most systems could not yet accomplish this process, some not at all, others not efficiently. Some systems offered to replace the dissolving and merging capability with shading and an external key to shaded areas, as shown in Figure 3.

This alternative may be completely satisfactory for some uses, though it does not reduce the data density and so make the map amenable to subsequent scale reduction.

Technical problems related to dissolving and merging stem from the fact that the process requires the manipulation of the graphic file and also tends to create large polygons (in fact, it sets out to do so). If the data volume in a graphic file is large relative to other systems, because of the chosen coding format or lack of data compaction, then the processing of that file tends to be slow and results in poor

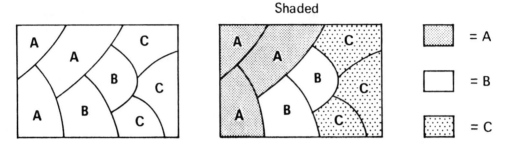

FIGURE 3. *An alternative method of generalizing data by addition of shading and an external key.*

times to run the dissolving and merging process. Greater efficiency will probably come from more compact data notation and more efficient logical data structures rather than faster computers. In some systems, the limitations on buffer size prohibited the merging of already large polygons and caused the programs to halt or abort completely. Buffer size can be changed, but this may affect other aspects of system processing and tie up the system unnecessarily. In this, as with other problems associated with large polygons and long line segments, the answer must come from a different approach to large element handling. At least two systems already handled minor polygons and also automatically partitioned, processed and reassembled large data elements without difficulty or operator intervention. *Scale change.* Scale change itself is a linear and orthogonal stretching or shrinking process. It is frequently used to change the scale of one map or photograph to match the scale of another, to reduce material from the detailed level of original survey to produce a smaller, more usable end product, or to reduce the size of a document further for formal presentation in printed reports.

In Saskatchewan the aerial photography for the forest inventory and the main series of forest inventory maps are at a scale of 1:12,500. At that scale, it takes approximately 1,650 map sheets to cover the commercial forest area. A further 150 map sheets at 1:50,000 cover the reconnaissance forest zone, for a total of 1,800 map sheets for the provincial forest. There is a requirement for summary maps and derived maps to be produced at medium scales around 1:50,000 and, for illustrative purposes, at much reduced scales (1:250,000 to 1:1,000,000).

Enlargement or shrinkage of a digital map image is a straightforward technical problem. Spatial data in digital form inside a computer are stored with reference to a mathematical coordinate system, and are to some extent independent of the dimensions of the piece of material on which the map is eventually plotted. In theory, a map can be plotted out at any scale. Problems arise in the legibility of reduced-scale plots, in the graphic enlargement or compression of shapes (and hence loss of information content), in the diminished size of map spaces resulting from scale reduction, and in problems of continuity of data between map sheets when many map sheets are encountered. Enlargement of a map sheet can similarly reduce synopsis, can go beyond the fidelity of the original data and result in lines which have a marginal information content and are graphically unacceptable. A measure of satisfactory digital scale change is the degree to which measures are included to overcome these ancillary problems.

The majority of systems could perform basic scale change, but few had resolved the ancillary problems. Those with capacity to reclassify, dissolve and merge could produce low-density data sets which were amenable to scale reduction. Only one system could carry out scale changes that handled a set of maps crossing a UTM zone. The majority of other systems could handle maps only on a sheet-by-sheet basis, which would be a laborious process if hundreds (or thousands) of map sheets were involved. Only one system carried out automatic line weeding during a scale change process. No system carried out line smoothing. The problem of labeling small-scale maps had not been solved adequately. Some systems allowed a choice of reduced character size, but had no control over character placement. Labels were overprinted and the small characters were illegible. One system produced a shaded map with an external key to the shaded areas. The resulting product was pleasing and produced a reasonable 1:50,000 map for greatly simplified material.

In general, an adequate overall capacity existed for small to modest automatic changes of scale. Scale reductions from 1:12,500 to 1:50,000 are carried out in a marginally satisfactory manner, the key factors being the amount of generalization and reduction in data density that has been carried out beforehand, and the degree to which the labeling problem has been resolved by a particular system. No system demonstrated a capacity to produce legible 1:250,000 scale maps from 1:12,500 source material.

Generation

The ability to generate new graphic elements (points, lines, or polygons) in the data base is used to define locations to which queries can be addressed (what is at a point, what is along a line, what is within a region) or to insert new data into a data bank at specific locations. Given a digital data base of Saskatchewan forest inventory maps, for example, points might be used to define the location of new sample plots, lines to define new roads, and polygons to define new forest regions, administrative districts, burn areas, or diseased areas. Other forms of polygons such as grid cells can be used to extract data in patterns amenable to statistical modeling, circles to gather data within so many kilometres of a pulpmill or sawmill, and corridors to define visual buffer zones of unharvested forest along roads and streams.

The definition of a figure and its topological insertion into the data base should be straightforward, with a minimum of operator involvement and a minimum of commands being invoked. Two approaches are commonly used. The batch approach requires the input of a coordinate listing, either one of known coordinates for simple figures, or one resulting from the automatic digitizing of more convoluted figures. The interactive approach allows both simple and complex figures to be drawn on the cathode ray tube and subsequently inserted in the data base. The use of either approach must result in the accurate spatial positioning of graphic figures which have a resolution that is adequate to the data being inserted or the questions being asked. Regular figures such as straight lines, grid cells, circles, or latitude and longitude lattices should be generated by calculation rather than digitization from drawn figures. There should be no constraints on the number of points, length of lines, or size and complexity of

49

polygonal shapes that can be generated. If a graphic data element is required for repeated use, one should be able to store it for future recall and have a name attached to it so that it can be referred to (Circle A, Circle B, Circle C, and so on).

Points. Very few systems can perform easily the seemingly trivial task of defining a new single point location and allocating it to a data base. Sometimes this simply reflects a lack of adequate software development, but sometimes it is because the structure of the data base inhibits single points being defined and entered. Most grid-cell-based systems cannot allocate points with a finer resolution than the size of the grid cell being used. Some polygon-based systems generate very small circular polygons as a surrogate for points. These then have to be inserted into the data base by an overlay process, which can be very slow and expensive and altogether a very poor method of generating points. Some systems can generate points on a cathode ray tube, some can even cause such points to be plotted out on new maps, but in neither case can the new points be inserted into the data bank and be used in conjunction with the other data therein. The fundamental differences of capability between current systems probably stem from their original design objectives. Some systems have been designed for the generation and updating of graphic drawings. (There is a large market for such systems for producing engineering drawings, utility company maps, and printed circuit masters.) These systems produce graphic data elements very easily and with great flexibility, but have placed much less emphasis on the building of spatial data bases and the subsequent manipulation of spatial data. The second category of systems reverses the emphasis, having originated with the aim of creating spatial data bases and providing data handling capability (particularly for natural resources analysis on a project-by-project basis). These systems appear to have placed less emphasis on data input and the generation of graphic data entities, and frequently labor under cumbersome procedures.

Very few systems in 1980 combined both types of capability and both will be required for efficient natural resources inventory data handling.

Lines. Most existing systems can define a line and insert it into the data bank, some with more ease than others. Again, some data base structures inhibit the feature generation process. Users of grid-cell-based systems may encounter low resolution and gaps in lines that are not aligned (either orthogonally or diagonally) with their grid cell structure. Some polygon-based systems treat all open-ended lines as error conditions and linear features must be represented as long, narrow polygons which results in cumbersome input procedures. Other systems have constraints on the length of line that can be input. There are limitations with respect to the number of coordinate points that can be used to define any one line segment. Long lines thus must be broken and the pieces separately coded. In general, there are no insurmountable technical problems for inserting lines into the data bases, but some procedures are considerably more clumsy than others.

Polygons. For simple five-sided polygons, the capability is that of generating a simple straight-sided figure from a given list of corner point coordinates, perhaps to define a new administrative region. The ease of entering the information into the computer is the main criterion for judgment. A five-sided polygon is a minimum requirement. It has to be topologically defined to the data base so that it delimits an area within which measurements can subsequently be carried out.

Some systems cannot create such a region directly from a list of corner coordinate points. Others are constrained to four corners (unless the area is digitized and overlaid on the data base as an irregular polygon). Some systems, however, have no difficulties. Both batch and interactive approaches are utilized effectively. The elapsed time needed to create the figure and insert it into the data base varies between 10 and 20 minutes. Systems without the ability to link maps in a uniform data base structure are, however, restricted to polygon definition within one map sheet or map file.

For irregular polygons with islands, the type of ability required is to define an irregular, highly convoluted polygon using 1,000 or more coordinate points. Within that polygon, it should be possible to define other minor polygons (islands) using 100 or more coordinate pairs each. The objective is to define a new region, perhaps simulating a newly burned area of a forest that comprises the interior of the larger polygon and excluding the interior of the smaller polygons (representing unburnt areas). The new region could then be used in conjunction with the data base to determine the type and extent of forest stands burned.

In the current state of development, most systems could generate the complex polygons required. The majority used a batch approach. The polygons were digitized separately (off-line) and the resulting coordinates were subsequently merged with the data base. Occasionally this was a two-step process, the large polygon being created and merged first and the smaller polygons created afterwards and removed from the region now defined by the large polygon. Some systems had the ability of weeding, after digitizing, to reduce the number of points involved in subsequent processing. This is judged to be a useful feature. Other systems showed good error diagnostics and reporting during the input process, functions that are particularly useful when complex features are generated. The average time taken using the batch input approach was approximately 45 minutes (36 minutes to 55 minutes). The interactive approach enabled the irregular polygon to be defined quite quickly (2 to 3 minutes), but the subsequent step of topologically inserting the figure into the data base appeared to be time-consuming (40 minutes).

Problems in both approaches centered on current system limitations on the size of polygons that can be entered. Most systems had not yet solved the problem of creating or handling large graphic data elements and were restricted to between 1,000 and 2,000 coordinate pairs per polygon. The input process frequently ran into this limit or the limit of coordinate pairs in a line segment mentioned above, and polygons then had to be arbitrarily divided into two or more parts by the operator and then redigitized. New programs were being written that would allow larger polygons to be handled, as indeed they must if typical natural resources inventory is to be handled in digital form. These new programs should not just increase buffer size to ameliorate the problem, but must recognize that large graphic data elements are a common occurrence in natural resources inventory, that it is extremely difficult to determine the maximum size of such graphic data elements before they are encountered, that some data manipulations even set out to create large data elements for natural resources management purposes, and that it is necessary in future programming to remove all constraints on the handling of large graphic data element size by a process of automatic partitioning and sequential processing of the pieces, without operator intervention.

In the case of circles, given a center point and a radius, a system should be able to define a circle by calculation and to insert it topologically into the data base. It should not be necessary to digitize the circle and digitally overlay it on the data base.

Some systems could not generate circles at all but had to digitize the circle and treat it as an overlay. Systems with well-developed interactive graphic capabilities generated circles easily, given center point coordinates and a radius expressed in miles or kilometres. Often the resulting circle could be moved around the screen to adjust its placing. This required the existing map data base to be drawn on the screen. Some systems did this very quickly. One, however, had no fast-draw capability and required an elapsed time of more than 90 minutes to create the temporary plot file of the data base on the screen before the circles could be drawn upon it. Insertion into the data base was generally straightforward; the size of the circle in the interactive system was, however, limited to the size of the area covered by the map file being handled.

Differences in the resolution with which circles were drawn should be noted. Some systems increased the number of points with the size of the circle. Others had a fixed number of points, for example, 200. One system created each circle in four quadrants, each with a variable number of points, the maximum being 12, for a total of 48 for the full circle. Large circles were thus increasingly polygonal, with a slight diminution of circular area. The purpose of reducing the number of points was to reduce the amount of effort required in the subsequent use of the circles, particularly in overlay programs. Given the inefficiency of most of the current overlay programs, this was a laudable aim, but perhaps detracts from the real need to write overlay programs to handle efficiently large graphic elements and large data volumes of the kind found throughout natural resources inventory.

To the task of drawing grid cell nets, systems using an interactive approach appeared to offer a straighforward answer. One interactive system, however, was constrained by the length of time needed to create the temporary plot file of the data base on the screen before locating the grid cell net. Another system set up the grid cell net very quickly but ran into buffer size constraints and halted when subsequently using the grid cell net to overlay and compute areas from the data base. Small parts of the test area were partitioned and handled separately, turning the grid cell net into a cumbersome and frustrating framework for extracting data from the data base. The problem here is, of course in the overlay procedure and not in the generation of the grid cell net itself. Other systems did not have the ability automatically to generate a net of grid cells that could be topologically related to the data base. It would probably be simple to write such a program, but new software development is required.

Latitude and longitude lattice generation was not a capability demonstrated by any of the systems. However, no technical problems were identified that would inhibit such software development in other than grid-cell-based systems.

Corridors are zones of selected width along both sides of a linear feature. The linear feature can either already exist in the data base (such as a road or a stream), or it may be defined by the generation of a new line (a new road location, for example). The linear feature may be of single line width (a trail, or a narrow stream), of varying width (irregular-width streams with two banks), or some combination of these (a narrow stream joining lakes). The linear feature may be highly convoluted, with some of the convolutions forming loops that almost close (oxbows on a stream). The linear features may extend over several map sheets. Corridors should be generated by calculation, given the desired width of the corridor and the identification and extent of the feature to be followed. Overlaps and inclusions in highly convoluted areas should be handled automatically. The resulting corridors should be topologically inserted into the data base, specifically to allow subsequent measurement of the entities that they contain.

No system had a satisfactory corridor generation process. At best the methods were cumbersome and slow. Many could not create corridors along highly convoluted linear features (such as a meandering stream). The most nearly automatic system used a batch approach. The command language was powerful, but corridors around linear features of a single line width and of varying width had to be generated separately and subsequently concatenated, which slowed the process down. Interactive systems had not yet been developed to the stage where they could generate corridors around convoluted linear features. Erroneous spikes were produced which had to be edited out on the screen, again a laborious process. The subsequent procedure to insert the polygon into the data base tended to be slow. Elapsed times for generating 15 cm of corridor fell between 45 minutes and 60 minutes, for both batch and interactive methods. Other systems relied on off-line digitizing of the corridor and could not generate a corridor from dimensional commands. In general, the ability of systems to generate corridors was poor and new or modified software is required. The best present procedures are functional but cumbersome.

Data Extraction

Data extraction is the process of retrieving data from a file for use in a decision-making process. The retrieval process may be the simple reading of the file or specified parts of the file to determine facts that are already written therein. Much of the information content of a map, however, is not written on the map but is inherent in its structures. Such factors as distance apart, size of areas, direction, spatial correlation of items, shortest route, nearest neighbor, line of sight, connectivity, and contiguity, have to be extracted from the data base by measurement and comparison. Data extraction can include the analysis of complex space-attribute-time correlations which are fundamental to understanding the environment and the nature of the interrelationships of its many components.

The capabilities required by Saskatchewan are described below.

Search and identification. A very basic capability of any data handling system is its ability to read the data it contains. An extension of this is the ability to read in an intelligent manner, to search and identify specific items of data, and, in the context of spatial data systems, to find out where the items are located.

Such a search may be prescribed by attribute or spatial parameters, or both. A typical task would be to ask the system to identify items with certain attributes, say, forest stands which contain only hardwoods, more than 80 years old, and more than 25 m tall, to list and summarize these forest stands, and also to plot a map showing where the specified forest stands are located. An additional task might be to measure the area of the stands and add that figure to the list. ("Measure" and "plot" are higher-level functions and are considered separately in this report, but are typically combined with forest stand search and retrieval.)

A system should make it easy to define the item(s) being sought. It should be possible to define ranges of values, and broad categories of items for inclusion or exclusion. One should be able to choose attributes from any part of the total attribute label of any entity. The profile of characteristics being specified should be able to incorporate the full range of Boolean commands. The search operation should be initiated simply, be performed on one pass, and produce the required lists and maps in a standard format or an easily prescribed custom format.

In general, a search and identification operation must be regarded as a fundamental, and often-needed, capability of an inventory type of information system. It should be extremely easy to command and quick to carry out. Current system capabilities ranged from the awkward to the competent. Nearly all had some limitations. With the exception of those that exhibited no capability or whose software persistently failed to complete a search, the least desirable were those with a poor ability to define the items of interest or those whose search capability was limited. Some systems required that every possible individual code for every entity that needs to be identified be specified in advance. This drawback can be overlooked if a system is tested only on its ability to identify one or two specific items, but becomes readily apparent when entities with a wide range of values and a complex set of sub-attributes are sought. Some systems could only summarize their internal attribute files and could not produce a list of individual entities of interest. Some systems could search attributes but could not identify the location of the individual data elements; they could not link back to the location identifiers after an attribute search.

The majority of systems had better capabilities, but with limitations that should be identified when they are being considered. The set-up process was either batch or interactive in the different systems, and it usually took less than 5 minutes to pose a question. The best took less than 1 minute. Some unfortunately had a limited set of Boolean commands. The search process for selected polygons (three attribute criteria) in one-quarter of a typical forest inventory sheet varied from less than 1 to more than 8 minutes of CPU time. The time depended to some extent on the computer utilized, but 8 minutes of CPU is very slow. Some searches were efficient in small areas, but because system characteristics gave increasingly slower access times as larger amounts of data were handled, they would be extremely slow over a wide area. (One was estimated to take 4 days of CPU if the 1,800 maps of the Saskatchewan Forest Inventory had to be searched.)

Some systems performed the search adequately but had minor problems in printing lists and plotting maps that resulted from the search. Some could not suppress unwanted characters in the lists but had to print the complete description of each selected entity and not just the desired attributes. Others could not plot maps that showed the selected attributes only. These are annoying rather than serious problems, and could be eliminated with minor programming changes.

In general, the process of search and identification does not involve any intractable technical problems. The logic is straightforward and the software should work easily and quickly. Where inadequacies exist in present software, it should be possible to make improvements with modest effort, but it is worth checking that the facilities required in search and identification are present before purchasing software.

Measurement. The need to measure maps; to count the number of specific data items, to measure distances, to calculate areas, is easily perceived. Its importance is hard to overestimate, for the bulk of spatial information carried in the traditional map form, the essential knowledge of the relative position of things, is contained in the arrangement of the data on the map sheet, in other words, the data structure. The extraction of this type of information from a map sheet requires visual examination and if quantitative values are required, involves measurement. The state of development of spatial data structures in digital systems available in 1980 is still rudimentary. In essence, they attempt to preserve the traditional map in digital form and are not data bases where spatial relationships are more explicit. There are several reasons for this. A newly developing technology often mimics the products of the earlier technology before realizing its own capabilities. Moreover, several fundamental questions concerning the nature of spatial query, and indeed the nature of spatial relationships, have to be answered before new forms of spatial data structure, with more explicit spatial relationships, can be designed effectively. Whatever the future development, the present systems rely heavily on measurement and quantitative comparison to determine quantitative spatial values. Some of the simpler forms of measurement were tested, as described below.

Measuring distances can be expressed as the ability to measure straight line distances from a specified point to another specified point and to measure the distance along a prescribed convoluted linear path such as a road or stream. It should be possible to carry out the measurements between points and along linear features that cross the boundaries of map sheets and map projection discontinuities. A further requirement may be to measure the length of a path through three dimensions: a road through mountains, a flight path, a bore hole. However, these latter were not required by Saskatchewan.

Distance measurement is not a complex manual procedure. The set-up for the computer version should be straighforward and quick. Certainly, it should not take longer than performing the measurement manually. Where point-to-point distances are measured, the points themselves should be specified precisely and marked temporarily. The ability to attach quickly and positively (snap closure) on to already specified points, for example road intersections, is desirable. Where distances along convoluted linear features are measured, the linear features should be identified for measurement with very few actions or commands. Measurements across map boundaries or map projection discontinuities should be transparent to the user. For data retrieval associated with natural resources inventory, the repeatable accuracy of measurement should be less than ±1% of distance. The results should be swiftly forthcoming, and should relate unequivocally to the place being measured. It should be easy to generate a written record of the results if desired, particularly in cases where numerous measurements are taken.

Several systems could not perform distance measurement and new software development would be required before they could do so. The ease with which such software could be written would depend on the existing internal file structures of the individual systems. In most cases there were no logical difficulties.

In systems which now have distance measurement capability, the point-to-point measure was performed most easily. Each used interactive pointing to specify the desired locations. Ten accurate straight line measures could be determined in 10 minutes or less. Results were displayed on the screen. It was not possible with the existing software to measure distances that crossed map file boundaries or map projection discontinuities.

Distance measurement along convoluted linear features could be accomplished, but the degree of effort required to identify the linear feature to be measured differed widely between systems. If the linear feature was topologically established as a line and could be pointed to as an entity, the process was straightforward. If the linear segment was part of a more complex topological entity, such as a polygon, and had to be unlinked before it could be added to other segments to form the desired linear feature, or if it was not topologically established in the system but was simply a graphic entity not continuous with other graphic entities, then the process of assembling the line to be measured, while quite possible, was more laborious. Similarly, if the editing during the input of maps to the system was incomplete, if there were breaks or gaps in the linear features, particularly as a result of partial or imprecise edgematching procedures, then these too had to be identified and the linear feature made whole before distance measurement took place. The degree to which such work needs to be done before distance measurement takes place translates into total elapsed time for the distance measurement procedure. If the linear feature could be readily identified, measurement took 1 or 2 minutes. If complexing work was necessary, then 3 km of road could take 7 minutes to identify and measure, 6 km of convoluted stream 31 minutes, and so on. The distance measurement could be accomplished, but only at the same or slower rates than the same process could be done from a map by manual methods.

Area size measurement is the ability to measure the area of the interior of a specified polygon or polygons and, by calculation based on map scale, to convert that measure to a statement of size of the polygon(s) on the ground, in some suitable units such as hectares or acres. The resolution and the accuracy of measurement must be adequate for the decisions that result from the use of the measurement. For the forest inventory maps used, the resolution required was 0.1 ha. The acceptable accuracy of the measure related to the forestry source data was less than ±4% of any shape of polygon greater than 100 ha, and of aggregates of polygons with like attributes whose combined area was greater than 100 ha.

In practice, the task of area measurement includes specification of the items to be measured, calculation of the area measures, and the process of listing the results and summaries of the results.

All current working systems could measure areas; most could do it with acceptable resolution and accuracy. Many systems actually calculated the area of every polygon during the phase of creation of the original data base, and stored the results of the measurement for subsequent retrieval. The set-up time to obtain the measurements for a specified region and produce two types of list (one a list ranked by size of individual polygon, one a list of summed areas of certain broad forest types) took between 2 and 24 minutes in the various systems, with a median time of approximately 13 minutes. This median time could probably be reduced to 2 to 5 minutes given standard, often-used list formats that might reasonably be set up in an inventory operation.

Minor problems were often encountered. One system apparently converted to areal units incorrectly and produced figures that were in error by a factor of approximately 4.2x to 4.5x. Another could not handle the number of items on the attribute list and omitted measuring polygons that contained the last of the specified attributes. Another could not print out the full label describing each polygon measured. This type of problem, while annoying, could be corrected with modest programming effort.

A similarly annoying set of results in some systems stemmed from the inability of their edgematch routines to piece together polygons that had been arbitrarily split during the input process. These typically involved real polygons that lay across the boundaries of maps or air photographs used during input. This splitting can create an unusually high proportion of small polygons which become very apparent when area measurement results are listed. These arbitrarily created polygons do not occur in nature and can give a wrong impression of the distribution of the natural resource when a list of measurements is examined. Moreover, the percentage error in individual polygon measurement tends to increase as the size of the polygon decreases, and numerous "sliver" polygons created by arbitrary boundaries represent a potential source of error. The answer to this particular problem is to modify the edgematching and data bank formulation software so that whole polygons are created during the input phase.

A somewhat more serious problem exhibited in two systems was the sensitivity of the measurement process and other similar software to the polygonization or complexing of polygons that takes place during input. The problem is serious because, unlike the creation of small polygons at the input stage, this input error can cause the measurement process itself to be in error; in fact, polygons can actually be recorded as having negative sizes, and occasionally the measurement program ends without measuring all polygons. This is particularly pernicious if the reporting format used does not include a listing of every individual polygon measurement, and instead summarizes or accumulates totals. The spurious polygon or polygons can be hidden within the summation and the total area measurements can be seriously wrong. Up to 30% accumulated error was noted in one part of one test; it could have been much more. Again, the problem is not in the measurement process itself but stems from the sensitivity to the input process, which must be totally error-free if measurement is to proceed correctly. Clearly such systems need a reliable arc-to-polygon routine that complexes polygons automatically and identifies errors during input. They might also benefit from area measurment programs that could recognize incorrectly complexed polygons that may cause measurement errors, and bring these to the attention of the user.

Comparison. Overlay is a well-understood term in manual cartographic practice. One map is brought to the same scale and map projection as another, is transferred to transparent material which is then physically overlaid and aligned on the second map, the two are then viewed in concert, and the spatial relationships of entities on the two maps are observed or measured. In a computer system, it is the ability digitally to overlay one map on another and produce a third digital map which is the concatenation of the first two, in short, an intersection of the data sets involved. It should be possible to perform all the other data handling capabilities of the system on the concatenated digital map.

One type of intersection is the point-in-polygon overlay. Systems were required to show that they could digitize a set of labeled points, overlay that set of points on a map of polygons already in the system, and count the number of points in each polygon. As an additional task, the systems were asked to calculate the number of points per hectare in the polygons where they occurred.

Most systems did not have a point-in-polygon overlay capability. One system could not handle points, but did offer to digitize very small circles as surrogates for points, and subsequently overlay the circles. This is a slow and expensive way of handling point data. Two minicomputer-based systems did overlay points on polygons. The set-up time took between 10 and 20 minutes; CPU times were between 9 and 36 seconds for this relatively straightforward task. No minicomputer system produced the density calculation requested. One system had a full point-in-polygon capability.

The point-in-polygon overlay is not a technically difficult data handling operation. Several well-established algorithms exist; in fact, the British government has even published a report examining their relative efficiency. However, this simple capability has not been incorporated into many spatial information systems to date.

A second type of intersection is the polygon-on-polygon overlay. This classic version of overlay is an often-requested but sometimes not needed operation. Many earth scientists trained in sieve-mapping analysis and well used to performing the process manually, albeit with an extraordinary amount of effort, require that it be part of any natural resources information system they consider acquiring. Not infrequently the results needed could be gained by the overshading of one data set on another without topological intersection, or alternatively the degree of intersection of one spatial data set with another could be measured by statistical sampling techniques. There are, however, occasions when polygon-on-polygon overlay is required and it is also axiomatic (or should be) that information systems provide capabilities that match the requirements and capabilities of those using them. Accordingly, systems were examined for this function. The test required that a small map (20 cm × 25 cm) containing 22 labeled polygons be overlaid on a map of labeled polygons already in the system. A plot of the concatenated map, with concatenated labels, was requested and a list of the new polygons and their size was required. This is quite a simple test; it does not involve multiple overlays, it does not cross map boundaries or map projection zones, and it involves about one sixth or less of a standard forestry map sheet.

The systems tested had varying degrees of difficulty in completing the operation. Two systems did not finish the test, one because it did not have the software in place to attempt it, and one because its software was not capable of handling the nested polygons that occurred on the test map and caused the program to stop prematurely. One system could carry out the process partially, in that it could graphically draw one map over another, but it could not create new polygon files, it could not carry out the topological intersection, and hence it could not produce the list of results or concatenated labels.

The two systems that could produce results had their own problems. One was extremely sensitive to data volume and during the first pass took over 6 hours of CPU time to complete the test (but did complete it). A second unobserved attempt was made after redigitizing and generalizing the test map to the greatest extent possible to reduce data volume and then carrying out the overlay on partitions of the map already in the system. The total CPU time reported was a very reasonable 24.5 minutes. The extreme sensitivity to data volume is clearly a problem, but there is some possibility that the software can be further developed to overcome this difficulty. The other system was functional but awkward. Nested polygons had to be identified and linked during input. The program was sensitive to the shape and orientation *vis-a-vis* the axis of polygons on the maps. The map input phase took 45 minutes elapsed time and the overlay itself took 2 hours and 21 minutes of CPU time. The concatenated labels on the resulting map were poorly placed (several were overprinted, causing them to be illegible), although that is a labeling problem.

Overall, the minicomputer-based systems did not perform overlay well, and software development is required if this is to be an economically performed process.

The process of overlay is a difficult programming task, particularly when large areas are involved and when the density of data is high. At least one large computer-based system has solved the problems and performs the task swiftly and economically. Software development in minicomputer-based systems faces the restrictions imposed by minicomputers themselves, but it is probable that they can and will be resolved in the near future.

Interpretation

Many of the above three functions, data manipulation, generation, and data extraction, can be used in concert, in a manner specified by a weighted algorithm, to arrive at interpretations of the data contained on maps. These can be used to identify optimal conditions, determine suitability for specific purposes, and determine conditions that are most desirable for a set of objectives. The nature of the interpretation depends on the set of prespecified rules. In this way, for example, the best route between two points, the most suitable land for reafforestation, the most desirable among several sites for a new park, and similar conditions, can be recognized and evaluated. The specified requirement was only for straightforward data handling functions in the data manipulation, generation, and data extraction categories, and no tests of interpretation capability were called for.

CONCLUSIONS

The present state of development of the systems represents a substantial capability for handling spatial data, one that in many cases already exceeds in economic terms the manual capability for carrying out the same operations. Development is progressing apace. It is an intensely competitive field and funds and resources are being expended continually to make existing hardware and software more reliable and efficient and to add new capabilities.

Conclusions about individual systems can be drawn only with respect to their ability to handle a certain set of requirements. In the case of the Saskatchewan forest inventory, these requirements were not met and the recommendation was made that none of the systems tested be purchased at that time. Much of the software needed by the forest inventory is under development but is not yet reliable. Some software exists, but is so slow that it would put an unnecessary and heavy load on the computing facilities; it should be rethought and rewritten before anyone buys it.

The software required by Saskatchewan (with the relatively minor exception of the block adjustment procedure) is all general-purpose spatial data handling software suitable for many users in the field and is under active development by many of the potential vendors. There is every evidence that this software will be forthcoming and will be suitable for use in the Saskatchewan forest inventory in the foreseeable future.

APPENDIX A

FUTURE REQUIREMENTS FOR DATA HANDLING, SASKATCHEWAN FOREST INVENTORY
(As estimated in October 1979)

A A NEW PHOTOGRAMMETRY SUBSYSTEM

The objective of this subsystem is to convert the geometric and interpreted forest stand boundary and attribute data directly from air photographs to a digital form, to perform digital aerotriangulation and transformation on those data, and to produce an acceptably error-free digital file for all Saskatchewan in a form amenable to subsequent digital cartographic manipulation and analysis (see Section c below).

1 *Input Capacity*

Maximum load: 13,000 stereo photo models per year (based on 300 maps) (1 print in each pair fully annotated).

Normal operational load: 9,600 stereo photo models per year (1 print in each pair fully annotated).

Typical number of coded photogeometric points per annotated print: 9.

Typical number of forestry polygons per annotated print: 75.

Typical number of alphanumeric characters and/or symbols per forestry polygon: 10 (maximum 25).

2 *Output Map Accuracy Relative to Source Data*

Horizontal error as expressed by the fitted accuracy (residual error) of the block adjustment for the benchmark area:

$< \pm 12.5$ m on the ground, $< \pm 1$ mm at a mapping scale of 1:12,500.

Relative positional errors between adjacent features resulting from digitizing, transformation, and edgematching of boundary line data: $< \pm 1$ mm at a mapping scale of 1:12,500.

Logical and/or topological errors: zero.

B A NEW MAP DIGITIZING SUBSYSTEM

Given that the photogrammetry subsystem is not implemented or does not produce a digital version of the Saskatchewan forest information, there will be a requirement to convert the manually produced forest inventory maps into an acceptably error-free digital file in a format amenable for subsequent digital cartographic manipulation and analysis (see Section c below).

1 *Input Capacity*

Maximum load: 300 map sheets per year.

Normal operational load: 220 map sheets per year.

Typical size of map sheet (100 km² @ 1:12, 500): 80 cm × 80 cm.

Typical number of forest polygons per sheet: 1,800.

Typical number of alphanumeric characters and/or symbols per forestry polygon: 10 (maximum 25).

Map projection: 6° Universal Transverse Mercator (3 zones involved).

2 *Digitized Map Accuracy Relative to Source Data*

a If boundaries are digitized, positional accuracy must be $< \pm 1$ mm at source map scale;

b If map converted to grid cell format (dot method), maximum size of grid cell = 5 mm² and all forest stands must be recorded;

c Logical and/or topological errors: zero.

NB Subsystems A and B must include the editing and reformatting steps necessary to create a digital cartographical data base that covers the whole Province of Saskatchewan, allows edgematching of map sheets within that cartographic data base, and allows source map sheet boundaries to be ignored when calling for areas to be analyzed, for example, to change scale and produce one map at, e.g., 1:50,000. All functions described in c below must be able to be carried out on any part or all the whole province-wide digital file.

C A NEW MAP DATA HANDLING SUBSYSTEM

The following capabilities are required to be available in the map data handling subsystem in the first year of operation. No further capabilities will be required for 5 years.

1 *Area Size Measurement*

Measurement of each source document polygon, summarized by attribute and by larger areas known at time of measurement, e.g., total of one stand type within each map unit. Accuracy relative to source data: $<\pm4\%$ of any shape of polygon greater than 100 hectares and of aggregates of like polygons whose combined area is >100 hectares.

2 *Updating*

Updating of digital data base with new polygons and polygon nets and/or new attributes to add to or replace previous regions in the data base. Previous polygons and attributes to be maintained separately in digital data base.

New polygon(s) location data to be derived from 70 mm Hasselblad photography annotated with new polygon(s) boundary and attribute data, and other sources.

Updating of 10% of total map area per year in anticipated.

3 *Derive New Maps*

Search and remeasurement of forest stand polygons by:

a newly defined area of any size and shape, and

b attribute or attribute profile ("and" and "or")

Generate lists, summaries and plot maps.

4 *Corridor Analysis*

This is a special case of 3a above and systems will be judged on the ease with which the corridors are defined to the system.

Measure size of forest stand polygons by attribute contained within a specified distance (e.g., 15 m, 30 m, and/or 90 m) of a linear feature (e.g., a stream, existing road, or new road location).

The system must allow the corridors to be defined to a resolution of $\pm20\%$ of corridor width.

6,000 cm of corridor analysis per year is anticipated.

5 *Generate*

Arbitrary regular areas and overlay on data base to define regions for retrieval:

a Circles – with center and radius only specified;

b Grid cells – with dimension and orientation only specified;

c Latitude and longitude net.

6 *Reclassify, Merge and Dissolve*

Reclassify attributes according to a set of rules within a specified region, and/or

Merge adjacent polygons with specified attributes, dissolve interior boundaries, and reclassify.

7 *Scale Change*

To be performed in computer rather than at plotting table (Weeding routines and/or look-ahead line smoothing will be incorporated in a scale reduction),

Range: 1:12,500 to and from 1:1 million (approx. f × 8).

8 *Overlay*

The ability digitally to overlay one polygon type map on another and produce a third digital map which is the concatenation of the first two. All capabilities listed above to be able to be performed on the concatenated digital map.

The ability similarly digitally to overlay a set of irregularly spaced points or a set of grid cells on a polygon map.

9 *Distance Measurement*

The ability to measure straight line distances from a specified point to another specified point.

Distance measurement along linear features.

Perimeter measurement of polygons is not required.

Automatic contained centroid allocation for polygons (to which distance measures could be taken) is not required.

10 *Background Data*

Establish a separate digital file of linear data such as streams and roads and 1 km grid cells and be able to plot that background data as output on top of polygon data on demand.

This data file to be updated with the capabilities noted in 2 above.

11 *CRT Browse*

Interactive search and retrieval with all of the capabilities listed above in Section C.

Color CRT(s).

One CRT at FIS, a second CRT at a remote location (Fire Protection Div.) Hard copy unit at each CRT; b & w adequate, color preferred.

12 *Line Plotting and Attribute Labeling*

Drum, roller or flat bed device;

Medium precision;

High-speed.

13 *Drafting Capabilities*

The following drafting capabilities are required:

Maps: Interior polygon labeling, exterior polygon labeling with arrow to polygon, cross-hatching of various densities, thick and thin lines, dotted and dashed lines, single and double lines, special symbols, external map titling, color pen plotting.

Diagrams: The ability to produce graphs, histograms, bar graphs from statistical data.

14 *Dicomed Color Film Plotter Output*

Given a map data handling subsystem based on grid cells (dot method), the ability to produce low-cost color grid cell pictures on 70 mm film would be required.

PART 2
LAND SUITABILITY STUDIES

Land use suitability mapping is a technique which can help find the best location for a variety of developmental actions given a set of goals and other criteria. The mapping technique is based on natural and human processes and analyses the interactions among three sets of factors: location, development actions, and environmental effects. The technique can yield three types of maps: a map showing what land use will cause the least change in environmental processes, a map showing qualitative predictions of environmental impacts of proposed developments, and given certain developmental actions to be carried out and specific environmental actions to be controlled, a map showing the most and least suitable locations for those actions. Two case studies are given in which suitability mapping was used as a tool to help local, state, and federal agencies formulate public policy regarding land use decisions in coastal Southern California.

Computerised Land Use Suitability Mapping

John Lyle
California State Polytechnic University

and ## *Frederick P. Stutz*
San Diego State University

Over the past twenty years, the methods and practices of rural land use planning, like those of many applied social sciences, have been engaged in a revolution. Progress on at least two fronts has been so dramatic as to shift the basic theories and attitudes of which land use decisions are based. The first of these fronts is the inclusion of environmental concerns, and the second is the application of more systematic methods.

Though land use planners have generally included such variables as soils, vegetation patterns, and microclimates in their planning processes for at least a century, most plans for urbanisation (where there were any plans at all) were based mostly on political and economic concerns. During the 1960s, this began to change noticeably with the sudden growth of environmental awareness among the general public. In the early '70s, the passage of the National Environmental Policy Act (NEPA) made environmental concerns an essential part of all planning processes related to federal lands and projects. NEPA's offspring in a number of states, most notably California and Massachusetts, imposed similar requirements on private efforts.

The specific requirement of these laws was for an Environmental Impact Statement (EIS). But thoughtful planners soon realised that the EIS could hardly serve its purpose if it were simply an addendum tacked onto the planning process. Developed as an integral part of the plan, the EIS could be a most useful tool. Planning decisions could be based on impact predictions made along the way. This meant that information concerning the environment had to be gathered and organised at the very beginning of the process and integrated with planning efforts.

Since for a sizeable project, the information was usually complex and voluminous, orderly techniques were needed to deal with it. So it was fortunate that the second front, that of systematic methods and computer cartography, had been progressing on a parallel course.

Systematic methods of land use planning draws heavily on the technical repertoire developed by the systems theorists since World War II. The steps in the planning process, all moving in the direction of clearly defined objectives, are carefully identified in advance. Information is organised into discrete categories and applied to the development of models. The models are used in defining alternatives, and choices among alternatives are made on the basis of their effectiveness in meeting the objectives. Various systematic tools and techniques, including flow diagrams and matrices, are usually used along the way. But the most important tool is the computerised map.

Such methods are useful, not only in keeping the information organised, but also in communicating with and convincing a larger audience. The complex social and political environment that has evolved from years of conflict over environmental issues demands planning processes that can be clearly explained and defended. Especially in governmental planning efforts, citizens groups usually participate, and effective participation requires an understanding of what is being done. Here, the simple computer map is invaluable. Roles must be defined with clarity, and information, attitudes, and values must be articulated by various groups and individuals and incorporated into the structure of the plan where each is appropriate.[1]

Furthermore, since most major governmental planning efforts are now challenged in court, it is important that they be able to withstand judicial scrutiny. The structural coherence and logic of the process are especially important in this context, as is the reliability of the information, the sources of the controlling values, and its simplicity.

At the heart of the environmentally sensitive, systematic land planning and in the pivotal position between analyses and the definition of alternatives, are mapped models of the type that is coming to be generically termed the suitability map. A suitability map assesses the ability of each increment of land under study to support a given use. As the technique is now developing, the assessment of suitability is commonly based on predictions of the results likely to come about if a certain development is placed on a particular piece of land. Thus, in suitability mapping, environmental impact analysis and land planning can be effectively merged.

The purpose of this paper is to demonstrate how this can be accomplished, specifically to show the use of a new and relatively simple suitability weighting and mapping process for rural land planning. Two case studies developed by the authors for government agencies in Southern California will be presented. The output of the land suitability analysis is a set of maps, one for each potential land use, showing the level of suitability attributable to each land cell. These maps, along with others, formed the bases for land use planning for the areas concerned.

LITERATURE REVIEW

The literature of rural land use planning methods is sparse. This is partly because, until recent years, relatively little work was done in the field that could be said to have a definite methodology. And much of what has been done, both by private consultants and university researchers, has not been published.

Nevertheless, the history goes back at least sixty years. Steinitz has traced the use of map overlays to develop a plan as far back as an ambitious undertaking by landscape architect Owen Manning which was published in 1923.[2] Manning developed a series of 363 data maps for the entire United States and then overlaid these in some way to shape a land use plan. But that was long before clearly defined methods were considered worthwhile, and Manning never explained exactly how he used his maps to derive a plan.

After that, applications of overlay techniques appear from time to time in the literature, but there are no notable advances until the development of sieve mapping in England.[3] This was an orderly process using map overlays to define land areas that, for various reasons, should not be developed. It was widely used in the large-scale planning of the British new towns and other development after World War II.

During the 1950s, the systematic analysis techniques that had grown out of World War II weapons research were finding widespread applications in a variety of fields, especially those related to business and engineering. But they were little used in land planning and design until 1964, when Christopher Alexander published his seminal work, *Notes on the Synthesis of Form*.[4] Alexander had worked with Marvin Mannheim on highway route selection studies using systematic processes for combining locational factors. In his book, he presented such a process for the planning of a village in India.[5] He did not use map overlays, but set out a specific set of criteria for locating various activities, and used a matrix to analyse their compatibilities and conflicts.

The clarity and logic of Alexander's process had obvious application to the increasingly complex issues of land use planning that were emerging during the 1960s. And perhaps more importantly, they offered convincing support for the results that the process produced, a feature that had considerable value in contentious times.

Ian McHarg's *Design with Nature* brought systematic land use planning using map overlays to a broad audience when it was published in 1969.[6] His examples had the virtues of being both systematic and simple enough to be easily understood and replicated.

Most of the criticisms of McHarg's approach were based on two major difficulties: his assumption, to use McHarg's own phrase, of 'ecological determinism', and the equal weight that he gave to all of the variables considered.[7]

Gold articulated the former criticism in his review of McHarg's work, arguing that the location of given activities must be established to some degree by the character of the activities rather than exclusively by the 'presumption for nature'.[8] Ivan Marusic describes McHarg's method as an 'absolutisation of values in an otherwise relative social and physical space.'[9] The second shortcoming of McHarg's work is largely due to the limitations of hand-drawn overlays. If we then assign different weights to the data on each map, the task can become virtually unmanageable.

During the late 1960s, in attempting to overcome these shortcomings, a number of researchers began working with computer-aided systems. The SYMAP computer mapping programme was coming into use at this time, and it was applied to a variety of land planning problems. Carl Steinitz and his colleagues developed a simplified version of SYMAP, which they called GRID,[10] and used it to comparatively analyse different systematic land planning methods that had been developed over the decade of the 1960s by various researchers, planners (including McHarg) and planning agencies.[11]

The Steinitz group went on to develop a modelling concept for urbanisation, based on resolution of 'attractiveness' factors, which make a site desirable for urban uses and 'vulnerability' factors, which deal with ecological impact.[12] Patri *et al.* later used ecological vulnerability as a basis for their 'early warning system', which identifies lands that are likely to be vulnerable to certain kinds of change if they are developed in certain ways. [13]

Some recent work in the field has focused on linking suitability models with more precisely quantified predictions of environmental impact. According to Hopkins, the models most commonly applied for this purpose are those already in use for other purposes.[14] These include the universal soil loss equation, runoff calculations and intervisibility models.

Largely because they are familiar, these tend to find ready acceptance, and they fit neatly with EIS requirements. But because they tend to break down the physical environment into discrete factors and to deal with these in isolation, there is a serious danger of losing the holistic quality, the expression of the environment as an interacting whole, that is the hallmark of the geographic approach and that distinguishes the more inclusive, though less precisely quantified methods.

A FOUNDATION IN THEORY

At this point, a decade of experimentation and practical applications has served to define the essential characteristics of suitability mapping, and their potentials and limits. The principal difference between the approach presented here and earlier deterministic work is the concept that map overlays should not lead directly to a plan. Rather, a series of modelling steps should be applied to define interaction among collections of data describing geographically distributed variables and weighted values related to given land use parameters. The resulting maps, along with other types of information, provide bases for the decisions which shape a plan.

In this approach, as in those described above, ecological factors are considered as fundamental determinants of land use patterns. That is, it is assumed that variations in ecological character render some areas of a landscape more suitable for supporting a given human activity than others, and that the difference can be very important to environmental quality. Indeed, there is strong evidence that ecologically unsuitable locations are responsible for a great many of the environmental dysfunctions that plague our society. This being the case, we can build a strong argument for assigning assessments of ecological suitability a primary role in any planning process. But rarely will they have an exclusive role.

Another distinguishing feature of this approach is the strong emphasis placed on organisation of information concerning three component sets of factors and analyses of interactions among them. The problem to which we are directing our attention, then, is finding locations most suitable for various land uses. Land use suitability, we propose, can be defined by analysing the interactions among these three factors: location, development actions, and environmental effects. That is, given certain developmental actions to be carried out and specified environmental effects to be controlled, we can map the most and least suitable locations for those actions.

This concept incorporates a close connection between suitability mapping and environmental impact prediction and at the same time casts both in a larger perspective. It uses the term 'effects' rather than 'impacts' because the latter carries with it a negative bias, a subtle implication that effects are necessarily adverse. The term 'developmental actions' is preferable to 'land use' because it is useful to know the specific activities that act on the land to bring about particular effects.

INFORMATION FOR MAPS

The suitability mapping process, then, begins with a collection of information forming coherent descriptions to these three component sets of factors, and a means of defining the connections among them. These three sets of information—describing locations, developmental actions, and environmental effects—form the three legs of the tripod on which the suitability mapping process is built (*Figure 1*).

It is important to recognise that we do actually need all three. Since suitability mapping has become more widespread, efforts to simplify and hasten the process have often resulted in failure to take into account either developmental actions or environmental effects. Too often, locational maps are simply overlaid without definition of either the uses or environmental factors being considered. This can lead to meaningless conclusions. It happens partly because the collection on locational information is the most visible, usually more

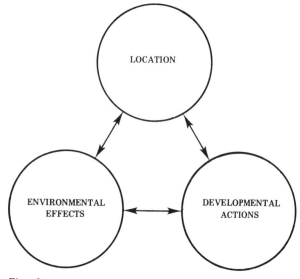

THE PLANNING SYSTEM IS BASED ON INTERACTIONS OF:

Figure 1.

extensive and complex than the other two. This collection is essentially a map file showing geographic distribution of the locational variables that interact with developmental actions and environmental effects. Generally, at the most basic level, these include the horizontal layers of the biosphere, starting with bedrock composition and other geological factors, and proceeding upward through soil types, hydrology plant and animal communities and microclimates. To this we often add human contributions to the environment such as existing land uses, transportation routes, accessibility, and political and social boundaries.

Any or all of these above land factors can be almost indefinitely expanded. Soil types can be grouped by various characteristics, such as bearing capacity, expansion potential, or porosity. Or they can be divided into capability classes according to the US Soil Conservation Service system. Hydrology might include surface water features only, or it might also include ground water basins and their capacity and recharge areas. Often, streams and tributaries are mapped to second, third or fourth orders. Sometimes rates of flow are included. Man-made factors can likewise vary in their level and type of detail.

The character, precision, and inclusiveness of the locational data base, then, varies according to its intended use. In general, we can distinguish between two types, which we can call the one-time, one-use type, and the many-time, many-use type. As the name implies, the one-time, one-use type is prepared for one application, to produce one map or set of maps. The many-time, many-use data file is assembled for a number of applications over a long period. The variety of specific uses is usually unpredictable when the data are assembled, so a broader range of factors is included. Organisation and mapping is done in a way that facilitates retrieval and use of the data.

Two basic types of techniques are available for computer mapping: the polygon and the grid. In the grid mapping process, the area is divided into square grid cells located by co-ordinates, and data are recorded for

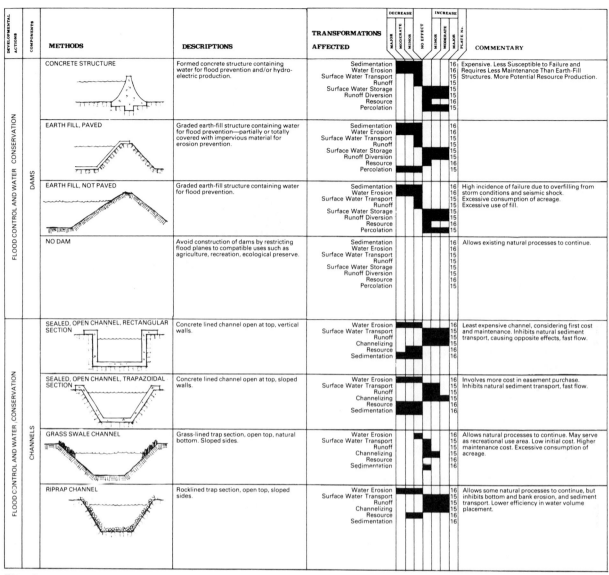

Figure 2.

each grid cell (*e.g., Figure 8*). This makes it possible to use the line printer as a drawing device. Since each grid cell can be considered as a unit, it also facilitates analyses of interactions among variables on a unit by unit basis. In any case, data recorded by use of a digitiser and stored by polygons can be easily converted to grid cell form. So, given the present day state of the art, it is probably best to record and store data in polygon form and then convert to grids for application to suitability models.

Data sources also exert some influence on choices among techniques. The first set of information needed is on land variables. Major efforts in the mapping of variables over the last two decades have made locational data far more available than before. The soil conservation service has mapped soil types over much of the United States, and the Geological Survey has made topographic and hydrologic data widely available. Multispectral remote sensing is likely to greatly expand both the amount and the availability of data in the future.

The second collection of information is that describing developmental actions, which we can define as human activities occurring on the land which may bring changes in the environment. It is often useful to divide them into two categories: capital and operational.

Capital actions are those that invest energy and natural resources in physical alterations and additions to the landscape. Operational actions involve human use of the environment, usually without specific intent to alter it. For example, construction of a road requires several capital actions including cutting, filling and paving. These occur once, bringing immediate major changes, which are likely to be more or less permanent. The use of the road after construction will involve several operational actions, of which driving motor vehicles and maintenance activities are the most common and obvious. These bring ongoing effects, which are continuously renewed. Often, they are cumulative in nature. For the San Diego Coastal Plain project, which will be more fully described in the case study and modelling sections of this article, developmental actions were listed and described in a series of 42 charts (*e.g., Figure 2*). These were developed to provide simplified comparisons of the environmental effects brought about by different methods for producing a developmental action. Forty-

69

cutting	energy generation
filling	energy consumption
excavation	sewage disposal
dredging	outdoor lighting
soil retention	roof drainage
retaining walls	site drainage
shoreline protection structures	fences
vegetation removal	paving
vegetation introduction	pest control
animal species removal	fertiliser application
animal species introduction	weed control
dams	groundwater extraction
channels	off-road vehicle use
walkways	active recreation
settling and debris basins	passive recreation
spreading grounds	irrigation
power transmission lines	aquaculture
power transmission structures	soil cultivation
demolition	automobile operation
building foundations	solid waste disposal
building superstructures	

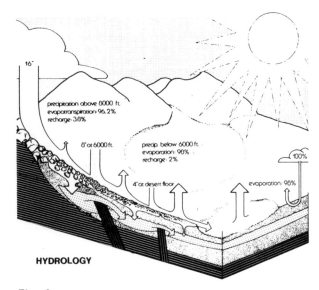

HYDROLOGY

Figure 3.

two developmental actions were included (Table 1).

For each developmental action, of course, we can identify a number of technical means. And environmental effects vary with the technology employed. In some cases, it is useful to take these variations into account. For this purpose, the developmental actions data may include a range of techniques. For example, several technical means are available for carrying out the developmental action of flood control channels (*Figure 2*). These are all standard methods commonly used in landscape construction. In addition to describing available techniques, it gives an indication of the intensity of effects likely to be brought about by each technique. This provides a practical linkage with the information concerning environmental effects.

The third collection of information, that describing environmental effects, may be assembled in an almost unlimited variety of ways. For one-time, one-use applications it is often adequate to use short lists of effects that are important to the situation at hand. When considering limited sets of actions within small areas of land, we can often determine the potential effects fairly easily on the basis of local experience.

At the other extreme, where we want to generate large numbers of suitability maps for an unpredictable variety of uses over a long period of time, more complete descriptions of the ecological systems operative within the area of concern are often useful. Such descriptions can be relatively simple (*e.g., Figure 3*). This represents the hydrologic system as it functions in a desert foothill study area that was the subject for a project carried out for the US Bureau of Land Management.[15] In this case, the modelling process had to be very simple for application under field office conditions.[15]

Far more elaborate is the flow diagramming technique used for the San Diego Coastal Plain project (*e.g., Figure 4*). Here, it was important to have a holistic system for including all the major elements of the subject ecosystem. The technique is based on the symbolic language developed by Howard T. Odum for tracing energy use.[16] Ecological processes are represented as flows of materials and energy. The symbols indicate specific ecological operations, which for convenience are grouped into six categories: inputs and outputs, storages, workgates, photosynthesis, animal respiration and

TABLE 2

Transformation: Percolation

INPUTS: Condensation   OUTPUTS: Seepage
Surface water import   Soil water export
Surface water storage   Soil water storage

Related Dev. Action	Related Land Var.	Relative Importance	Attribute Sensitivity Range High	Low
Vegetation Removal and Introduction	Vegetation Type	1	Grass	None
	Soils (run-off)	2	High	Low
	Slope	3	0–5	50+
	Plant Climates	4		
	Rainfall	5	35″	10″
Structures	Soils (run-off)	1	Low	High
	Slope	2	0–5	50+
	Rainfall	3	35″	10″
Site Planning Foundation Systems	Soils (run-off)	1	Low	High
	Slope	2	0–5	50+
	Rainfall	3	35″	10″
Walks	Soils (run-off)	1	Low	High
	Rainfall	2	35″	10″
	Slope	3	0–5	50+
Roads and Parking	Soils (run-off)	1	Low	High
	Rainfall	2	35″	10″
	Slope	3	0–5	59+
Cut	Soils (run-off)	1	Low	High
	Slope	2	0–5	50+
	Vegetation	3	Grass	None
	Existing Land Use	4	Open	Urban
	Rainfall	5	35″	10″
Fill and Compaction	Soils (run-off)	1	Low	High
	Slope	2	0–5	50+
	Vegetation	3	Grass	None
	Existing Land Use	4	Open	Urban
	Rainfall	5	35″	10″
Excavation	Soils (run-off)	1	Low	High
	Vegetation	2	Grass	None
	Existing Land Use	3	Open	Urban
	Rainfall	4	35″	10″
Soil Retention	Soils (run-off)	1	High	Low
	Slope	2	50+	0–5
	Vegetation	3	None	Grass
	Rainfall	4	35″	10″
Retaining Walls	Soils (run-off)	1	High	Low
	Slope	2	50+	0–5
	Vegetation	3	None	grass
	Rainfall	4	35″	10″

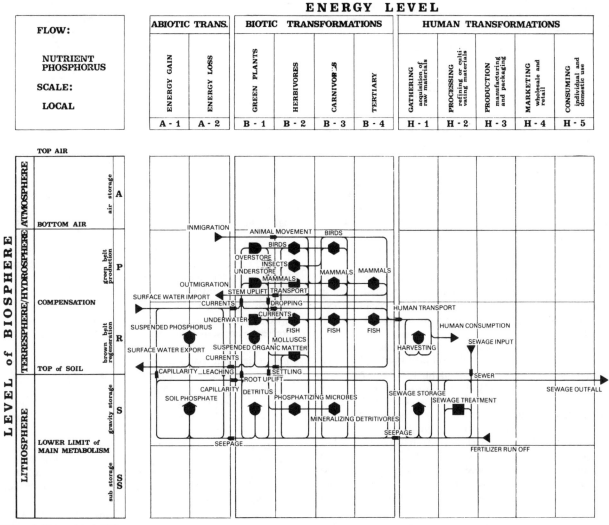

ENERGY LEVEL

FLOW: NUTRIENT PHOSPHORUS SCALE: LOCAL	ABIOTIC TRANS.		BIOTIC TRANSFORMATIONS				HUMAN TRANSFORMATIONS				
	ENERGY GAIN	ENERGY LOSS	GREEN PLANTS	HERBIVORES	CARNIVORES	TERTIARY	GATHERING acquisition of raw materials	PROCESSING refining or culti- vating materials	PRODUCTION manufacturing and packaging	MARKETING wholesale and retail	CONSUMING individual and domestic use
	A - 1	A - 2	B - 1	B - 2	B - 3	B - 4	H - 1	H - 2	H - 3	H - 4	H - 5

Figure 4. Biogeochemical Processes.

economic enterprises (*Figure 5*).[17] The lines define the interrelationships among them.

In this case, such a diagram was constructed for each major flow, which provided a means of identifying and summarising the important ecological operations. Once identified, these could be analysed in relation to developmental actions and locational variables.

Once the three basic collections of information are established and organised, it remains to identify the interactions among them. That is, how does a factor in one category influence factors in the other two categories? This is accomplished by means of the Transformation Key Chart. For each transformation, the Key Chart lists the developmental actions with which it interacts, and gives other related information. The example is the Key Chart for water percolation, as developed for the San Diego Project (Table 2). From this, we can see that sixteen different developmental actions are likely to affect surface runoff, and for each of these, between two and six land variables will influence the extent of change.

These key charts can be included in the computer data base. Beginning with the information in any column, a sort program can be used to identify the related factors in any other column.

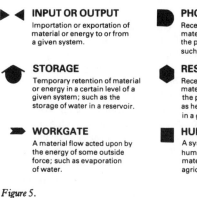

INPUT OR OUTPUT
Importation or exportation of material or energy to or from a given system.

STORAGE
Temporary retention of material or energy in a certain level of a given system; such as the storage of water in a reservoir.

WORKGATE
A material flow acted upon by the energy of some outside force; such as evaporation of water.

PHOTOSYNTHESIS
Reception and processing of materials and energy through the process of photosynthesis; such as in green plants.

RESPIRATION
Reception and processing of materials and energy through the process of respiration; such as herbivores and carnivores in a grazing food chain.

HUMAN ENTERPRISE
A systematic or purposeful human activity, transforming material and energy; such as agriculture or manufacturing.

Figure 5.

TWO EXAMPLES OF SUITABILITY MAPPING USED BY GOVERNMENTS

To demonstrate the application of these tools and the modelling process in practice, we will present two case study projects in which these methods were actually used to help create a land use plan. The first is the San Diego Coastal Plain project and the second is the Vandenberg Air Force Base project.

71

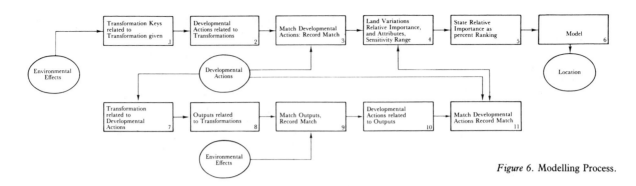

Figure 6. Modelling Process.

The San Diego Coastal Plain project was undertaken as a component of San Deigo's Integrated Regional Environmental Management (IREM) programme, which was funded by the Ford Foundation. IREM's purpose was to develop new techniques and processes for the management of the physical environment on a regional scale by local government. The 345-square mile study area covered a semi-arid, sparsely vegetated, gently rolling stretch along the coastline of the Pacific Ocean north of the City of San Diego, in the direct path of urbanisation.

Vandenberg Air Force Base covers an area of about 98,000 coastal acres some 300 miles to the north of San Diego, midway between Los Angeles and San Francisco. It is located in the ecologically important transition zone between southern and northern California. Because the Air Force programmes require large areas of vacant land to absorb the effects of missile launching, about 65 *per cent* of the land area is still in its natural state. The situation here required a collection of information and the process for using it as a basis for land use decisions and environmental impact prediction related to development of the base, particularly with respect to the new Space Shuttle facilities.

In each of these cases, while locating and predicting the effects of immediately foreseeable developments were matters of particular concern, the larger purpose was to institute an ongoing land management process. Thus, both are examples of the many-time, many-use system, using the computer for storing and processing data.

In order to provide for both very broad and very detailed determinations, these projects use a modular organisation of grid cells. The largest module represents the regional scale. Maps at this scale are generally used in making broad policy-related decisions, such as determining general areas in which urbanisation should be encouraged, or pointing out directions for expansion of support facilities, or in formulating growth policies. For these purposes, maps are intended to describe general patterns rather than precise location. So, accuracy with respect to points or lines on the ground was not a major concern at this scale. Data did not need to be very precise. The 1000 foot square grid cell was accurate enough for maps of these types, but not significantly more accurate than needed.

There are some situations, however, in which maps are used to establish suitable locations or to evaluate proposed locations for specific developments. In these cases, it is important to establish the boundaries of areas defined by maps with respect to property or jurisdictional boundaries or to actual physical features. For this purpose the 1000 foot grid cell is too crude. So provision is made for subdividing the larger cell into nine equal parts, which yields cells 333.3 feet square, or 2.54 acres in area. Since this is accurate enough for land use planning applications, we called it the planning unit scale. For purposes of the Coastal Plain project, planning units coincided with watersheds, which were areas of fifty to one hundred square miles.

Also included in both projects were a number of site specific issues; especially important were those related to tolerances of sensitive wetlands. For these, the planning unit cell was still too large. So it was subdivided again into nine equal parts to produce a 0.28 acre cell, 111.1 feet square. This was accurate enough for site planning, and for virtually any planning application.

The major difference between the two projects is in the character of the information base. The unusually wide range of future uses anticipated for the San Diego Coastal Plain required an unusually extensive collection of geographic information. At Vandenberg Air Force Base, the future uses were more restricted, but the ecological vulnerability was greater, and the available data was less complete. So, in this case, a major effort was expended on data gathering and analysis. Quantitative description of the plant and vertebrate species present in each vegetation community was determined by establishing 34 permanent quadrants, which coincided with the planning unit scale grid cells. Populations in these quadrants were sampled at regular intervals during the study as a basis for describing existing community structure and predicting future changes.

Besides vegetation cover, the locational data base for Vandenberg includes soil types, slopes, exposure, elevation, and marine meteorological, and geological data. Much of this information was interpreted from aerial photographs, using over 150 sites for ground truth verification. Maps showing distributions of soil and vegetation types as well as for a dozen other variables were constructed.

WEIGHTING THE MAPS

As already explained, these tools may be used in several different ways to derive at least three types of models. If the location and proposed developmental actions are known, then potential environmental effects can be qualitatively listed. If the location and environmental effects that are to be limited are known, then the permitted developmental actions can be determined. Finally, most important and useful is the suitability model; that is, when the proposed developmental actions and the environmental effects that are to be limited are known, we can map the most and least suitable locations. The suitability model, then, is derived from the criteria that are established at the outset, the information contained in the tools, and the judgment of the model builder.

In situations like those of the San Diego Coastal Plain and Vandenberg projects, where suitability mapping is to serve as an ongoing management tool, it is important that the weighting process be described in a precise, step-by-step manner (*Figure 6*). Consistency and continuity are essential. The suitability modelling process for these projects is briefly described as follows:

GIVEN: (1) Environmental effects to be minimised (stated in terms of transformations to be controlled) in priority order.

(2) Developmental actions to be carried out.

FIND: Most or least suitable locations.

STEPS		SOURCE TOOLS
1A	List transformations related to given developmental actions.	Developmental action key
1B	List developmental actions related to given transformations.	Transformation key
1C	List effects (outputs) related to transformations in 1A.	Transformation key
2A	Match list of developmental actions in 1B with those given and record those that match.	
2B	Match list of effects (outputs) in 1C with those given and record those that match.	
3A	List locational variable, relative importance and attribute sensitivity range related to development actions in 2A.	Transformation key
3B	Determine where interactions occur between locational variables and transformations in 2A and 3B.	
3C	Assign a range to each variable according to its interactions; ranges must total 100 *per cent*.	Scoring matrix
3D	Assign a score to each attribute based on its sensitivity range.	Transformation key
4	Print weighted map based on these scores.	Data Base

The most complex and difficult of these steps is, of course, 3D, the assigning of scores. These scores control the suitability distribution on the map. Since land

Figure 7. Agriculture Model

ENVIRONMENTAL EFFECTS

RANGES

LAND VARIABLES	Sediment-Transport	Nutrient Transport	Pesticide Transport	Erosion	Productivity	Disruption of Wildlife Habitats		Total
Vegetation Types						5	=	5
Slope	5	4	5	5			=	19
Fertility Classes					30		=	30
Water Features	2	3	5	2			=	12
Rainfall					10		=	10
Runoff Potential	3	8	10	3			=	24
Total	10	15	20	10	40	5	=	100

SCORES

Vegetation Types	Attribute Code	0	2	3	4	5	6	7	8	9	10	11	12	13	14	
	Model Value	0	1	0	2	3	2	3	3	5	5	0	1	4	4	
Slope	Attribute Code	0	1	2	3	4	5	6	7	8						
	Model Value	0	1	3	5	8	11	13	15	19						
Fertility Classes	Attribute Code	1	2	3	4	5	6									
	Model Value	30	24	18	12	6	1									
Water Features	Attribute Code	0	2	3	4	5	6	7	8	9	10					
	Model Value	0	6	12	6	12	12	6	12	12	12					
Rainfall	Attribute Code	1	2	3	4	5	6									
	Model Value	10	10	10	5	5	1									
Runoff Potential	Attribute Code	1	2	3	4											
	Model Value	1	8	16	24											

73

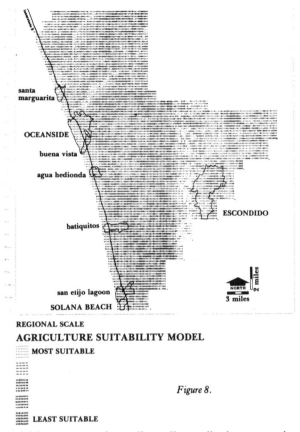

REGIONAL SCALE

AGRICULTURE SUITABILITY MODEL

MOST SUITABLE

Figure 8.

LEAST SUITABLE

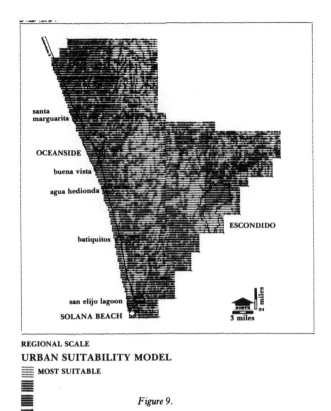

REGIONAL SCALE

URBAN SUITABILITY MODEL

MOST SUITABLE

Figure 9.

LEAST SUITABLE

variables are not, in reality, all equally important in determining environmental effects, the model will be considerably more reliable if it can incorporate varying degrees of importance or weights, than if all variables are assumed to be equally important. This is a major advantage that computer mapping holds over manual techniques. Manipulation of different weights with manual overlays is so cumbersome as to be generally impractical, while the computer can manipulate even very complex weighting systems with ease.

However, the assigning of land variable weights requires judgments that are often difficult to support with precise data. Suitability modelling involves a variety of factors far too complex, given the present state of the art, to be described entirely by absolute values. This can lead to arbitrariness and subjectivity in relation to weighting processes if they are not carried out in an orderly and thoroughly consistent way. With a controlling framework and responsible, knowledgeable participants, the process becomes one of systematic decision making. It is important that the process always be explained and that a clear framework be provided for converting professional judgments into numerical weights in an orderly sequence.

Among the techniques used for reaching agreement among groups of knowledgeable people developing scores is the delphi technique. Where this technique is used, each of the people involved is asked to derive a recommended score independently. These are then plotted on a straight line, and usually they cluster very tightly. When this happens, the average value may be used as a virtual consensus score. Sometimes, however, it happens that the scores are widely distributed between two extremes. In such cases, the respondents are all told

all of the scores and the reasoning behind them. Then they are given some time for deliberation and asked to submit another score, which may be the same as their first one or not. Usually the result is a tighter cluster, which is not always in the middle but often tends to one extreme or the other, depending on the arguments (some experts alter their position on the basis of new or more reliable evidence). This process is repeated several times until the cluster of responses is tight enough to constitute general agreement. The result is an interval scaled set of land variables for each environmental effect for each potential land use.

The second technique used to assign importance scores to each land variable is the method of paired comparisons. This technique is useful when environmental experts are not able to assemble together for discussion. The method of paired comparisons requires respondents to choose between pairs of land variables. Each variable is matched with each other sequentially. If there are m land variables, then there are $m(m-1)/2$ unique pairs. With some manipulation, the process yields an intervally scaled set of land variables that is similar to results of the delphi technique.

At some point in the future, more elaborate simulation techniques may advance to the point where absolute quantification is more feasible. One absolute quantification technique is factor analysis and this is sometimes suggested to avoid the human judgmental factor. However, the inability to use the results of factor analysis in actual scoring has been well documented.[18] Since ecosystems vary so widely among and within study areas for which suitability modelling can be used, it seems unlikely that suitability modelling processes will soon be

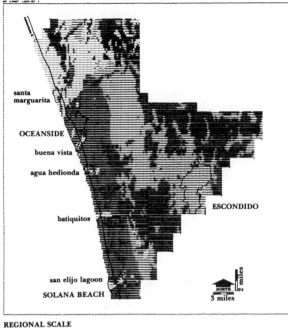

REGIONAL SCALE
FIRE HAZARD MODEL

::::::::: UNDETERMINED IMPACT

▦▦▦▦ SLIGHT IMPACT

■■■■

Figure 10.

■■■■ SEVERE IMPACT

able to abandon the human judgmental factor. Meanwhile the scoring matrix provides a convenient device for dealing with relative numbers in a clear and orderly way (*Figure 7*).

In using this matrix, the land variables determined in step 3A are first listed in the left hand column and the environmental effects determined in step 2B are listed across the top. These effects are then assigned weights, based generally on the priorities established in the criteria for the model and derived more precisely by the process described above. These weights are expressed as percentages, all of them adding up to 100 *per cent*. They are written along the bottom or "total" line of the matrix.

For the agricultural suitability model for the San Diego Coastal Plain project illustrated here, productivity was given the greatest weight among the six environmental effects under concern. It was assigned 40 *per cent* of the total value of the model. Pesticide transport was second with 20 *per cent*, nutrient transport third with 15 *per cent* and so on. It is important to remember that these values are related to this specific situation and are not intended as measures of the relative importance of these factors for agriculture in general.

Next, interactions between land variables and environmental effects are analysed and a weight is assigned to each interaction depending on the importance of that variable in controlling the effect in question. The weights in each column must add up to the total already given on the bottom line. The row of scores for each land variable are then totalled in the column on the right hand side. These totals give the relative importance of each land variable in providing the related environmental effect in

the model, and they also add up to 100 *per cent*. In this agriculture model, the most important land variable is the soil fertility class with 30 *per cent*, while next is runoff potential with 24 *per cent*.

The final step is then to distribute these scores among the attributes within each land variable using the second matrix illustrated (*Figure 7*). All the vegetation types constitute the attributes of the land variable: vegetation type. The basic information on which this distribution is based is contained in the last two columns of the

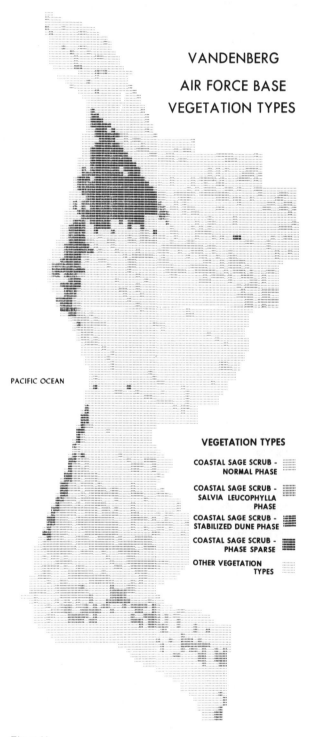

VANDENBERG

AIR FORCE BASE

VEGETATION TYPES

PACIFIC OCEAN

VEGETATION TYPES

::::::: COASTAL SAGE SCRUB -
NORMAL PHASE

::::::: COASTAL SAGE SCRUB -
SALVIA LEUCOPHYLLA
PHASE

▦▦ COASTAL SAGE SCRUB -
STABILIZED DUNE PHASE

▦▦ COASTAL SAGE SCRUB -
PHASE SPARSE

::::::: OTHER VEGETATION
TYPES

Figure 11.

Transformation Key Chart, which are labelled "Attribute Sensitivity Range". Once this is done, the scores are simply inserted in the modelling programme to be totalled for each grid cell and the results printed by the computer. The printout that resulted from the matrix described above is shown (*Figure 8*). In this case, the purpose was to to accurately identify specific lands with agricultural potential in a single watershed within the San Diego Coastal Plain where urban development is progressing rapidly and where agricultural uses are in direct competition with residential and recreational uses. The scores were applied to this 85 square mile area at the planning unit scale; that is, each cell on the printed map represents a subdivision of one ninth of the 1000 foot grid cell, or an area 333.3 feet square. As mentioned earlier, this cell is small enough to provide reliable determinations of specific pieces of land.

Models for urban suitability and fire hazards at the regional scale were derived by similar means (*Figures 9* and *10*). In these examples, each cell represents the full 1000-foot square. The urban suitability model is broad and inclusive, encompassing the consideration of a range of variables, while the fire hazard model is more limited in scope. The urban suitability map can provide a basis for urban growth policies and infrastructure locations and other long-range planning decisions. The fire hazard map can plan a role in these also, but it can also be used for such practical purposes as locating fire stations or areas where shingle roofs should be banned or fire retardant plants planted.

For Vandenberg Air Force Base, erosion was considered among the most serious environmental problems. Heavy construction, including runways and launching pads will ideally be concentrated in areas that are less subject to erosion. For this reason the model shown in *Figure 11* showing erosion potential based on soil types and vegetation communities, is one of the most important early warning models developed for that project.

CONCLUSION

These examples constitute a sampling of applications for suitability mapping, and demonstrate clearly that it is a practical and useful planning tool. However, it is also a planning tool in the infancy of its development. The techniques, especially those related to quantification, remain relatively crude. It remains to be seen to what degree models can become both holistic and precisely quantified. Much of the data used in suitability mapping also remains crude. More accurate data and better means of interpretation will result in more reliable models.

Communication of the meanings of models to the diverse audiences required by public participation processes is also a continuing problem. Computer generated maps are mysterious to most people, including some professional planners, and the concept of suitability is often difficult to explain.

So a great deal of research and development remains to be done. Since it involves so many different skills and areas of knowledge, suitability mapping presents a rewarding opportunity for interdisciplinary effort. But it is not likely to be a large, productive area for pure research. The large-scale studies of the past were made in the process of carrying out actual planning efforts related to real issues, and this is likely to remain so in the future.

NOTES

1. The needs and requirements of applied geographical research for state and local governments have recently been explored and three groups have been identified that must function smoothly in order to produce usable research results: the local agency staff, the elected officials, and the university researchers. University researchers are reluctant to take an advocacy role which the public planning process requires, especially in view of the traditional University rewards system. Yet state and local government agencies have a need for short and long term research. In-house staff and consultant companies can best fill the short term needs while intermediate and long run needs can best be filled by universities because of better quality research at lower expense. See Stutz, F. P. (1980). 'Applied Geographic Research for State and Local Government: Problems and Prospects,' *The Professional Geographer*, **32**, 393–399.
2. Steinitz, C. (1979). *Defensible Processes for Regional Landscape Design*, Landscape Architecture Technical Information Series. Washington DC, American Society of Landscape Architects.
3. Tyrwhitt, J. (1950). 'Surveys for Planning' in *Town and Country Planning Textbook*. London, The Architectural Press.
4. Alexander, C. (1964). *Notes of the Synthesis of Form*. Cambridge, Harvard University Press.
5. Alexander, C. and Mannheim, M. (1962). *The Use of Diagrams in Highway Route Location: An Experiment*. Cambridge, MIT Civil Engineering Systems Laboratory.
6. McHarg, I. (1969). *Design With Nature*. New York, Natural History Press. A number of students at San Diego State University have made improvements using a modified McHarg approach. See for example, Hartsook, T. (1973) 'Determination of a Least Cost Corridor for Route 52,' unpublished Master's Thesis, San Diego State University; Hobbs, G. (1974) 'Determination of a Least Cost Corridor for Anza Road,' unpublished Master's Thesis, San Diego State University; and White, T. (1979) 'Determination of Least Social Cost Corridors in Chaparral Land Management Planning', Master's Thesis, San Diego State.
7. A short elaboration and critique of McHarg's method can be found in Stutz, F. P. (1976) *Social Aspects of Interaction and Transportation*, Association of America Geographers Resource Paper, 76–2. Washington DC, AAG, 59–60.
8. Gold, A. J. (1974). 'Design With Nature: A Critique,' *Journal of the American Institute of Planners*, **40**, 284–286.
9. Marusic, I. (1980). 'Landscape Planning Methods in the USA—An Outside View,' *Landscape Research*, 5, 4–7.
10. Sinton, D. and Steinitz, C. (1969). *GRID*, Cambridge, Laboratory for Computer Graphics at Harvard University.
11. Steinitz, C., Murray, T., Sinton, D. and Way, D. (1969). *A Comparative Study of Resource Analysis Methods*. Cambridge, Department of Landscape Architecture Research Office, Harvard University.
12. Steinitz, C. and Rogers, P. (1970). *A Systems Analysis Model of Urbanization and Change*. Cambridge, MIT Press.
13. Patri, T., Streatfield, D. C. and Ingmire, T. J. (1979). *Early Warning System: The Santa Cruz Mountains Regional Pilot Study*. Berkeley, Department of Landscape Architecture, University of California.
14. Hopkins, L. (1980). 'Landscape Suitability Analysis: Methods and Interpretation,' *Landscape Research*, 5, 8–9.
15. Lyle, J. T., Jokela, A. and Bess, D. E. (1979). *Land Tenure Strategy: Planning for Private Use in the California Desert*. Pomona, California State Polytechnic University.
16. Odum, H. T. (1971). *Environment Power and Society*. New York, Wiley Interscience.
17. For a more complete treatment of the use of material and energy flow diagrams for planning purposes, see Van Hook, R. T. (1971), 'Energy and Nutrient Dynamics of Population in a Grassland Ecosystem,' *Ecological Monographs*, **41**, 1–26.
18. This problem has been reported on Hopkins, *op. cit. (note 14)* and in Stutz, F. P., 'Assessing Social Costs for Transport Route Location,' *High Speed Ground Transportation Journal*, **8**, 134–138.

ACKNOWLEDGMENTS

Thanks are due to federal, state and local government agency staffs as well as graduate students at our universities who provided and gathered the bulk of the data on which these analyses were based. The assistance of Mark von Wodtke, Charles Cooper, Arthur Jokela, David Bess and Richard Reilly is gratefully acknowledged. The financial support of the County of San Diego, New Environmental Research Corporation, and The Center for Regional Environmental Studies at San Diego State University and the USAF is gratefully acknowledged.

Implementing LESA on a Geographic Information System—A Case Study

T. H. Lee Williams
Department of Geography, University of Kansas, Lawrence, KS 66045

ABSTRACT: The USDA Soil Conservation Service has developed the agricultural Land Evaluation and Site Assessment (LESA) system to determine the quality of land for agricultural uses and to assess agricultural land areas for their economic viability. The system is implemented manually on a site-by-site basis. However, many possible uses of LESA involve area-wide evaluations or are concerned with modeling the potential effect of changes such as rezoning, and are difficult to perform using the manual approach. A study was carried out to evaluate the feasibility of implementing LESA on a computer-based Geographic Information System (GIS). The study demonstrated that the GIS approach extends the utility of LESA into area-wide modeling and allows a more flexible site analysis. Furthermore, the recent advances in microcomputer hardware and software now make possible the development of an affordable county-level computer-based LESA system.

INTRODUCTION

THE U.S. DEPARTMENT OF AGRICULTURE Soil Conservation Service (SCS) has developed the agricultural Land Evaluation and Site Assessment (LESA) system to determine the quality of land for agricultural uses and to assess sites for their agricultural economic viability (SCS, 1983; Wright *et al.*, 1983; Dunsford *et al.*, 1983). The LESA system was designed to assist farmland protection policies and is developed at the governmental level at which it will be used (state, county, township, or town). Many of the data inputs to the LESA model are derived from remote sensing imagery (e.g., land use and surface hydrology), but many are drawn from other map and tabular sources (e.g., zoning, sewer lines, soils data). Integration of the many varied data sources for a particular site is therefore a tedious and inefficient manual process. A demonstration study was carried out to evaluate the feasibility of implementing LESA on a Geographic Information System (GIS).

There is no clear definition or agreement between cartographers and remote sensing researchers about what the term GIS actually covers. Cartographers have developed and researched digital geographic data bases as a means of automating map production and storing maps in a digital form. Their work has emphasized the efficient encoding, storage, retrieval, and display of the digital cartographic data. Each map element or theme in the data base, e.g., contour lines or hydrology, is independently retrievable, and a selected set of themes may be simultaneously displayed on a CRT screen or printed on a paper map product. The digital data are processed in a multiple-independent form in which the various themes are functionally separate. Remote sensing researchers, however, view the digital data base as a mechanism which will allow the map themes (land use, elevation) to be overlayed and combined in various ways to simulate a particular resource management analysis procedure. For example, soils, land use, and hydrology information may be used in concert to locate areas of potential soil erosion (Campbell, 1979). This use of the digital data base involves multiple-related *analysis* of the data base map themes, as distinct from the multiple-independent use of the data base as a map *storage* and retrieval device. Both approaches are included in the common definition of a GIS, although it is the former use of the digital data base that is of concern in this study.

The techniques used for analyzing the multi-theme data in a GIS involve application of arithmetic and logical operations on the data base themes in a structured procedure termed "cartographic modeling" (Tomlin and Berry, 1979). While the basic analysis functions and general modelling procedures have been available for a number of years, GIS techniques in the resource management field have been limited largely to national and regional programs. However, powerful microcomputers are now available at a cost within the reach of local (i.e., county level) programs. The advances in hardware and software have brought powerful 16-bit, 1 megabyte microcomputers, 10 megabyte disks which can handle county-level data bases, and digitizing tables into the $7,000 price range. GIS and image processing software is now available for microcomputers at prices starting around $1,000. It is apparent, there-

PHOTOGRAMMETRIC ENGINEERING AND REMOTE SENSING,
Vol. 51, No. 12, December 1985, pp. 1923-1932.

0099-1112/85/5112-1923$02.25/0
© 1985 American Society for Photogrammetry
and Remote Sensing

fore, that more county-level GIS programs will develop in the next few years.

The work reported here presents a case study in which a GIS was used to automate and facilitate a county-level LESA model. The choice of LESA for a county-level demonstration is appropriate as LESA represents a national program that is mainly implemented at the county level. The study does not introduce new ideas on database design or natural resource models, but presents a practical example of using GIS technology to automate a tedious manual process.

LESA

The LESA system consists of two parts, land evaluation (LE) and site assessment (SA):

- *Land Evaluation.* In agricultural land evaluation, soils of a given area are rated and placed into groups ranging from the best to the worst suited for a stated agricultural use, i.e., cropland, forest land, or rangeland. A relative value is determined for each group: the best group is assigned a value of 100 and all other groups are assigned lower values. The land evaluation is based on data from the National Cooperative Soil Survey.
- *Site Assessment.* Site assessment identifies important factors other than soils that contribute to the quality of a site for agricultural use. Each factor selected is stratified into a range of possible values in accordance with local needs and objectives. This process provides a rational, consistent, sound basis for making land-use decisions.

(SCS, 1983, pp. 600-1)

The SA factors and LE are combined to give a LESA score for the site under consideration. The score is scaled to give values between 0 and 300, with high scores corresponding to areas of high agricultural value. The LESA scheme is implemented manually: the LE and SA scores are derived and compiled for each site individually using a worksheet. In order to compare several sites or search for possible alternate sites, the LESA score must be compiled separately for each site. The scheme is appropriate for use in a reactive mode, i.e., responding to a request for consideration of a specific site. However, many potential uses of LESA involve proactive area-wide evaluations (to identify regions of low or high value) or modeling the effect of changing conditions (e.g., the impact of rezoning on agricultural value), which are time-consuming and awkward using the manual approach. A study of the typical SA factors and the scheme for combining the SA scores and LE scores shows that the scheme is well-suited to a GIS approach. Most of the SA factors are derived from raw data available in map or spatially-defined form, and the factors are derived from the raw data using simple spatial algorithms.

STUDY AREA

Douglas County in northeastern Kansas, the home of the University of Kansas, is experiencing many of the development pressures that threaten agricultural land throughout the U.S. In the past 15 years it has lost 40,131 acres of productive land (13 percent of its total area) to Clinton Lake and Park, rural residential development, and industrial expansion. The number of rural dwellings increased 12 percent in three years and 32 percent in the past decade. The city of Lawrence is among the fastest growing cities in the U.S., and the rural townships have already passed their projected population for the year 2000. The newest industrial development is full and another 400 acre industrial park is planned (Lawrence Journal-World, 1982; University Daily Kansan, 1981). At the time of this study, Douglas County was developing the LESA scheme and was therefore a suitable choice as a study area.

A 6 by 5-km area between Lawrence and Clinton Lake was chosen for the study (Figure 1). The area is experiencing development pressure from Lawrence due to the Clinton development. It also includes all the major physiographic types (floodplain, bluffs, and rolling uplands) and land uses (urban, cropland, pasture, etc.) found in Douglas County.

LESA DATA BASE DESIGN

Some design considerations for the data base are (1) will it be vector-based, raster-based, or a hybrid vector-raster system? (2) to what coordinate system will the data base be referenced? and, (3) what will be the minimum mapping unit (for a vector system) or grid cell size (for a raster system)?

A vector-based system explicitly records the boundary of each area unit as a string of x,y coordinates, and provides efficient storage for the data (Monmonier, 1982). It is favored in GIS systems that emphasize digital map storage and retrieval. Comparison or 'overlay' of different themes (e.g., land use versus soils data) is, however, difficult as the areal units do not usually match (e.g., a given land use will cross several soil boundaries). The vector-based system is not favored, therefore, in GIS systems where map overlay analysis is required, unless exact polygon boundaries need to be preserved. The raster-based system divides the study area into uniform cells. Each cell is encoded with an attribute value for each theme (e.g., a land-use value). Overlaying different themes is a simple process of comparing attribute values for each grid cell. The raster system is, however, a relatively inefficient method of storing spatial data. The trade-off between the vector and raster systems is, therefore, one of data storage versus the ease of overlay analysis. Hybrid data structures use a combination of raster and vector representations in order to preserve the data storage efficiency of a vector database and the theme overlay capabilities of a raster system. One hybrid approach encodes and stores the data in vector format, but displays and analyzes the data in raster

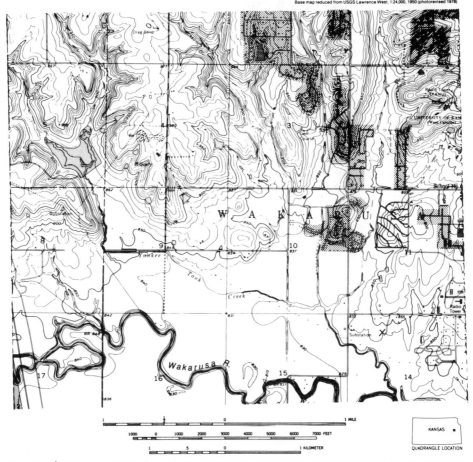

Fig. 1. 7½' topographic map of the 6 by 5-km study area on the west side of Lawrence in northeastern Kansas. Note the Wakarusa river and its floodplain in the lower portion of the study area and the new developments (shown with an overlay dot pattern) in the eastern portion of the area.

format. Rapid vector/raster conversion is achieved using custom software routines and the standard screen painting routines available on microcomputers. Another approach uses a true hybrid data structure that contains elements of both vector and raster structures. This 'vaster' approach has been proposed for very large databases where the map data are digitized using raster flatbed or drum scanners (Peuquet, 1983). A hybrid structure is superior to either the vector or raster approach, and is the system of choice for microcomputers. However, the decreasing cost of mass storage devices makes the data storage limitations of a raster system less critical for moderate data bases of reasonable resolution, such as county-sized areas of about 1000 by 1000 cells. A raster-based system already available at the University of Kansas was adopted for the county-level LESA case study discussed here.

The second design consideration involves the choice of coordinate system to which the data base will be referenced. Although there are many potential options, two options were considered to be feasible: the Universal Transverse Mercator (UTM) and the Township-Range-Section (TRS) systems. The TRS system matches the pattern of land holdings (i.e., it is a cadastral system) and is familiar to the public and land resource agency personnel, but the systematic adjustments and random errors in the section boundaries make it unsuitable for geodetic purposes. The UTM grid is preferred by cartographers for its geodetic fidelity, but the coordinate system does not correspond to any pattern of land use and, hence, grid cells will be more likely to fall across land feature boundaries than those of the TRS system. However, as many data sources are available for a GIS in digital grid form (e.g., Landsat) and are easily referenced to the UTM grid (e.g., USGS and Defense Mapping Agency digital elevation data), the UTM scheme was selected.

The choice of grid cell size in a raster system is a trade-off between the large data volumes associated with small cells and the loss of detail incurred by the use of large cells. The particular considerations for this study are the minimum parcel size that will be considered by the LESA scheme, the errors in distance measurement (± 1 cell width) acceptable in the LESA scheme, and the minimum mapping units of the source maps. The minimum parcel sizes will relate to residential parcels, which are zoned to be a minimum of 5 acres for rural areas, 3 acres for designated suburban growth areas, and 2 acres for designated rural growth areas. The distance criteria for the LESA SA factors are in units of $^1/_8$ mile. The limiting factor, therefore, is the 2-acre rural parcel, and a cell size of 100 by 100 metres (2.5 acres) was selected. (Note that a hybrid data structure allows variable cell-size, limited only by the x,y precision of the vector data input.) For an average mid-western county size of 30 by 30 miles, this corresponds to a 480 by 480 cell array for the county, which is a convenient size relative to the 512 by 512 CRT displays commonly available.

LESA COMPONENTS

The national LESA handbook (SCS, 1983) describes in detail the design and uses of a LESA scheme. The general outline and procedures are fixed, but the details of a particular scheme (e.g., SA factors, relative weights for each factor) are designed by local working groups formed within the jurisdiction being considered (state, county, township, city). At the time of this study, Douglas County was in the process of defining the LESA factors and weights, but the process was not completed. The factors and weights finally used in the study were a combination of those defined by the county working group, supplemented by typical values derived by the author from the national LESA handbook.

LAND EVALUATION (LE)

The Douglas County Soil Survey was used to identify eight soil productivity groupings. Each group was then rated according to the yield of an indicator crop (grain sorghum was selected for Douglas County), taking into account the economic costs associated with the crop, e.g., necessary soil improvements. The outcome is a relative value rating for each group, with the highest group set to a value of 100 and the others prorated. Figure 2 shows the soil survey and relative value (LE) maps for the study area.

SITE ASSESSMENT (SA)

The SA rating incorporates physical, economic, social, and cultural factors in an assessment of the suitability of the site for agriculture, based on factors other than soil productivity. Table 1 shows the SA factors determined for Douglas County. Each factor is quantized and assigned values on a scale of 10 to 0, where a value of 10 indicates high suitability for agriculture, and 0 indicates low suitability. The quantization and value ratings for Factor 1, "Percent of Land in Agriculture within 1½ miles," are shown in Figure 3.

Each factor (i) is assigned a relative weight (w_i), which indicates its importance relative to the other

SA factors (see Table 1). The SA rating (SA_j) for a site (j) is then given by

$$SA_j = \Sigma_i w_i v_{ij} \qquad (1)$$

where v_{ij} is the value for factor i on site j. The values of w_i are adjusted (scaled) so that the maximum possible SA rating for a site is 200 (Table 1).

The LESA rating for a site is then given by

$$LESA_j = LE_j + SA_j = LE_j + \Sigma_i w_i v_{ij} \qquad (2)$$

A maximum possible SA rating of 200 and maximum LE value of 100 gives an SA:LE relative weight of 2:1, which was found to be reasonable in LESA test cases (SCS, 1983, pp. 601-12). The ratio can be adjusted for specific purposes by rescaling the LA rating or the LA factor weights.

PROCEDURE

SOURCE DATA

Table 1 lists the source data from which each SA factor is derived. Note that one factor has no source identified because the data were not available in a spatially-defined form. Factor 5, "Agrivestment in area," includes onsite investments (e.g., barns, conservation measures) and the agricultural support system (e.g., farm equipment suppliers, grain dealers). It is perhaps feasible to derive an estimate of the agricultural support system for a study area, but the onsite investments would have to be assessed on a site-by-site basis. A complete site inventory for the study area is infeasible. It is apparent, therefore, that there is at least one factor that will not be incorporated into the data base. It is likely that other study areas will have similar factors that may not be included because of availability, cost, or rate of change of the data. The approach taken in this study was to compute a partial LESA score based on all available factors, and to add in the remaining factor for a particular site when it is selected for study.

The data were coded onto transparent grid overlays on the source maps, and entered to the data base using run-length encoding (Wehde et al., 1980).

MANIPULATION

The Map Analysis Package (MAP) developed at Yale University (Tomlin and Berry, 1979) is used in GIS courses at the University of Kansas. As much of the database construction for this project was carried out in a GIS course, MAP was adopted for the project. MAP allows a wide range of arithmetic and logical operations to be applied to single or multiple map themes. Five basic operations are used singly or in combination to generate the LE and SA ratings in this study:

- Renumber. Assigns new values to the categories of a map. It is used to combine map categories and

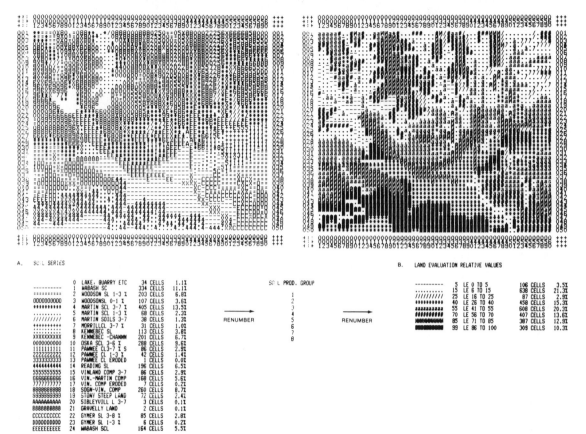

FIG. 2. The soil series (a) are grouped into eight soil productivity units which are then renumbered to show the land evaluation (LE) relative values (b), which can vary between 0 and 100.

rescale measurements. Figure 2 demonstrates how the LE relative value ratings are derived—map categories (soil types) are combined into new categories (productivity groups) and the categories rescaled (to show the relative value of each group). The renumbering function is used for all the SA factors to rescale or combine categories.

• Neighborhood. Computes the average value of all cells that lie within a specified radius of a site, or counts the values for the four adjacent grid cells. Figure 3 shows how SA factor 1, "Percent of Land in Agriculture within 1½ miles of Site," is derived from the land-use map. The neighborhood functions are used in SA factors 1, 2, 4, 6, 7, 9, and 13 (Table 1).

• Intersection. Identifies cells with a specified set of values on two or more source maps (logical AND). It is used in SA factor 9 to locate sites that are both non-farmland and have poor soils.

• Distance. Computes the geographic distance from a cell to a line network, area, or point. Figure 4 shows how SA factor 20, "Distance from Central Water," is derived from the source map of water lines. The distance function was used in SA factors 18 to 21.

• Arithmetic. Computes arithmetic combinations of two or more maps on a cell-by-cell basis. The SA factor maps were multiplied by their relative weights and summed, to give the overall SA rating, which was added to the LE map to give the LESA scores (Figure 5).

Note that the scores are incomplete as they do not contain Factor 5, which is not in the data base. As discussed earlier, this factor would be evaluated and added to the LESA score of an individual site to give its final value.

RESULTS

As expected, location and acquisition of the source data and the entering of the data were the most time- and personnel-intensive parts of the study. Several of the source data items were not available in a standard form. For example, the utilities maps (water, sewage) were available on single maps within the city limits, but outside the city were available only on several different maps and tabular listings that had to be combined onto a common base before they could be coded. The mechanism of manually encoding the data is also inefficient but is suitable for such a small study area. The use of an electronic digitizer is recommended for encoding county-size areas. The initial construction of a geographic data base is personnel-intensive. County agencies will probably opt to contract the database construction, but will perform the necessary periodic updates in-house.

The final partial LESA map (Figure 5) corresponds well to the study area map in Figure 1. As expected, the Wakarusa floodplain had the highest scores,

TABLE 1. THE PRELIMINARY SITE ASSESSMENT FACTORS FOR DOUGLAS COUNTY, THEIR RELATIVE WEIGHTS, AND SOURCE MAPS

Site Assessment Factor	Source Map	Relative Weight
Land Use/Agricultural		
1) Percent of area in agric. within 1½ miles	Land use	10
2) Land in agriculture adjacent to site	Land use	7
Agricultural Economic Viability		
3) Farm size	Parcel size	2
4) Average parcel size within 1 mile	Parcel size	4
5) Agrivestment in area	—	3
Land Use Regulations		
6) Percent of area zones agric. within 1½ miles	Zoning	8
7) Zoning of the site and adjacent to it	Zoning	6
Alternative Locations		
8) Availability of land zoned for proposed use	Zoning	6
9) Availability of non-farmland or less productive land as an alternative site within area	Zoning/Soils	6
10) Need for additional urban land	Land use/City lim.	8
Compatibility of the Proposed Use		
11) Compatibility of proposed use with surrounding area	Land use	7
12) Unique topographic, historic or ground cover features or unique scenic qualities	Unique areas	3
13) Adjacent to land with unique topographic, historic or groundcover features or unique scenic qualities	Unique areas	2
14) Site subject to flooding or in a drainage course	Surface hydrology	8
15) Suitability of soils for on-site waste disposal	Soils	5
Compatibility with Adopted Master Plans		
16) Compatibility with an adopted comprehensive plan	Master plan	5
17) Within a designated growth area	Growth area	5
Urban Infrastructure		
18) Distance from city limits	City limits	6
19) Distance from transportation	Transportation	5
20) Distance from central water	Water lines	4
21) Distance from sewage lines	Sewer lines	4
		114

$$\text{Scale by } \frac{200}{(114 \times 10)} = 0.175$$

while the urban Lawrence area and new developments in the northeast have the lowest scores. The undeveloped dissected bluffs and upland areas in the northwest show intermediate values. Note that the variations in LESA scores follow closely the ridges, slopes, and valleys in this area.

DISCUSSION

The LESA map gives an area-wide view of agricultural value, and identifies the location and size of regions of relatively low or high value. Unfortunately, the LESA scheme has not been completed or adopted by Douglas County, and therefore the results of this study cannot be critically evaluated.

However, the national LESA handbook (SCS, 1983) discusses extensively the uses of the LESA scheme, and provides a suitable framework for a general evaluation of the potential uses of the digitial LESA database.

The digital data base format allows rapid manipulation of all components of the LESA scheme, including the factors to be used, factor quantization and scores, factor weights, and the relative LE:SA weighting. This flexibility is central to the utility and application of the LESA scheme as it allows modeling of the effects of changing parameters and assessing the effects of specific actions, e.g., the installation of a new sewer line. The design and re-

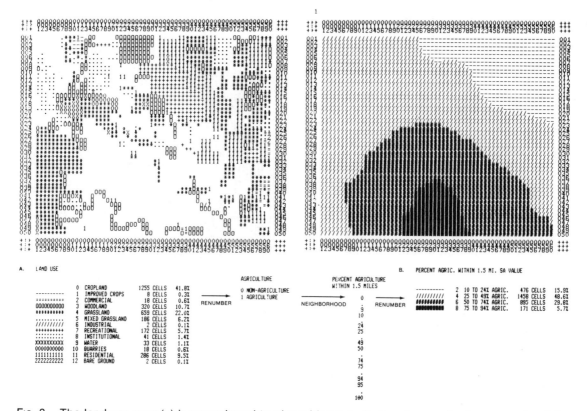

A. LAND USE

	0 CROPLAND	1255 CELLS	41.8%
----------	1 IMPROVED CROPS	8 CELLS	0.3%
==========	2 COMMERCIAL	18 CELLS	0.6%
0000000000	3 WOODLAND	320 CELLS	10.7%
++++++++++	4 GRASSLAND	659 CELLS	22.0%
....	5 MIXED GRASSLAND	186 CELLS	6.2%
//////////	6 INDUSTRIAL	2 CELLS	0.1%
++++++++++	7 RECREATIONAL	172 CELLS	5.7%
::::::::::	8 INSTITUTIONAL	41 CELLS	1.4%
XXXXXXXXXX	9 WATER	33 CELLS	1.1%
0000000000	10 QUARRIES	18 CELLS	0.6%
1111111111	11 RESIDENTIAL	286 CELLS	9.5%
2222222222	12 BARE GROUND	2 CELLS	0.1%

AGRICULTURE

0 NON-AGRICULTURE
1 AGRICULTURE

RENUMBER →

NEIGHBORHOOD →

PERCENT AGRICULTURE
WITHIN 1.5 MILES

0
.
.
9
10
.
24
25
.
49
50
.
74
75
.
94
95
.
100

RENUMBER →

B. PERCENT AGRIC. WITHIN 1.5 MI. SA VALUE

	2 10 TO 24% AGRIC.	476 CELLS	15.9%
----------	4 25 TO 49% AGRIC.	1458 CELLS	48.6%
//////////	6 50 TO 74% AGRIC.	895 CELLS	29.8%
##########	8 75 TO 94% AGRIC.	171 CELLS	5.7%
██████████			

FIG. 3. The land-use map (a) is renumbered to give a binary agriculture/non-agriculture map. A neighborhood operation on this map produces a map showing the percent of agriculture within 1.5 miles of each site, and the map is then renumbered to give the site assessment (SA) Factor 1 map (b), which has possible values between 0 and 10.

finement of the original scheme parameters (e.g., the LE:SA weighting) are based on field tests of various parameter combinations (SCS, 1983, pp. 602-13), which is a time-consuming process using the manual LESA scheme. The scheme is also use-specific (e.g., cropland to residential use conversion),

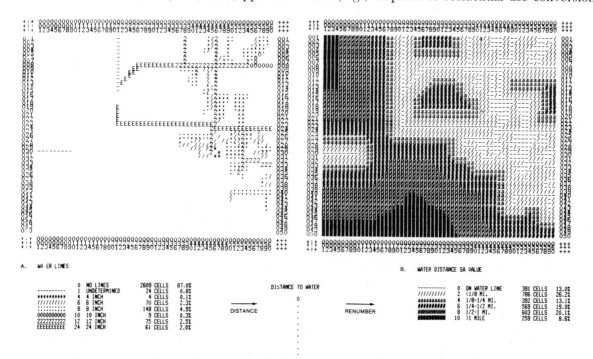

A. WATER LINES

	0 NO LINES	2609 CELLS	87.0%
----------	1 UNDETERMINED	24 CELLS	0.8%
++++++++++	4 4 INCH	4 CELLS	0.1%
//////////	6 6 INCH	70 CELLS	2.3%
....	8 8 INCH	148 CELLS	4.9%
0000000000	10 10 INCH	9 CELLS	0.3%
2222222222	12 12 INCH	75 CELLS	2.5%
EEEEEEEEEE	24 24 INCH	61 CELLS	2.0%

DISTANCE TO WATER

0
.
.
.
.
.

DISTANCE →

RENUMBER →

B. WATER DISTANCE SA VALUE

	0 ON WATER LINE	391 CELLS	13.0%
----------	2 <1/8 MI.	786 CELLS	26.2%
//////////	4 1/8-1/4 MI.	392 CELLS	13.1%
aaaaaaaaaa	6 1/4-1/2 MI.	569 CELLS	19.0%
##########	8 1/2-1 MI.	603 CELLS	20.1%
██████████	10 >1 MILE	259 CELLS	8.6%

FIG. 4. A distance operation is applied to the water lines map (a) to produce a map showing the distance from each site to the nearest water line. The distance map is renumbered to give the site assessment (SA) Factor 20 map (b), which has possible values between 0 and 10.

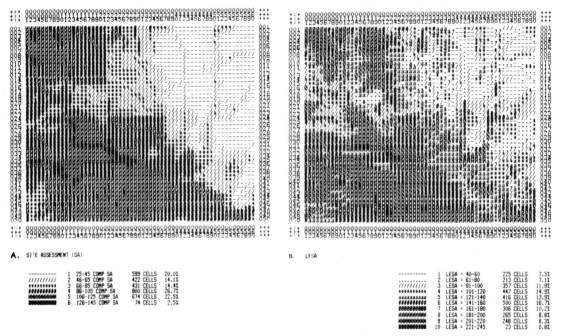

A. SITE ASSESSMENT (SA)

----------	1	25-45 COMP SA	599 CELLS	20.0%
//////////	2	46-65 COMP SA	422 CELLS	14.1%
++++++++++	3	66-85 COMP SA	431 CELLS	14.4%
++++++++++	4	86-105 COMP SA	800 CELLS	26.7%
##########	5	106-125 COMP SA	674 CELLS	22.5%
##########	6	126-145 COMP SA	74 CELLS	2.5%

B. LESA

----------	1	LESA = 40-60	225 CELLS	7.5%
....	2	LESA = 61-80	213 CELLS	7.1%
//////////	3	LESA = 81-100	357 CELLS	11.9%
++++++++++	4	LESA = 101-120	447 CELLS	14.9%
aaaaaaaaaa	5	LESA = 121-140	416 CELLS	13.9%
##########	6	LESA = 141-160	500 CELLS	16.7%
##########	7	LESA = 161-180	306 CELLS	10.2%
##########	8	LESA = 181-200	265 CELLS	8.8%
##########	9	LESA = 201-220	248 CELLS	8.3%
##########	10	LESA = 221-245	23 CELLS	0.8%

FIG. 5. The final site assessment (SA) rating map (a) is produced from a weighted linear combination of the individual SA factor maps, using Equation 1. It is added to the land evaluation (LE) relative value map (Figure 2b) to give the final LESA map (b).

and several versions of the scheme are required to represent the range of proposed uses.

Many of the potential applications cited for the LESA scheme involve the use of selected parts of the scheme (e.g., the LE ratings), which is easily accomplished using the stored LE and SA factor maps. Agricultural value assessment and the size of a site to meet income requirements can be derived from the LE value ratings (SCS, 1983, pp. 603-5, 6). The acquisition of development rights involves both the complete LESA rating and individual SA factors (e.g., distance to city, flooding, scenic easements, and adjacent land use) (SCS, 1983, pp. 603-6, 9). The relative LE:SA weighting is also adjusted for different development situations (SCS, 1983, pp. 603-7).

The LESA map in Figure 5 was derived from a cell-by-cell (100-metre area) analysis, whereas in practice the LESA scheme is applied to complete sites which are generally more than one cell in size. The LESA rating for a site can be computed as the average of the individual cell values, up to a site size of 100 acres (SCS, 1983, pp. 601-26). Alternately, the rating for an extended (multi-cell) site can be derived directly from the database.

A significant limitation of the scheme is related to the linear arithmetic structure of the LESA rating (Equation 2) where each factor is evaluated independently and the results added (SCS, 1983, pp. 602-13). The linear combination of factors does not account for interrelationships between factors. For example, Factor 15 rates the soils for septic tank suitability, but the importance of the factor is in practice related to the availability of a central

sewage system (Factor 21). The relationships between factors can be expressed as simple arithmetic formulae or as complex symbolic decision rules. Expert system models that allow both arithmetic and symbolic reasoning are now being applied to remote sensing and natural resource problems (Mooneyham, 1983; Tinney et al., 1983). Expert system techniques are being used in ongoing work at the University of Kansas to refine the LESA scheme (DeMers, 1985).

MICROCOMPUTER IMPLEMENTATION

For convenience, the test study was conducted using the University of Kansas mainframe computer. It is likely, however, that any implementation of an automated LESA system at the county level would use a microcomputer system. The question arises, therefore, as to whether and at what cost a microcomputer system has sufficient central processor size and mass-storage capabilities to handle a county-level data base, and whether appropriate GIS and image processing software is available for such microcomputers. The answer to each of these questions is encouraging: 16-bit microcomputers with one megabyte central processors are available for around $3,000; 10-megabyte hard disks are available for around $2,000; suitable digitizing tables cost around $2,000; and GIS software packages are available at prices that start at $1,000.

A county data base with a 100-metre grid cell for a 1000 square mile county (typical of the mid-west) is 509 by 509 cells in size, or 259,081 cells. If each attribute value for a cell is coded into one byte (thus allowing 2^8 or 256 possible values), each map theme (e.g., land use) will use 253K bytes of storage. A 10-megabyte disk could hold 39 map themes, which is more than adequate for the LESA scheme. Note that a more complex raster data base structure would allow more efficient use of space by encoding certain map themes in few bits per cell. For example, the city boundary theme (used in Factor 18, Table 1) could be encoded in 1-bit form (i.e., 0 = outside city, 1 = inside city), thus using 253K ÷ 8 = 32K bytes of storage. Also note that a hybrid vector/raster database structure would be considerably more efficient in data storage.

A one-megabyte computer is easily large enough to hold both a complex GIS program and two full map themes in core simultaneously. As all GIS operations can be broken down into sequences of operations on two map themes, the input/output problems associated with large data sets on small microcomputers (64K to 127K in size) will not be experienced with the larger microcomputers. The display screen resolution on microcomputers is also improving rapidly, and the cost of 512 by 512 multicolor displays is dropping rapidly. Systems with limited display sizes can still be useful, as a full county will usually be displayed to view general patterns rather than specific detail about any one area. Therefore, the effect of the resolution limitations of lower-cost microcomputers ($3,000) can be minimized by displaying a sampled version of the county (e.g., selecting every third cell on every third row will give a 170 by 170 display) to view general county-wide patterns, and zooming in to smaller areas at full resolution. Reasonable quality hardcopy output can be effectively produced on matrix printers costing about $500.

A few vendors currently are advertising image processing and GIS software for microcomputers. The effective operating speed of the hardware/software package is of concern in most GIS applications, due to their computation-intensive nature. However, LESA uses a simple linear arithmetic algorithm, which allows recomputation of the LESA scores to be performed by recomputation of only the affected part(s) of the model. Thus, once the initial LESA model is obtained, subsequent operations will not require rapid data processing or data transfer capabilities.

It is apparent from this short discussion that existing hardware and software are available to implement the LESA scheme at the county level on a microcomputer. The market is also changing rapidly, and improved capabilities at a lower cost are being introduced frequently. However, one should look for a hardware/software package that is appropriate to the intended use and requires minimal development.

CONCLUSION

The case study demonstrated that a GIS approach to implementing the LESA scheme provides a flex-ible framework within which many of the design considerations and applications of LESA can be performed more easily than in the manual approach. Most of the site assessment factors can be implemented, although some factors may be either unavailable in spatial form or change too rapidly to be entered into the database. The digital LESA scheme computes, therefore, a near-complete LESA score, with any additional factors added later on a site-by-site basis. A preliminary evaluation of available microcomputer hardware and software indicates that a microcomputer-based implementation at the county level could be accomplished at around $9,000 hardware/software cost. No estimate was made of the cost of database development.

ACKNOWLEDGMENTS

The data base construction and initial work on the project were conducted in a seminar course in the Department of Geography at the University of Kansas. The author would like to thank John Ackerman, Brian Brisco, Tshow Chu, Mike DeMers, John Hutchinson, Bo Jung, Randy McKinley, Bob McMaster, Emily Roth, and Bob Yoos for their assistance and input into all phases of the study.

REFERENCES

Campbell, W. J., 1979. An application of Landsat and computer technology to potential water pollution from soil erosion. *Satellite Hydrology.* American Water Resources Association. pp. 616-621.

DeMers, M. N., 1985. The formulation of a rule-based GIS framework for county land use planning. *Proc. 16th Annual Pittsburgh Modelling and Simulation Conference.* Pittsburgh, Pennsylvania.

Dunsford, R. W., R. D. Roe, F. R. Steiner, W. R. Wagner, and L. E. Wright, 1983. Implementing LESA in Whitman County, Washington. *J. Soil and Water Cons.* 38(2):87-89.

Lawrence Journal-World, 1982. Planners consider measures to regulate residential development in county. *Lawrence Journal-World,* Lawrence, Kansas, September 21, 1982. p. 8.

Monmonier, M. S., 1982. *Computer-Assisted Cartography, Principles and Prospects.* Prentice Hall. 214 p.

Mooneyhan, D. W., 1983. The potentials of Expert Systems for remote sensing. *Proc. 17th Intl. Symp. on Remote Sensing of Environment.* Ann Arbor, Michigan: 51-64.

Peuquet, D. J., 1983. A hybrid structure for the storage and manipulation of very large spatial data sets. *Computer Vision, Graphics and Image Processing.* 24(1):14-27.

SCS, 1983. *National Agricultural Land Evaluation and Site Assessment Handbook 310-VI, Issue 1.* U.S. Department of Agriculture Soil Conservation Service, Washington, D.C. January 1983.

Tinney, L. R., C. Sailer, and J. E. Estes, 1983. Applications of Artificial Intelligence to remote sensing. *Proc. 17th Intl. Symp. on Remote Sensing of Environment.* Ann Arbor, Michigan: 255-269.

Tomlin, C. D. 1980. *The Map Analysis Package.* Yale

School of Forestry and Environmental Studies, New Haven, Connecticut. Draft copy, 81 p.

Tomlin, C. D., and J. K. Berry, 1979. A mathematical structure for cartographic modelling in environmental analysis. *Proc. American Congress on Surveying and Mapping,* 39th Annual Meeting. Washington, D.C. pp. 269-284.

University Daily Kansan, 1981. Industry threatens farmland. *University Daily Kansan,* University of Kansas, Lawrence, Kansas, December 1, 1981.

Wehde, M. E., K. J. Dalsted, and B. K. Worcester, 1980. Resource application of computerized data processing: the AREAS example. *J. Soil and Water Cons.* 35(1):36-40.

Wright, L. E., W. Zitzmann, K. Young, and R. Googins, 1983. LESA—agricultural land evaluation and site assessment. *J. Soil and Water Cons.* 38(2):82-86.

(Received 24 December 1984; revised and accepted 5 April 1985)

M. K. Beard
N. R. Chrisman
University of Wisconsin-Madison
Madison, WI 53706

T. D. Patterson
Southeastern Wisconsin Regional Planning Commission
Waukesha, WI 53187

Integrating Data for Local Resource Planning
A Case Study of Sand And Gravel Resources

ABSTRACT. This study examines trends in
resource planning and their implications for
geographic information system development, and
reports on these as they apply to sand and gravel
resource management. The objectives of the
project were to determine the requirements of a
comprehensive resource analysis, and demonstrate
system capabilities for accomplishing the
analysis. A set of data records in different
scales and formats were integrated to identify
limitations on sand and gravel supplies and to
define possible areas for reservation. The
project demonstrates software functions essential
to the manipulation and analysis of disparate
spatial data at a level of detail required for
local planning.

INTRODUCTION

Dangermond makes the case that the urgency of our
resource situation is driving the evolution and acceleration
of GIS development (3). If this is so, current trends in
resource management should be examined to determine what
they may require of GIS technology. Some trends have been
well articulated and are only reiterated here. The need to
combine diverse information, to examine a broad range of
alternatives, and to replace tedious and cumbersome manual
techniques, have been well documented and have become
established system capabilities (15).

Trends which have received minor attention but will
likely direct system development in the near future are:

1.) Legal status and enforcement of resource boundaries (13). Explicit and objective data will be required to support resource policy decisions and uphold legally binding resource legislation in cases of litigation.

2.) Abdication of the Federal government role and expansion of local government resource responsibilities. As a growing share of land and resource related responsibilities fall to local governments their need for GIS technology will increase. Due to access and economies of scale, GIS technology to date has catered to the federal agencies (9). Now small systems able to perform a variety of tasks should become available for local government use at reasonable costs. Production of digital resource data should reflect the local need for more detail, and system design reflect the need to process and manage this level of detail.

The two above trends provide much of the impetus for the third.

3.) The need for higher accuracy and documentation of data quality (1). Both factors influence the confidence with which resource decisions can be made. Resource decisions with legal implications require the best available and well documented information. Within the parochialism of local politics, accurate and well documented data can give officials tenable grounds for supporting and enacting controversial legislation (9). Requirements for accuracy and documentation will be followed by a need to maintain these through all phases of data processing.

This study examines the requirements for managing sand and gravel resources, and investigates the functions of GIS technology for meeting these requirements. Development of a sand and gravel management plan embodies many of the trends discussed above. The issues are predominantly local, potentially urgent, highly controversial, and likely to result in some form of legal protection and enforcement.

SAND AND GRAVEL RESOURCE ISSUES

Sand and gravel are resources we take for granted and presume to be abundant and ubiquitous. Sand and gravel deposits are not ubiquitous, but occur in specific geographic locations. Supplies are abundant but not necessarily available or useable relative to demand. Economic, political, and physical limitations combine to create local shortages. Several urban centers including New York City, Washington D.C., Atlanta, Denver, and several counties in California are currently experiencing shortages of sand and gravel (6;12;14;18).

Economically, supplies are limited by the low per unit
value and the high costs of transporting these materials.
Most of the demand for sand and gravel is in urban areas
(6). Due to the high transportation costs, deposits remote
from these locations are not economical to mine. A paradox
is that due to the low per unit value of these materials,
operations can not compete with higher valued land uses and
are forced to locate in more remote locations. Politically,
supplies are limited by restrictive regulatory legislation.
Zoning restrictions and environmental legislation cause vast
acreages to become inaccessible (6). Physically, supplies
are limited by their fixed location and quality. Permanent
development over deposits precludes their extraction. In
urban areas as building accelerates and urban areas
coalesce, natural deposits are covered over (6). Sand and
gravel must meet strict construction specifications, and low
quality deposits or deposits which require substantial
processing are not considered viable commercial sources.

REQUIREMENTS FOR SAND AND GRAVEL RESOURCE MANAGEMENT

The problems encountered in achieving a sand and gravel
resource management plan include:

1.) Convincing the public of the desirability of
reserving deposits for future use. The public is
currently more concerned with strictly regulating the
industry than with protecting the resource.

2.) Reserving deposits to maintain an adequate low cost
supply, yet minimize conflicts.

A comprehensive planning approch for resolving these
problems requires three information components: spatial,
non-spatial and temporal. These components must be combined
to determine the need for protection, convince the public of
the need for protection as well as regulation, and if
necessary develop an appropriate protection strategy.

The non-spatial components are all those attributes
that describe qualities and quantities pertaining to a
spatial entity. Attributes include descriptive and
statistical tabular data such as demand, supply, resource
quality, and cost figures.

The temporal component in this case is the lifespan of
an extractive operation. The duration of the operation
defines the planning timeframe: How long a supply is
available at a particular location, over what time period
conflicting land uses must be restricted from an area, and
how reclamation and sequential use of a site should be
scheduled.

The spatial component includes such information as deposit locations and extent, road networks, land use, land cover, zoning districts, and population centers. These phenomena are broken into geometric entites - points, lines and areas - which can be described by coordinate locations. The spatial data may exist as maps at various scales, photographs, or as digital data.

The primary spatial determinant is the location of deposits, but an inventory of potential deposits is not sufficient. Given a pool of identified deposits, their availability and potential for reservation must be further defined by proximity to points of use, access to transportation corridors, adequate separation from conflicting land use, and the extent and location of other natural resources. The spatial relations among these factors must be analyzed and the spatial conflicts resolved.

Ideally deposits near population centers deserve preferential protection. While this strategy increases the potential for land use conflicts, these could be minimized by careful planning. Despite the land use conflicts, the strategy offers substantial advantages. Transportation costs would be reduced, stabilizing material prices. Reduction of road haul distances would reduce road deterioration, road maintenance costs, and public exposure to truck traffic. Higher land values in urban locations could increase the incentives for reclamation and subsequent use of extractive sites. The success of the strategy will depend on the ability to combine and analyse all the relevant components so that public needs and values are satisfied.

GEOGRAPHIC INFORMATION SYSTEM CAPABILITIES

What does a geographic information system offer in terms of meeting these requirements? Currently many systems provide the means to combine and analyse information in diverse forms (17). Several sets of information compiled at different scales and map projections can be readily transformed and combined. Graphic, electronic, and tabular data can be integrated.

Geographic information systems allow more explicit and objective analysis and more rapid examination of alternatives than can be accomplished by manual methods (15). The process is repeatable and not subject to the different interpretations which may occur with each step in manual overlay and analysis. Results are less tenuous and less vulnerable to the litigation which may result from the more rigorous scrutiny given resource issues today.

Although a more explicit, objective and comprehensive analysis can be undertaken the quality and accuracy of the information has not neccessarily improved. Explicitly recording map data as numbers creates an illusion but does not generate greater accuracy than existed in the original data (1). Until recently the quality and accuracy of the digital data has not been an issue. Gordon in a study of systems in the Pacific Northwest, discovered a "noticeable lack of concern for data accuracy and data documentation."(8, p.263) If current trends in resource planning prevail, data accuracy must be improved and data quality documented not just presumed.

OVERLAY ANALYSIS

Resource analysis is commonly done with map overlay. Used for over half a century, the process continues as a fundamental concept (16) and has been adopted by GIS technology. Early computer applications used grid data, and this form is still recommended and assumed to be the most efficient and effective form for computer overlay (4). Grid form perhaps simplifies the process, but generalizes the data.

Detailed representation or legal definition of resource boundaries presents a problem. Resource boundaries by their nature are uncertain, highly variable lines. At some points they may be quite distinct and at others, indeterminant. A grid data structure is incapable of representing this variable quality in the individual lines. Since this information cannot be included in the structure it is not incorporated in the analysis and thus not accounted for in policy decisions.

Resource planning at the local level requires combining uncertain resource boundaries with detailed parcel level information. Overlay of these two sets of information has created concern for the erroneous assignment of information to specific parcels of land. A "judicious choice of grid cell size" (11, p.103) has been suggested as a solution to this problem. A large cell size is recommended to assure correct allocation of cells to areas. For detailed planning this solution fails since large cells further generalize the data and lose small features. Although large cells may be defensible for the resource data, they are too coarse to tie to the legal and fiscal network.

Detailed resource planning requires a data structure that captures the integrity of the item of interest and data processing capabilities which maintain this integrity. Vector format can capture a desired level of detail from source documents, and polygon overlay preserves this level of detail. A particular advantage of vector data and polygon overlay for resource planning is that resource boundaries can be represented and maintained as individual lines. Where the uncertainty of the boundary varies, it can be explicitly expressed as an error band around the line (2). An error band localizes uncertainty where it exists instead of using a coarse grid over the whole area. The advantages of this approach have been known, but no effective polygon overlay program was available for many years.

THE STUDY

This study employed an assortment of maps typical of a local resource planning application. A small scale sand and gravel resource inventory identifying known and potential deposits was combined with detailed land use and zoning district information to identify limitations imposed on supplies and to identify deposits for reservation. The study area covered the four northwestern townships of Walworth County, Wisconsin.

To process, combine, and analyse the information, the study used a state of the art geographic information processing system, ODYSSEY (10), developed at Harvard Laboratory for Computer Graphics. The system maintains vector form through all phases of processing and is able to perform efficient polygon overlay with a relatively complex data set (5).

Data Sources

The sand and gravel resource inventory was the product of a PhD thesis by R. Fausett (7) in cooperation with the Wisconsin Geological and Natural History Survey. Fausett integrated several existing land records including soil surveys, well logs, glacial and geological records, topographic maps and some field checks to identify sand, gravel, and bedrock deposits. These deposits were delineated on a mylar 1:62,500 scale base map. At the 1:62,500 scale, the line width delineating deposit boundaries occupies approximately 140 feet on the ground.

The Southeastern Wisconsin Regional Planning Commission (SEWRPC) provided 1975 land use data in digital form. Land use categories were interpreted on 1:4800 scale aerial photography. Land use classes were coded in 79 categories at a level of detail in which individual lots and farmsteads could be distinguished. The interpreted photos were digitized on SEWRPC's Calma system using State Plane Coordinates. The digital data for 144 sections were transferred on tape in separate files as linework and land use symbols, and were restructured for compatibility with ODYSSEY software.

The Walworth County Planning Office provided the zoning district information for portions of the study area. The zoning districts were compiled on SEWRPC's 1:4800 scale base aerial photography with line width occupying approximately 20 feet on the ground. A four section area was converted to digital form. Information on whether extraction could occur in a zone was listed as an attribute for each zone.

Consistent coordinates were established to prepare for the overlay analysis. The aggregate inventory and zoning districts were converted to the State Plane Coordinate system to match the land use data originally compiled in this system. The transformation used SEWRPC's records for the surveyed location of remonumented Public Land Survey System (PLSS) section corners and estimated locations for corners not yet remonumented.

The first overlay combined the land use data and the sand and gravel inventory to detect deposits covered by conflicting uses. The land use data file for the four township area included approximately 8600 polygons and the resource inventory included approximately 1200 polygons. The overlay of the two files was completed in 30 minutes of CPU time on a VAX 11/780. While this is not a trivial amount of time, the cost is well within reason.

Certain boundaries, lake shorelines in particular, were common to both source maps, and where there were large discrepancies between the two versions, slivers and gaps occured in the overlay. The program eliminates small slivers and gaps within a tolerance (a distance within which points may be considered the same) (5). Specification of a tolerance distance maintains a desired level of detail, yet will remove trivial detail and reduce the size of the overlay file. In the case of this overlay, 243 slivers were removed at a five foot tolerance, leaving 9857 polygons.

The overlay program produces a cross reference file which describes new polygons created in the process and associates these with the polygons and attribute data of the original coverages. The program also calculates perimeter and area figures for each new polygon. The cross reference file provides the raw material for any overlay question. In this case land uses restricting sand and gravel extraction (including all built structures, wetlands, lakes, recreation and special agriculture) were grouped and compared to deposits. Tables 1 and 2 show results of this analysis.

TABLE 1

OVERLAY RESULTS
SAND AND GRAVEL DEPOSITS/LAND USES (in acres)

Land Use Categories

Deposit Quality and Status	Extractive	Residential	Agriculture	Forest	Lakes
High					
Existing	70.0	28.8	100.1	26.5	23.6
Potential	1.8	221.8	635.2	515.1	170.2
Intermediate					
Existing	26.1	54.1	729.3	285.1	32.8
Potential	44.9	456.8	5173.2	3827.1	162.8
Marginal					
Existing	24.6	13.8	1021.1	231.1	13.0
Potential	33.4	75.1	753.0	261.3	13.7
Lakes	0.0	117.3	13.2	25.2	2147.4

Discrepancies between the extractive land use category and existing deposits and between the lake boundaries on the two coverages can be attributed to both time and scale differences. The 117.3 acres of lakes from the deposit map found to be actually in residential land use shows that boundary errors pick out structural dependencies (2) such as the shoreline development in this area.

TABLE 2

LAND USE RESTRICTIONS ON DIFFERENT QUALITY DEPOSITS

	Restricted	% Restricted	Available	Total
High Quality				
Existing	140.9	39.5	215.5	356.4
Potential	837.9	40.0	1246.1	2084.0
Intermediate Quality				
Existing	220.3	16.9	1081.0	1301.3
Potential	2311.3	19.8	9311.4	11622.7
Marginal Quality				
Existing	148.5	10.2	1296.8	1445.3
Potential	217.2	17.1	1052.6	1269.8
Totals	3876.1	21.4	14203.4	18079.5

Figure 2 shows that high quality deposits are more likely to be restricted than intermediate or marginal quality deposits. Of the deposits in the area, 21.4 percent are restricted. Residential land use composes the largest percent (21.9) of the restrictions.

The zoning information was added to determine political restrictions on deposits, and to test the program's ability to combine several layers. A four section area was used for this test. The overlay involved two steps, first combining zoning and deposits then adding in the land use map. This last step took 4 minutes of CPU time on a VAX 11/780.

The cross reference file created for this overlay related the new polygons to each of the input coverages. With this file deposits could be checked for coincidence with particular zoning districts, land use classes or both. Figure 1 shows the composite overlay.

Restrictions On Deposits

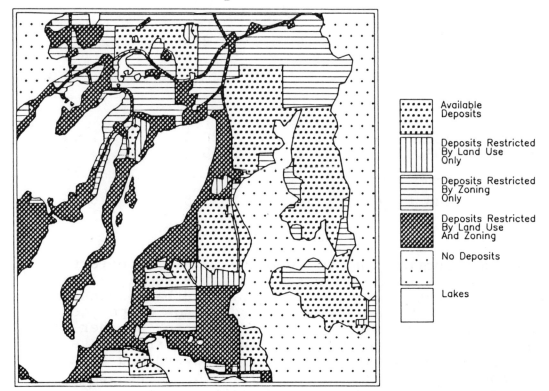

The total area of Figure 1 is 2827 acres. The sand and gravel inventory indicates 1533 acres of proven or potential deposits within this area. Area calculations from the overlay indicate that 1042 acres of deposits are restricted by zoning, land use, or both. Given these restrictions, only a third, 490 acres, of the identified deposits may actually be available for extraction. Land uses restrict only a small percent whereas zoning restricts over one third of the deposits.

TABLE 3

SUMMARY OF COMPOSITE OVERLAY
(in acres)

	Total Deposits	Restricted By Land Use	Restricted By Zoning	Restricted By Both	Avail Depos
High Quality					
Existing	14.6	0	7.3	7.3	0
Potential	705.7	18.5	253.0	325.2	109
Intermediate Quality					
Existing	108.8	8.5	10.7	13.8	7
Potential	613.4	22.6	242.5	109.3	23
Marginal Quality					
Existing	9.8	.6	2.4	0	
Potential	80.8	1.7	19.6	0	5
Totals					
Existing	133.2	9.1	20.4	21.1	8
Potential	1398.9	42.8	515.2	434.5	40
	1533.1	51.9	535.5	455.6	49

For this application, polygon overlay is an effective planning tool for identifying the available supply of sand and gravel and the supply precluded by suburban expansion and zoning restrictions. A further advantage of polygon overlay is that one layer can be used to check another (1), and in situations in which boundaries on different layers correspond the more precise boundary can overule a less precise boundary. In the resource inventory, the lake boundaries were substantially generalized. The corresponding boundaries in the land use file were assumed to be the more precise representation, and were made the definitive boundary in the composite overlay. In this situation the ability of the overlay process to maintain the detail of the land use data improved the quality of the less precise resource data.

CONCLUSION

Resource boundaries will remain imprecise but decisions regarding these boundaries will still be made. An efficient polygon overlay process is an improvement in computer

assisted resource planning and a step toward meeting the
requirements of a more rigorous resource planning
environment.

REFERENCES

(1) Chrisman, N. "The Role of Quality Information in the
    Long-Term Functioning of a Geographic Information
    System", Proceedings AUTO CARTO 6. Vol 2 p. 303-312
    1983.

(2) Chrisman, N. "A Theory of Cartographic Error and its
    Measurement in Digital Data Bases", Proceedings AUTO
    CARTO 5. p. 159-168 1982.

(3) Dangermond, J., L. Harison, and L.K. Smith. "Some Trends
    in the Evolution of Geographic Information System
    Technology", Harvard Library of Computer Graphics. Vol
    15 p. 27-34 1981.

(4) Deuker, K. "Land Resource Information Systems: A Review
    of Fifteen Years Experience", Geo-Processing, 1: p. 105
    - 128 1979.

(5) Dougenik, J. "WHIRLPOOL: A Geometric Processor for
    Polygon Coverage Data," Proceedings AUTO CARTO 4. p.
    304-311 1979.

(6) Dunn, J. and H.L. Moffet. "Mineral Exploration in the
    U.S.: Problems and and Answer," Mining Congress
    Journal, Vol 61 No 11. p. 22-26 1975.

(7) Fausett, R. "An Interdisciplinary Approach to Resource
    Inventory," PhD Thesis, University of
    Wisconsin-Madison, 1982.

(8) Gordon, K. "Computer-Assisted Environmental Data
    Handling: Issues and Implications," Proceedings AUTO
    CARTO 4, p. 259-266 1979.

(9) Little, C. "The Microcomputer and the Politics of Rural
    Land," American Land Forum Vol 3 No 4. p. 14-28 1982.

(10) Morehouse, S. and M. Broekhuysen. "ODYSSEY User's
     Manual," Harvard Laboratory for Computer Graphics,
     Cambridge, MA. 1981.

(11) Natinal Research Council. Procedures and Standards for
     a Multipurpose Cadastre, Washington,D.C. 1983.

(12) National Research Council. "Surface Mining of Non-Coal
     Minerals Appendix 1: Sand and Gravel Mining, and
     Quarrying and Blasting for Crushed Stone and Other
     Construction Minerals," Washington, D.C. 1980.

(13) Neimann, B. "Westport Land Records Modernization Project: Results and Implications," Paper presented at Harvard Computer Graphics Week, 1982.

(14) Sheridan, M. "Urbanization and its Impact on the Mineral Aggregates Industry in the Denver, Colorado Area," Information Circular 8320, U.S. Bureau of Mines. 1967.

(15) Smith, L.K. ed. "Review and Synthesis of Problems and Directions for Large Scale Geographic Information System Development," ESRI Redlands, CA 1983.

(16) Steinitz, C., P. Parker, L. Jordan. "Hand-Drawn Overlays: Their History and Prospective Uses," Landscape Architecture 66 (5): 444-455 1976.

(17) Tomlinson, R., R. Boyle. "The State of Development of Systems for Handling Natural Resources Inventory Data," Cartographica Vol 18 No 4 65-95 1981.

(18) Werth, J. "Sand and Gravel Resources: Protection, Regulation, and Reclamation," Planning Advisory Service Report No 347 1981.

PART 3
URBAN STUDIES

Jack Dangermond and Carol Freedman
Environmental Systems Research Institute
380 New York Street
Redlands, California   92373
(714) 793-2853

FINDINGS REGARDING
A CONCEPTUAL MODEL OF A MUNICIPAL DATA BASE
AND IMPLICATIONS FOR SOFTWARE DESIGN

Abstract

This paper offers a new perspective on the components and inter-relationships among land and geographic-related data which are used in a municipality. The various categories of land-related data and how they interact to form a municipal data base are described. In this paper, these various data components are identified, as well as the municipal tasks which generally make use of them, the types of generic software procedures which make them usable in an automated form, and the interrelations among them which must be understood as a basis for municipal data base design. The appendix to this paper provides a description of ARC/INFO, a software package which has been developed in response to and in support of this conceptual model and its implementation.

Introduction

Numerous jurisdictions have attempted to automate segments of their operational and procedural functions over the past several decades. These wide-ranging efforts have employed a variety of tools from the data processing community. One category of these attempts is associated with the automation of land and geographic related information.

Various communities have attempted to automate particular types of land and geographically-related information for individually selected municipal operations. For example: public works departments have attempted to automate plan and profile drawings, or maintain engineering drawings; planning departments have made attempts to automate parcel maps; environmental planners have attempted to automate environmental overlays for land suitability/capability; tax assessors and other municipal agencies that use parcel level data (such as zoning and building inspections and so forth) have attemped to automate a municipal tabular file of land related data; and a variety of agencies working with street networks have attempted to work with geographic base files to abstract the street network for address matching, census mapping and related kinds of applications.

To date, the implementors of these various kinds of automated municipal applications have been generally handicapped by a lack of a full understanding regarding a municipal data base model which generically describes the various components commonly found in a municipal data base. This paper provides a description of these various categories of data and how they interrelate with one another to comprise a municipal data base. The paper summarizes the results of an in-depth study which applied a rigorous scientific methodology and approach to the identification of a municipal data base model. The findings, which were intuitively understood in the past, are now substantiated on the basis of the scientifically documented methodology which was undertaken. The purpose of this presentation is to provide a tool for designers and implementors of such systems so they can better understand how various hardware and software tools, as well as data base concepts, relate to the overall data model that exists for the typical municipality.

In describing this municipal model an effort has been made to isolate it from any single municipality; therefore, the model is presented without regard to organizational structure, institutional context, or the way that entities are implemented within an existing municipality.

Background and Basic Concepts

This conceptual model has evolved over a number of years through research conducted in the City of Los Angeles, California, the City of Anchorage, Alaska, Santa Cruz County, California, Clark County, Nevada, and most recently through an extensive scientific research and application project sponsored by the Provincial government of Ontario, Canada. It was through this last effort that the conceptual model became clear and its form crystallized as it appears in this paper.

The Province of Ontario experience involved extensive interviews by a research team which examined four selected municipalities within the Province of Ontario. For each municipality, all of the departments which dealt with land-related information were interviewed extensively in terms of their mandates, their organizations and responsibilities, the tasks which they perform, the systems and data bases that they utilize, their existing data sources, and also their potential for use of geographic information systems. The Ontario effort involved numerous on-site interviews with each of the relevant agencies, as well as analysis of agency functions related to various ways that hardware, software and data bases could be used to their maximum potential.

From this basic research in Ontario, an extensive list of categorical tasks were identified, then generalized into a set of generic tasks common within municipalities. For example, generic municipal tasks include street addressing, building permitting, zoning subdivision inspections, and so forth. Table 1 lists thirty-three observed generic municipal tasks.

Municipal functions or tasks, and the types of data which support them comprise the vital elements of municipal operations. An examination of these generic tasks, together with the manual and automated data bases which are used in support of these functions, provide the fundamental framework upon which to build a "conceptual model" of geographic data entities and their relationships in a municipality . This understanding may then be directed into data base and information system design, and implementation planning.

Modern principles of data base management suggest that building a data base model may help to minimize or eliminate redundancies of data and operational function, and establish data entities and items which inter-relate with each other (so that transactions which occurring in the data base automatically update not only basic data entities but also the related information on which the data base is dependent).

A data entity, as it is termed in this paper, is a data source, such as a parcel map or a parcel attribute file. Each data entity is comprised of a number of data items such that, for example, one item in the parcel attribute file might be the area of a parcel. A descriptive summary of these components and their interrelationships is provided in the sections which follow.

Table 1

OBSERVED GENERIC MUNICIPAL TASKS

On-Going Procedural

1. Acquire and Dispose of Property
2. Process and Issue Parcel-Related Permits
3. Perform Inspections
4. Provide Legal Notification
5. Issue Licenses
6. Conduct Street Naming
7. Review Site Plans
8. Review Subdivisions
9. Create Street Addresses
10. Perform Event Reporting
11. Conduct Dispatching
12. Perform Vehicle Routing
13. Conduct Traffic Analysis
14. Conduct Facility Siting
15. Administer Area Districting
16. Administer Zoning By-Laws
17. Conduct Land Use Planning
18. Conduct Engineering Design
19. Conduct Drafting
20. Conduct Title Searches
21. Perform Tax/Fee Billing Collection

On-Going Managerial

22. Create and Manage Mailing Lists
23. Allocate Human Resources
24. Perform Facilities Management
25. Perform Inventory Management
26. Perform Resource Management
27. Conduct Weed Control
28. Perform Map Management
29. Conduct Drawing Management
30. Perform Data Base Management
31. Conduct Development Tracking
32. Disseminate Public Information
33. Respond to Public Inquiries

Municipal functions or tasks, and the types of data which support them comprise the vital elements of municipal operations. An examination of these generic tasks, together with the manual and automated data bases which are used in support of these functions, provide the fundamental framework upon which to build a "conceptual model" of geographic data entities and their relationships in a municipality . This understanding may then be directed into data base and information system design, and implementation planning.

Modern principles of data base management suggest that building a data base model may help to minimize or eliminate redundancies of data and operational function, and establish data entities and items which inter-relate with each other (so that transactions which occurring in the data base automatically update not only basic data entities but also the related information on which the data base is dependent).

A data entity, as it is termed in this paper, is a data source, such as a parcel map or a parcel attribute file. Each data entity is comprised of a number of data items such that, for example, one item in the parcel attribute file might be the area of a parcel. A descriptive summary of these components and their interrelationships is provided in the sections which follow.

<u>The Conceptual Model</u>

In a conceptual sense, there are ten data components in the municipal data model. Each of these may be conceived of as containing a series of data entities, such as maps or tables related to map coordinates. The ten basic components are as follows:

1. Base Map
2. Environmental Overlays
3. Engineering Overlays
4. Plan/Profile Drawings
5. Parcel Maps
6. Parcel/Street Address Tabular Data
7. Area Tabular Data
8. Street Tabular Data
9. Street Network File (geographic base file)
10. Area Boundary Maps (administrative boundaries)

Each of these different components contain data about geographic variation and status, which are commonly used to carry out municipal activities. However, some of these are used at much higher rates than others. Findings from municipal studies to date strongly indicate that the majority of municipal, transactions are associated with the land parcel or street address tabular data. For the broadest range of municipal applications, the tabular format is more efficient. Other data components, such as plan and profile maps, are used only occasionally in some of the engineering-related functions.

Figures 1 and 2 illustrate the interrelationships between these 10 different classes of data. The figure also indicates ways in which geographic information tools would be invoked to establish these relationships.

<u>Description of the Data Components</u>

1. <u>Base Map</u>. The basic elements of the engineering base map are control data and topographic elevation contours. These provide a framework and reference for integrating all of the other manuscripts and overlays for a particular geographic location. Control data are typically provided in the form of geodetic control. This gives the absolute location in the form of x,y coordinates, whereas contours are provided in the form of line drawings of topographic elevations and altitudes.

2. <u>Environmental Overlays</u>. These are typically maps of soils, geology, vegetation, land form, hydrography, slope, and related geographic features. In the typical municipality, these overlays are used to associate the suitability and capability of land for various types of uses, such as potential mineral or forest resources, as well as possible reference to natural hazards such as floods, earthquakes, landslides, areas unsuitable for given types of development, and so forth.

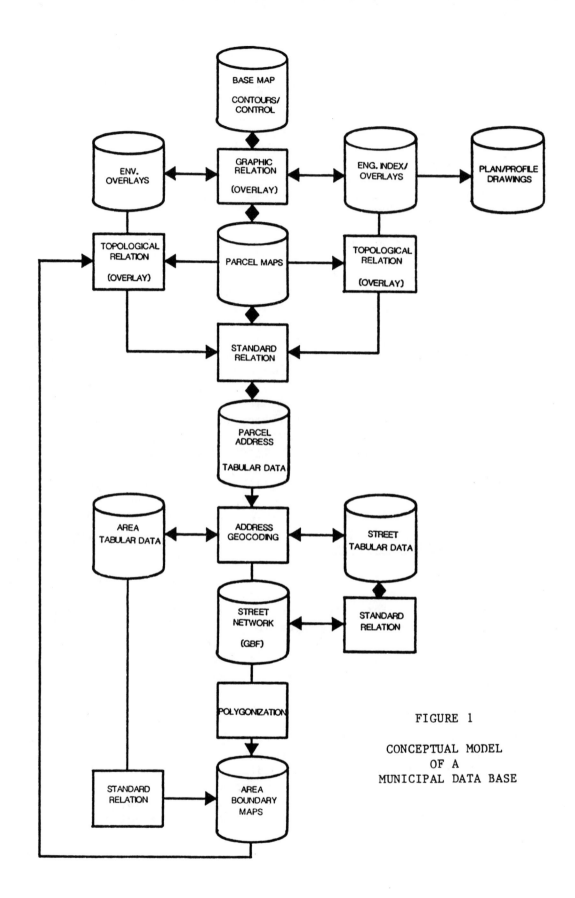

FIGURE 1

CONCEPTUAL MODEL
OF A
MUNICIPAL DATA BASE

FIGURE 2

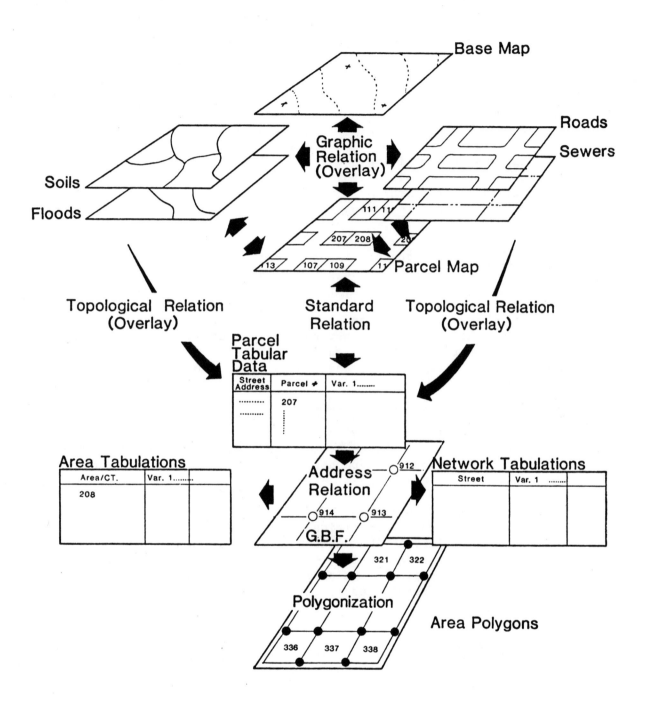

3. Engineering Overlays. These are maps drawn in reference to the base map, which indicate data such as locations of roadways, utilities and infrastructure, lay-outs (i.e., lines for water, sewer, cable television, gas, electric, hydro, sidewalks, fire hydrants, utility poles, etc.). These maps are typically used for reference in grading and construction of facilities.

4. Plan Profile Drawings. These are the engineering drawings generated as part of the engineering design process and typically performed by a "cad type" interactive graphic system. This type of system not only assists in design work, but also can be used as a manuscript or records manager for filing and retrieval of basic engineering drawing and plan/profile drawing information. These highly detailed engineering records are not considered to be GIS maps, but are normally considered in evaluating cartographic activities within the municipality.

5. Parcel Maps. These maps provide the basic drawings of land ownership boundaries for all public and private lands. Parcel maps are typically maintained at a variety of scales, and may frange from very precise to very general "cartoon maps" (which show only schematic layouts of the basic land cadastre). Parcel maps are commonly related in some manner to legal description documents which typically constitute the record for describing boundary ownership for a given parcel (e.g., metes and bounds descriptions).

6. Parcel/Street Address Tabular Data. This data typically includes a variety of tabular data sets associated with the parcel. In cases where buildings are actually constructed, these parcels are typically assigned to street addresses or site addresses which correspond with the tax block and parcel type designation commonly used by the tax assessor.

This data file is typically one of the central data bases used for supporting operational activities within a municipality. Against this file, a variety of activities occur (for example, in the areas of subdivision of land, zoning, building permits, and so forth). Therefore, it is typically maintained as a current file, often implemented with distributed processing terminals to a variety of agencies.

7. Area Tabular Data. This data entity is typically comprised of a variety of tabular data bases about blocks, census and immigration areas, and a host of other administrative areas summarized into aerial unit tabular form. These data could include police statistics on events, number of students by district area, and statistics from the census on relations, population and housing, health statistics, and so forth.

8. Street Tabular Data. The street tabular data entity includes a variety of statistics associated with the street link, (typically, a link running between two assigned nodes or street intersections). This tabular data may relate to road conditions, size, or maintenance quality; event data such as accidents or crime; or population and housing characteristics associated with properties which abut the street links (such as, the number of single family homes on a street face and their maintenance quality, the number of students living on a particular street block, or the amount of garbage requires pick up).

9. Street Network File (Geographic Base File). This is typically a topological street network file providing intersection numbers, street names, link numbers of each street, address ranges of the streets, and the polygons on the right side and left side of each street. This street base file provides a linkage between address data and street faces that surround a municipal block. By using the concept of address geocoding, parcel data may be matched with the street network to create street tabular data.

10. Area Boundary Maps. Area boundary maps are coordinate boundary data bases which define the polygon outlines of blocks, enumeration areas and various administrative boundaries such as school districts, police districts, census summary districts, zip codes, and so forth. These maps correspond directly to the area tabular data described above.

Table 2 presents an accounting of the ways in which particular generic municipal tasks are observed to use the ten identified data entities in the municipal model.

The types of data which municipalities generally require typically exists in one or more existing agencies. However, studies conducted by ESRI have found that often data desirable in one agency (e.g., street network data for ambulance dispatch) resides in and is maintained by another agency (e.g., roads maintenance or planning). Such data is not readily available in the format or location necessary for reuse or data sharing. Thus, duplication of effort in data collection, storage and maintenance are found. The integration of land-related data entities for a given study area into a single unified system will eliminate duplications of effort and cost, and result in greater economies and efficiencies for the jurisdiction(s) involved.

Description of the Basic Procedural Techniques Required

To create the necessary interrelationships among the data entities just described, five basic types of procedural software techniques are employed. As illustrated in Figure 1, these basic tools are: graphic overlay, topological overlay, address geocoding, polygonization, and relational matching.

TABLE 2
Page 1 of 3

DATA REQUIREMENTS

POTENTIAL USES	BASEMAP	ENVIRONMENTAL OVERLAYS	ENGINEERING OVERLAYS	PLAN/PROFILE DRAWINGS	PARCEL MAPS	PARCEL/ADDRESS TABULAR DATA	AREA TABULAR DATA	STREET NETWORK FILE (GBF)	STREET TABULAR DATA	AREA BOUNDARY MAPS
1 Acquire and Dispose of Property		X			X	X	X			X
2 Process and Issue Parcel-Related Permits		X			X	X				X
3 Perform Inspections	X	X		X	X	X		X		
4 Provide Legal Notification	X				X	X		X		
5 Issue Licenses	X									
6 Conduct Streetnaming										
7 Review Site Plans										
8 Review Subdivisions	X	X	X	X	X	X		X		
9 Create Street Addresses					X			X		
10 Perform Event Reporting		X			X		X			X
11 Conduct Dispatching					X	X				
12 Perform Vehicle Routing		X			X		X	X	X	X
13 Conduct Traffic Analysis		X	X					X	X	

TABLE 2
Page 2 of 3

DATA REQUIREMENTS

POTENTIAL USES	BASEMAP	ENVIRONMENTAL OVERLAYS	ENGINEERING OVERLAYS	PLAN/PROFILE DRAWINGS	PARCEL MAPS	PARCEL/ADDRESS TABULAR DATA	AREA TABULAR DATA	STREET NETWORK FILE (GBF)	STREET TABULAR DATA	AREA BOUNDARY MAPS
14 Conduct Facility Siting	X	X	X	X	X	X	X	X	X	X
15 Administer Area Districting					X	X	X	X		X
16 Administer Zoning By-Laws		X	X		X	X	X	X	X	X
17 Conduct Land Use Planning		X			X	X	X	X	X	X
18 Conduct Engineering Design	X	X	X	X	X	X	X	X	X	X
19 Conduct Drafting	X	X	X	X	X	X		X		
20 Conduct Title Searches										
21 Perform Tax/Fee Billing & Collect					X	X	X			
22 Create & Manage Mailing Lists					X	X	X			
23 Allocate Human Resources							X			X
24 Perform Facilities Management			X	X	X	X		X	X	
25 Perform Inventory Management	X	X	X	X	X	X	X	X	X	X
26 Perform Resource Management		X			X	X	X			X

111

POTENTIAL USES	AREA BOUNDARY MAPS	STREET TABULAR DATA	STREET NETWORK FILE (GBF)	AREA TABULAR DATA	PARCEL/ADDRESS TABULAR DATA	PARCEL MAPS	PLAN/PROFILE DRAWINGS	ENGINEERING OVERLAYS	ENVIRONMENTAL OVERLAYS	BASEMAP
27 Conduct Weed Control									X	X
28 Perform Map Management	X			X	X	X		X	X	
29 Conduct Drawing Management		X	X			X	X	X		
30 Perform Data Base Management	X	X	X	X	X	X			X	X
31 Conduct Development Tracking					X	X				
32 Disseminate Public Information	X			X	X	X			X	
33 Respond to Public Inquiries	X	X	X	X	X	X				

Graphic overlay involves overlaying several manuscripts atop one another to plot selected map elements which coexist at the same geographic locations.

Topological overlay involves the joining of geographic elements from two or more separate files to create an "integrated" cartographic file. This new file consists of intersecting of common elements of the selected data files (e.g., the joining of cartographic elements via polygon overlay to form new tables of attributes).

Address Geocoding is a procedure through which street address data and street network data are associated. Using a GIS matching capability, street addresses may be matched with a range of addresses assigned to street links.

Polygonization involves the chaining together of link segments (such as street network links) to form district boundaries comprised of related polygons (i.e., administrative districts for schools, police, census or other applications).

Relational matching is the GIS procedure used to relate two entities for functional purposes. For example, parcel tabular data may be relationally matched with parcel maps, or area tabular data with area boundary maps, and so forth.

These generalized categories of software procedures are representative of an expansive range of tabular and cartographic GIS procedures available for data entry, analysis, manipulation, display and management.

Distinctions must be drawn between automated geographic information systems and computer aided design-type systems. An automated geographic information system is a computerized system composed of hardware, software and a data base, all of which are designed to efficiently store, analyze, display and present geographic data. Systems built around the interactive display technology of the interactive CRT (typically accompanied by hard copy devices, digitizers and so forth) are considered graphic systems, many of which can create, store and display map graphics. The emphasis of these systems is on relationships visible to the user, resulting in implicit understandings of the user about how data interrelate. This contrasts with the emphasis of GIS systems upon data and topological relationships, which results in explicit data relations and data base contests.

The procedural techniques which spring from the municipal data model require a relational data base management structure within a GIS, and a broad array of tabular and cartographic procedures. The array of GIS software tools which are now available allow different data sets in mapped or tabular form to be correlated, interrelated and interpreted in a variety of ways. The GIS geometric tools are particularly important for translating implicit relationships held on maps into explicit relationships which may be stored in tabular form for use in diverse ways in conjunction with cartographic elements. Some of the more important of these tools used to establish these explicit relationships include polygon overlay, point-in-polygon, line-in-polygon, buffer and network analyses. By incorporating relevant data into usable tables, various manipulations may be conducted mechanically, thereby eliminating time-consuming manual searches for data, manual interpretation and reporting efforts. For example, searches to identify the distance of a sewer line to a given parcel or the distances between fire hydrants and a land parcel which they serve may be queried instantly, with the findings automatically produced.

Appendix 1 provides a description of software procedures which would be typically invoked to operate on the generic model and support the typical application functions common in a municipality. It is important to note also that this software system was designed to support both the basic procedural tools discussed here, as well as provide an expansive range of capabilities for managing geographic information in an efficient and readily usable system.

Interpreting Interrelationships among Data Components

Historically, the principles of data base management have been applied to the development of a GIS for a single-function aplication, such as the automation of tabular records on tax assessment. However, by conceiving of the entire municipal data model and each of its relevant inter-relationships and uses, it is possible to apply data base management principles to the entirety of land-related municipal operations. A multi-purpose GIS capable of responding to a broad range of municipal project needs in a systems orientation may be thus derived.

In examining the graphic representation of the data model (Figure 1), one can see the interrelationships between these generic data types and how, by interrelating them, one can create new data or convert original data into another form.

1. Interrelations using topological overlay. Environmental data may be transferred to the parcel address tabular file. In this case, environmental polygons, lines or points may be overlaid on top of the parcel maps, and a topological overlay technique may then be used to calculate areas with certain soil or flood-prone characteristics based upon the environmental overlay. The new data may then be stored in the actual parcel tabular file. An example of an environmental overlay on top of a parcel map could involve the overlay of flood plains on parcel maps, with a product being the designation on the parcel/address tabular data base of those parcels which contain flood hazards.

Another procedure involves the topological overlay of engineering maps on top of parcel maps to form tables of attributes by parcel concerning engineering data. For example, such a procedure could show the relationship of a sewer line in the street and a parcel that runs across the front of it, by tagging each parcel that has sewers.

This would be accomplished by running the polygon overlay of the parcel map on top of the sewer infrastructure overlay.

2. <u>Interrelations using graphic overlay</u>. The graphic overlay function can be used to overlay engineering indexes or engineering drawing data on top of parcel maps (i.e., to show the graphic relationship of utility overlays to specific parcels). The base map with contours and control can be used as a basic underlay for the composite of engineering and environmental or parcel maps within the municipality. This type of graphic function, although visually of value, is not a topological overlay as described previously in which maps are intersected and joined to create actual extensions of the tabular files.

3. <u>Interrelations using address geocoding</u>. By taking the address tabular data base and the address geocoding software, one can associate address tabular data to the street network file, thereby creating a street tabular data base. For example, by associating police statistics collected by street address to the street network, tabular data by street face regarding crime may be created.

4. <u>Interrelations using polygonization</u>. The street network file (geographic base file) may be converted into polygon boundaries by employing polygonization techniques. Specific software is available which may be used to polygonize the various street links into polygon boundaries.

5. <u>Interrelations using relational matching</u>. The relational matching procedure is another technique commonly used to relate various data entities for functional purposes. It is commonly used to relate parcel maps with parcel address tabular data, parcel tabular data with area or street network tabular data bases, network map with network tabular data bases, and area map with area tabular data bases.

Data Interrelationships Outside the Municipal Model

Beyond the basic model, a typical municipality has other kinds of data base linkages which should be recognized. For example, in the Province in Ontario, the basemap, environmental data and a large volume of parcel data is provided by Provincial agencies, much of the area tabular data is provided by Statistics Canada, and tabular data about street faces is provided by various transportation agencies also outside the municipal realm. In other words, there may be important relationships between the municipality and various outside organizations which are vital for the information they provide into the basic data model. In the U.S., the U.S. Census Bureau may provide area tabular data to municipalities, for example. Yet sometimes related base maps are provided by the municipalities themselves, which in turn rely upon control by an outside organization with geodetic monumentation. These types of data interrelationships must be carefully considered and understood in applying the principles of the conceptual model.

Findings and Implications of the Conceptual Model

The municipal data model described in this paper has broad applicability for other cities of varying sizes and configurations. The essential concepts discussed here should be relevant to a range of user situations and applications. What is important is the specification of essential data elements, user functions, and the tool box of software procedures which are fundamental for fulfilling the requirements of the urban municipality today. It is important to recognize that software packages are currently available to support the data model which has been described.

There are several basic implications of the municipal data model. Firstly, it provides a description and illustration of the basic municipal data types and their relations. Using this information, a system designer can better understand how these components interrelate and thereby design linkages which accurately account for data dependencies and independencies. Lacking such an understanding, the designer is likely to reiterate basic flaws and problems which have affected municipal data inventory design in the past.

This model should also be useful as a base for designing application procedures which respond to user needs. It is felt that this model provides a basic and meaningful context for the development of application software to be used in particular departments or applications. In addition to these two basic uses, the model has value for a data base administrator, for whom the model shows the relationships between individual data entities that must be solidified in order to have optimal interrelations of the data base.

It is important to note that a variety of single purpose functions are typically carried out using the same or similar data within a given municipality. When these commonalities are identified, and a broad-based set of data entities are established, redundancies of data and user function may be reduced or eliminated.

This model may also aid a non-municipal agency to better understand the data model, and ways in which such an agency may provide data to a municipality in a format which relates to the currently operational data bases within the municipal environment. For example, using the model, a survey organization can determine the best formats and geocodes in which to deliver data products which may become part of a municipal data base. In other words, the data model is intended to present a common vernacular which helps in communication regarding ways in which a particular data item being collected fits into the municipal model. The models also show the way in which information flows from very specific data which is collected (i.e., address information about a parcel) to more generalized summaries of data (i.e., tabular data summarized to the street face or to a specific aerial unit).

To integrate this conceptual model in an actual functional model, several steps are required. First, a detailed description of the data items which exist within each of the general data classes must be developed. In getting clear on this specification, it is necessary to identify for each data item the agencies from which the data originates, is updated and is utilized. Then, the linkages or interrelationships between these various data items and entities may be understood, as well as the ways in which various municipal transactions affect particular data in the system. Using this information as a base, the design of an efficient and sensible automated data system may proceed.

Description of
Generic Software Procedures

The generic software procedures described here comprise a detailed range of user-oriented functions observed as needed in a broad range of user applications. The generic procedures are categorized according to the types of data which are handled, and the focus of the procedures invoked.

Tabular Data Management Procedures

DATA ENTRY is a process of key entering tabular data into a structured file. This is typically accomplished by using a CRT terminal with a formatted screen.

UPDATE is a process of revising stored tabular data on an automated system. This is best performed using an alphanumeric screen editor with well-defined software procedures to control the process.

DATA MANIPULATION includes the abilities to perform various arithmetic and logical operations on tabular data, as well as to sort and relate different data sets to each other. This is normally done within the context of a model, process or procedure. There are various types of data manipulation: file and record processing, and logical/arithmetic manipulation.

File Processing is concerned with the operations of a set of entities within a file. Record processing is concerned with the retrieval of specific records from a file and the subsequent manipulation of fields within the record using arithmetic or logical functions.

Logical or arithmetic manipulation involves the manipulation of individual values for a particular purpose (e.g., addition of numbers in the rows of a table, or the use of logical procedures to model or analyze facts to obtain a reasonable response).

VECTOR CELL/CELL TO VECTOR CONVERSION is a process for the conversion of X,Y coordinate data expressed as vectors, points, lines and polygons into a cellular structure and also the conversion of cellular data into an X,Y coordinate format. The grid cell is a tile of an arbitrary two-dimensional ruling of the surface of the earth. The vector conversion requires the transferral of cartographic files from an X,Y coordinate form to a cellular form. Cells can be of any size, orientation or shape.

POLYGON DISSOLVE is the capability to merge two or more polygons which share common attributes by eliminating the shared boundaries.

POINT-IN-POLYGON is the procedure which determines if certain points lie within a given polygon. This capability can be enhanced so that the attributes of the polygon are assigned to points that fall within it.

Transformations are mathematical expressions used to convert coordinate data within one frame of reference to coordinate data in another frame of reference. It is used for a variety of applications including changing from one map projection to another map projection, or converting from one set of coordinates captured on a digitizer to UTM ground coordinates.

BUFFERING OR CORRIDORING is a process whereby a polygon or zone of fixed size is created around a point, line or polygon. These polygons may be created based on consistent or variable attributes of given cartographic features and produce continuous buffer polygon area..

WINDOWING is a function allowing the user to define a specific geographic area which is used to delimit a piece of a map or cartographic area. This process involves maintaining the attributes with the cartographic features. The cartographic features and the associated attributes can be retrieved using a window for subsequent display, analysis and manipulation.

PROXIMAL ANALYSIS is a procedure which involves the creation of polygons or areas around randomly spaced point locations. It is carried out by dividing equally the distance between paired points, then generating perpendicular lines to these mid-way points which are then extended to intersect and form areas. This process is used for creating polygons, mapping, and analyzing qualitative data where continuous coverage map is desired, but where contouring is not appropriate. It is possible to obtain pictorial displays of discontinuous data using this method.

COORDINATE FILTERING is the process of weeding out superfluous coordinates. It is used to generalize maps and to reduce requirements for computer storage.

LINE LENGTH CALCULATION is a procedure for computing the distance between successive coordinate pairs forming a straight line or following an irregular path.

AREA/PERIMETER CALCULATION is a procedure for geometrically computing the area or perimeter of polygon features.

SPATIAL AGGREGATION/DISTRICTING is a procedure which allows the user to semi-automatically determine districts. It provides the capability to aggregate polygonal data into larger units of polygonal data. Polygon dissolve is used as an operation of polygon aggregation. Districting is the process whereby areas may be interactively aggregated by utilizing toplogical boundary information around a set of entities (i.e., the ability to spatially aggregate polygons into larger polygons including the aggregation of the attributes by using topological identification of boundary nodes and segments, or by using a graphic function such as polygon overlay).

DISTANCE ACCUMULATION is the process of aggregating spatially distributed data within a specified distance from a feature. This procedure is used to summarize or enumerate geographic features such as points, lines or polygons and their attributes within a user specified distance of a particular geographic location. This is typically done using a grid cell-by-cell summarization with user specified distances.

CONTIGUITY ANALYSIS is concerned with adjacency relationships between any given polygon and its neighbors. Typically this involves summarizing and relating the attributes of neighboring polygons to the polygon being examined.

Cartographic Data Network Analysis Procedures

OPTIMAL PATH SELECTION is the process of using a network data base of road and road intersections to select the optimal path based on specified time/distance relationships.

FLOW SIMULATION is the ability to dynamically simulate movement of a collection of entities through a network. The technique is typically used to simulate vehicle flow on street, pedestrian movement through a pedestrian network, water moving through a pipe network, electricity moving through a concrete network, and so forth.

TIME/DISTANCE DISTRICTING is the ability to measure time/distance and accumulate time/distance data through a network radiating out from a given point, and associating these time/distance values to the points through which the measurements are taken. This procedure is used to define contour distances away from known points, and to determine districts which are created from a variety of such processes. It may be considered as proximal mapping on a network time/proximity/districting rather than distance bases.

Cartographic Data Query Procedures

SPATIAL QUERY allows the extraction of cartographic data based on user-defined windows, such as a circle, or other regular and irregular shapes.

ATTRIBUTE QUERY offers the ability to select data items from a file system based on the values of specific attributes or combination thereof, defined by arithmetic, relational and logical expressions.

Cartographic Data Display/Composition Procedures

QUERY/WINDOW defines the logical partition or bounds of the data base which determine the range of a query.

GENERALIZATION is the process of weighing coordinates and unnecessary data, such as in the deletion of polygons based on mineral size, elimination of other minerals, and so forth. This process concerns maintaining information content consistent with the skill with which it is being represented.

MAP JOIN/EDGE MATCH is the process of removing inconsistencies at the edges of maps from compiled map data so that the features match across map sheets. Edge match, map join and merge all constitute one function. Edge match can be considered a tie, map join is putting them together, and merge implies a global joining of a group of maps. Map joining compiles the physical and logical joining of map entities (features) from one map sheet to the adjacent sheet. While features may not be physically connected, they are logically connected. For example, a utility crossing several map sheets may not physically connect to itself at map edges, but its segments will be logically connected through pointers in a digital file. In a manual system, map joining simply requires ensuring that lines and features match at sheet edges where map sheets join. Map sheets, however, may also be physically connected as in the continuous map.

ADDRESS GEOCODING is the process of assigning alphanumeric locational identifiers (such as the municipal address or a physical location) to spatially related information. For example, an address may be matched to an address range on a street segment, or a given spatial area (i.e., the limits of a polygon, a line segment, a point along the segment, or an absolute point that has been coordinated). The process implies a geographic base file which can be used to pass addresses in order to find out characteristics about the geometry. This capability allows addresses to be matched to: 1) a specific point between street intersections; or 2) a complete line segment such as the street plan of a block, census tract, school district or other unit.

Geocoding can also be considered the encoding of something that can be related to position. It may involve matching an address to a place on the earth's surface, such as a specific property, a street face or block to a census tract or any other polygon. It may also involve digitizing of points, lines and polygons or legal descriptions to X,Y coordinates.

Cartographic Data Geometric Analysis Procedures

POLYGON OVERLAY is a function which allows the processing of one or more sets of polygons which have been overlaid to form a combined set of polygons. This results in a common set of values for the sets involved. There are different combinations of polygon overlays. One is used in update where it only takes on the atrributes of one of the previous sets, and is a process for cartographic analysis. Polygon overlay can be broken down into several other techniques such as line intersection, point-in-polygon, and cellular overlay. These are all techniques used to carry out the polygon overlay function.

STATISTICAL ANALYSIS is concerned with the aggregation of data according to well-defined statistical procedure in order to provide the user with a better perspective on the collection of data. The parameters prescribed regarding data collection itself are very meaningful for interpreting the data in the aggregate rather than individual data items. Statistical analysis includes a number of clearly defined functions, such as standard deviation, mean, average, regression, correlation, factor analysis, principal component analysis, and others.

In each of these examples, the data is viewed as a collection rather than as individual entities. Statistical analysis is usually defined in terms of a library of functions. It is important to understand what is being done by the functions in order to be able to properly interpret the results of the analyses.

ON-LINE QUERY is the process by which items of interest are selected according to certain criteria. The criteria are normally based on arithmetic and Boolean connectives. This process typically results in a list of data items which can be displayed by a report generator or stored in another file to be operated on by other functions. The output is a set of items which satisfy the condition or conditions specified.

REPORT GENERATION is a mechanism by which information is identified, extracted, ordered and displayed in tabular form, based upon access from a single file or a set of files. The user specifies output format on the file or files. A user may also define operations to be performed on the data items such as summarization, averages or any logical operations in a standard form.

DIGITIZING AND EDIT comprise the digital encoding of compiled map data including points, lines and polygons. Editing of digitized data is simply the elimination of errors created during digitizing, and the replacement or displacement of line segments to the proper position. Such errors are generally cartographic in nature.

COGO (Coordinate Geometry) is set of mathematical tools and functions for encoding and converting bearings, distances, angles, and so forth into coordinate information. This operation is normally done on an alphanumeric screen into which data is entered and the geometry is determined analytically.

SCALE CHANGE provides the ability to plot or display data at a user defined scale. With this capability, scale is not confined to the input scale. This procedure may involve generalization techniques.

COSMETIC ENHANCEMENT offers the capability of adding text and cartograph symbology, by automatic or manual means, to improve the presentation of a graphic product. This includes legends, names, surround material, etc.

GRAPHIC SUPERIMPOSITION is the integration of two or more graphics in a consistent manner over the same area.

Cartographic Data Display/Presentation Procedures

POLYGON MAPPING is a cartograhic display of regularly or irregularly shaped polygons and their attributes. Typically, this capability includes shading, symbology and numeric labeling, as well as a variety of other map cosmetic functions for generating alphanumeric labeling of polygons.

POINT MAPPING is the depiction of attributes with graphic symbols or alphanumeric values as per user specifications.

LINE MAPPING provides cartographic display of linear features by means of X,Y coordinate pairs.

SURFACE MAPPING is the display of three dimensional information in terms of 2-D or 3-D forms. This can be achieved through contour mapping of 3-D relief displays, or color, shading, hatchering or other techniques.

ANNOTATION/TEXT is the positioning of actual textual data including graphical symbology for place names, feature names, surrounding data, etc.

IMPLEMENTATION
OF
LAND RECORDS INFORMATION SYSTEMS
USING THE
INFORMAP
SYSTEM

Charles H. Drinnan
Assistant Manager of Marketing
Synercom Technology, Inc.
500 Corporate Drive
Sugar Land, Texas

Presented at

URBAN AND REGIONAL
INFORMATION SYSTEMS ASSOCIATION

August, 1982

IMPLEMENTATION OF LAND RECORDS INFORMATION SYSTEMS USING THE INFORMAP SYSTEM

Introduction

The Land Records Information System (LRIS) implemented by most registry authorities is a combination of detailed maps and attribute information maintained separately. The INFORMAP Land Records Management System developed by Synercom offers the proven capability to maintain these records on a single turnkey computer system. The maps rather than individual plat maps organized by deed book and page numbers are combined to become a Continuous Digital Map covering the entire taxing authority. Both the map data and the attribute data are immediately available at a graphics display station for the entire operating area. The Continuous Digital Map is maintained on an accepted coordinate system such as the State Plane coordinate system so that each lot may be located accurately geographically.

For each parcel, a centroid is defined and attribute information is stored about the parcel. This attribute information stored in the Continuous Digital Map may become the permanent maintained parcel level data base or may be the result of interchanging information from a data base maintained on a main frame computer. In either case it may be accessed on line to produce a display or plot relating parcel oriented information to geographical information directly.

The Continuous Digital Map provides an accurate base for all types of map related activities including planning, public works, public utilities, natural resources management, flood plane analysis, etc. Thus the Continuous Digital Map becomes a valuable resource to the management of a county, municipality or state government. It is a resource which should be shared by all map users.

However, the tendency to perceive the sharing of map data as a one way process from the taxing authority to other users ignores the fact that planning, zoning, utility availability, natural environment and public hazards affect the evaluation of public property. A vertically integrated system combining information naturally maintained using smaller scale mapping with taxing authority maps at larger scales offers the potential to include in the evaluation of a particular parcel gross effects as well as detail. Systems such as the ESRI Geographic Information System have been used to evaluate the most favorable property to develop based on effects such as population density and affluency, accessibility, zoning, topography, etc. If the developer can evaluate property suitability, it seems only natural that the tax accessor should also systematically evaluate these parameters to assure a fair assessment.

The planned integration of the ESRI Geographic Information System and the Synercom INFORMAP system will provide a turnkey system capability to vertically integrate small scale and large scale mapping. Thus on a single system with a single data base analysis of many factors may be immediately combined with detailed land ownership maps and viewed directly on a graphics CRT.

This paper categorizes Geographic Related systems and then relates the system categories to Land Record Information systems which may be developed by the Registrar of Deeds, City Recorder or Tax Assessor and/or Public Works Director. It suggests that vertical integration of small and large scale mapping offers new potential for the mapper and the planner.

Examples are drawn from Forsyth County, City of Houston and Orange County data.

Categories of Geographic Related Systems

To develop a framework for the paper a categorization of types of automated systems is necessary. Geographic systems differ in level of accuracy, content and purpose. These differences are often orders of magnitude. The following categorization of systems is from low accuracy general analysis systems toward high accuracy detail systems:

A. Geographic Base File Systems with street networks and the areal units which they define (the U.S. Census Bureau is the best example).

B. Image Processing Systems with land sar and related satelite image data.

C. Generalized Thematic and Statistical Mapping Systems including environmental planning, census mapping, resource management, suitability analysis, etc.

D. Cartographic Mapping Systems producing maps such as the USGS Quad type maps, topographic base maps etc. Usually these are very high quality cartographic products obtained from stereo compilation.

E. Engineering Mapping including public works, gas, electric, and telephone mapping and corresponding system modeling.

F. Cadastral Mapping for city mapping including lots and blocks at a large scale.

G. Property and Parcel Information Systems to maintain detailed attribute information about each parcel.

This general categorization follows closely that of Dangermond (Reference 1).

These categories represent orders of magnitude in levels of accuracy, cartographic quality, and size of the resulting data base. However there is considerable overlap between the various categories. Many systems in the market place cover several of the categories.

Another method of differentiating these categories is by the types of basic algorithms required by each category:

Category	Algorithms
Geographic Base File	Topological Networking
Image Processing	Pattern Recognition Satellite Data
Thematic and Statistical	Polygon Processing Grid Cell Capability Analysis of Multi Layered Data

Cartographic Mapping	Stereo Compilation
	High Quality Output
Engineering Mapping	Facility Modeling and Maintenance
Cadastral Mapping	Survey Data Entry
	Coordinate Systems Based on Monuments
Property and Parcel Information	Large Scale Geographic Data Base Capability

The Synercom INFORMAP system combined with the ESRI Geographic Information System will provide the complete vertical integration of all these categories. (Note image processing is supported by interface to other systems). As such it offers a unique opportunity to integrate such diverse mapping applications as census bureau, planning, topographic, utility and assessment data into a single Continuous Digital Map.

Registrar of Deeds Mapping Requirements

The minimum mapping requirements for most Registrar of Deeds, Tax Assessors, and Recorders are mandated by law. The mapping requirements are usually to maintain maps at a large scale (typically 1" to 100') in a standard coordinate system (state plane) with property lines, lot dimensions, lot numbers, block numbers, subdivision names and street names. Sometimes origianl lot lines as well as occupied lot lines are included. Typically the developer submitts a plat map of a subdivision which is then approved, stored in a plat book (referenced by volume and page) and becomes the registrar's map. Despite efforts to obtain high quality maps, these maps generally do not fit together well over a large geographic area so that composite maps are hard to achieve. If the recorder does not redraft these maps into a composite, the relation between adjacent subdivision and even blocks within subdivisions is difficult to perceive.

In addition to mapping requirements the registrar is mandated to maintain deed records, chains of title, ownership, mortgage, deeds of trust, etc. This data is usually maintained by year in large books with cross index books. A title search of parcel ownership over fifty years often requires considerable effort in a manual system. In more modern land records systems this information is organized into a data base and maintained on a main frame computer. This becomes a Property and Parcel Information system.

Cadastral System Requirements

If the taxing and registry offices have access to a Cadastral Mapping System, the plat maps are replaced or augmented by digital maps. Typically small scale maps such as Figure 1 are maintained as index maps. These maps contain the major arteries of the city and other general approximate location data. The next level of maps are typically of a medium scale (1" to 200' or 400') and merely provide more detail. Figure 2 (which is the rectangle shaded in Figure 1) is typical and includes block lines, block numbers and subdivisions. Maps which are developed from census data using Geographic Base File Systems may be used for index and medium scale maps. Figure 3 is the DIME file map of Manhatten.

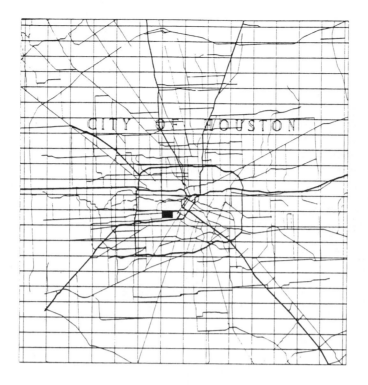

Figure 1

Small Scale Index Map

Figure 2

Medium Scale City Map

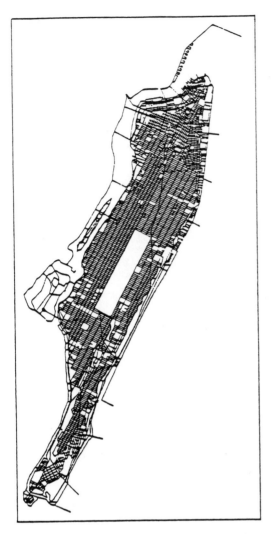

Figure 3

DIME File Map of Manhattan

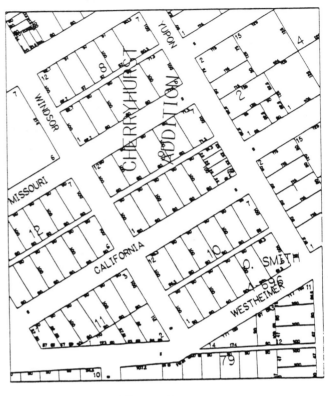

Figure 4

Land Registry Map

Land registery requires the addition of lot numbers and lot dimentions and thus larger scale mapping. Figure 4 is the shaded area in Figure 2 and is typical of land registry maps. However there is still a requirement to obtain plat maps from the composite maps. This may be done by defining a polygon for the boundaries of the plat and retreiving only the data within the polygon. Figure 5 is a good example of a plat map from the high quality maps being developed by Forsyth County, North Carolina.

The functional capability outlined above yields the system capabilities of modern Cadastral Mapping Systems. The map data base is maintained in a recognized coordinate system and polygonal retrieval is automatically supported such that original map boundaries are transparent to the user. Windowing from an index map to the actual desired area is supported. The data base is divided into logical overlays so that different selections of data may be made for maps at different scales.

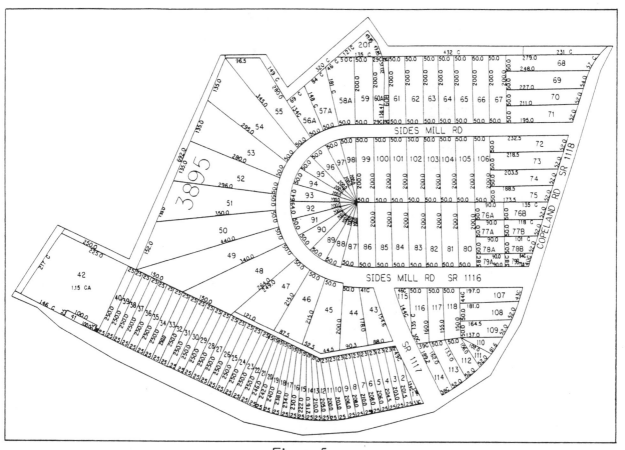

Figure 5

Plat Map Retrieved from the Continuous Digital Map

Furthermore the user sets the relative scale of symbols and annotation at the time the display or plot is made. Display time justification of annotation is an important feature to support multiple scale maps. Justification of annotation is defined as the location and scale of annotation. For example most systems support justifications such as centered, lower left, upper right etc. at the time of entry. This means that the annotation may be centered in the middle of the street between two intersections at entry. However if the scale of the annotation is changed relative to the scale of the map, the annotation may not remain centered. The INFORMAP system determines the justification at the time the display is made. Figure 6 shows the entry options and the results with the annotation scale doubled and halved. For particularly crowded situations even the height and spacing may be changed automatically.

The source for the development of the data base will be maps at different scales, orientation and area. Portions of even the same map sheet may be developed from different sources. Most users combine accurate construction of the property lines with careful digitizing of existing map sources. The Synercom Coordinate Geometry Capability (reference 2) provides the ability to develop a subdivision directly from survey or deed description data on an inexpensive alphanumeric terminal and have the result appear as a drawing in the data base. Corrections are then made at a full display station. The Coordinate Geometry logic may be used to directly interface with existing data such as to divide an existing block line into 10 lots. Macros may be written to proportion the lots along block lines for unevenly spaced lots.

131

Figure 6

INFORM Annotation Justification Determined at Display Time

To enter data by digitizing the system with user input must define a transformation from the source document to the coordinate system of the permanent data base. This transformation may be orthogonal based on the corners of the map sheet and taking into account paper shrinkage. Alternatively the transformation may be non linear based on known monument points on the source map or recognizable common points on the source map and in previously coordinated data (Figure 7). INFORMAP does the transformation from the drawing placed on the digitizer to the data base coordinate system as the data is entered. If the cursor puck is over the drawing on the digitizer, its location is displayed on the CRT in data base coordinates relative to existing map elements.

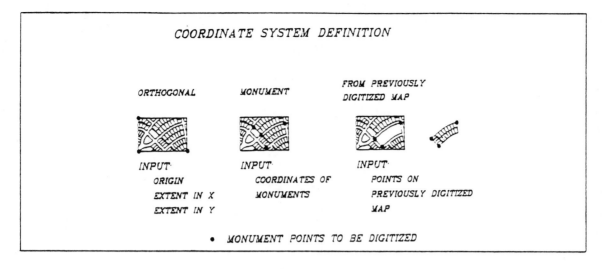

Figure 7

Coordinate System Definition Alternatives

Property and Parcel Information System

The mapping requirements are only a part of the requirements of a Land Records Information System. The other aspect is maintaining parcel oriented information such as ownership, title chains, improvements, sale value, etc. This information may reside on a host computer using standard data base systems. Geocoding the approximate parcel centroid and using the coordinates as a unique key has been used to fix the classical data base to approximate geographic location. However modern systems integrate the Cadastral System with a geographic data base which meets Property and Parcel Information System Requirements. These systems add a new perspective to the attribute information. Geographic relationships are perceived immediately. The user can point to the element he wants further information about and immediately obtain the attributes. Figure 8 is an example of parcel data.

Reports such as Figure 9 may be obtained immediately from the geographic data base. The elements to be included in the report are selected by polygonal area and criteria. The criteria may be any combination of tests made on the attribute data. An example might be all property in the Memorial area sold in the last two years whose value is greater than one hundred thousand dollars. Narrative reports are combined with graphic plots to provide the total picture. For example Figure 10 has symbology presented for the number of bedrooms in different classes of property. It is much more informative than any narrative reports.

If a Geographic Based File system such as the Census Bureau DIME files is integrated into the system, access may be by street address, street intersection or street address range. This is particularly useful in dealing with the public since they may not know any of the keys to their parcel but they do know the street address.

NON-GRAPHICAL RECORD FOR DATA TYPE PARCEL
BLK   =4278
LOT   =029
OWNNAME =HAWKINS BOBBY W & MARGARET
PROPADR =5901 GREENHAVEN CT
CITY   =
ZONE   =
LANDUSE =11
CENSUS =0
OCCUP   =0
STYLE   =RANCH
TWP   =5
STORIES =1
TYPE   =
BEDROOM =3
BATHS   =2
ERECTED =0
BASEMENT=4
ATTIC   =0
FIRECNT =2
GARAGE =0
NON-GRAPHICAL RECORD FOR DATA TYPE PARCEL
SQFT   =1980
HEAT1   =8
HEAT2   =0
HEAT3   =0
BASE   =0
REVDATE =0
REVAMT =50410
CURTAX =0.00
REAL   =50410
PERSONAL=0
FIREDIST=0
RSVD   =

Figure 8

Parcel Attribute Data

AUDITOR'S OFFICE LOT INFORMATION
INFORM REPORT
SYNERCOM TECHNOLOGY INC.

BOOK	PAGE	ITEM	LOT	LOT VALUE	TOTAL VALUE	LOT SIZE
862	07	001	111	14400.00	71600.00	8640.00
862	07	002	110	14400.00	68400.00	8640.00
862	07	003	109	14400.00	94500.00	8640.00
862	07	004	108	15900.00	69600.00	8640.00
862	07	005	107	14400.00	69600.00	8640.00
862	07	006	106	14400.00	68400.00	8640.00
862	07	007	105	14400.00	80300.00	8640.00
862	07	008	104	14400.00	68700.00	8640.00
862	07	009	103	14400.00	67600.00	8640.00
862	07	010	102	14400.00	80800.00	8640.00
862	07	011	101	14400.00	67600.00	8640.00
862	07	012	100	14400.00	70400.00	8640.00
862	07	013	99	14400.00	68400.00	8640.00
862	07	014	98	14400.00	70400.00	8640.00
862	07	015	97	14400.00	72700.00	8640.00
862	07	016	96	14400.00	71600.00	8640.00
862	07	017	147	14400.00	72600.00	8640.00
862	07	018	146	14400.00	64600.00	8640.00
862	07	019	145	14400.00	66500.00	8640.00
862	07	020	144	14400.00	66400.00	8640.00
862	07	021	143	14700.00	77600.00	8784.00
862	07	022	142	14200.00	77000.00	8496.00
862	07	023	141	14400.00	71300.00	8640.00
862	07	024	140	14400.00	77000.00	8640.00
862	07	025	139	14400.00	68400.00	8640.00
862	07	026	138	14400.00	77700.00	8640.00

Figure 9

INFORMAP Report

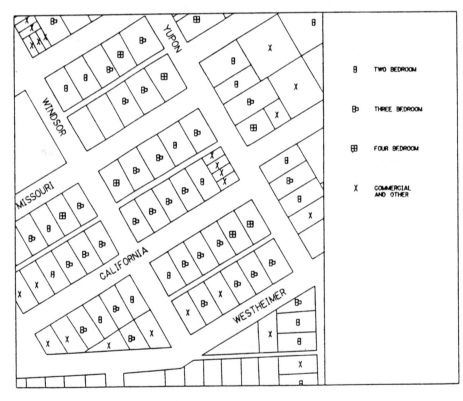

Figure 10

Number of Bedrooms on Each Lot

Engineering Mapping Systems

The cost of capturing and maintaining this Land Records Information Data Base is high. The quality of the maps in general is very good. Thus these maps make an ideal base for public works and utility companies. However the level of detail is much greater than that required by public works. In fact the detail gets in the way of the facilities, thus in many cities and counties a public works base map series and a tax map series are maintained manually.

With an automated system meeting Cadastral requirements, the level of detail may be chosen by the user at the time a display or plot is made. Figure 11 is a waste water engineering map made from the same data base as Figure 4 with some of the cadastral data not displayed and the waste water facilities added. Engineering maps have the same varying scale requirements that cadastral map systems have. See Figure 12 for an example.

Engineering Mapping Systems generally require modeling of the facility to become really cost effective. For example there are relationships between man holes, sewer mains and terrain. With the addition of attribute information about the sewer mains and equipment flow analysis of the sewer may be accomplished. Reference 3 explores the utility requirements for a geographic data base.

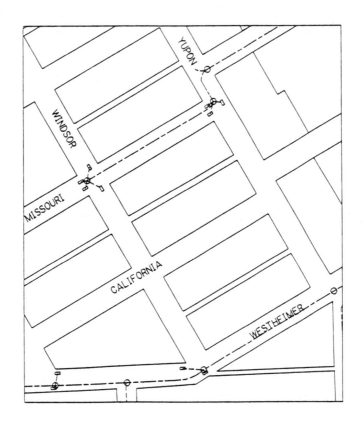

Figure 11

Waste Water Facilities on Cadastral Base

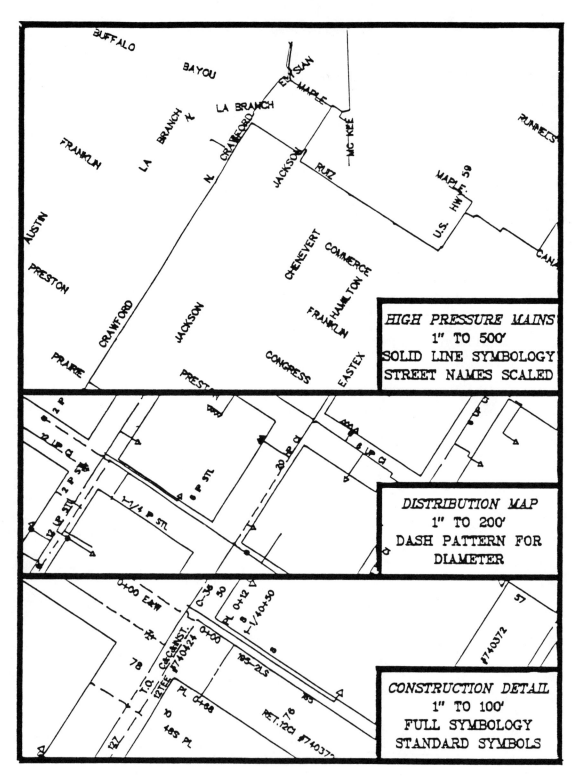

Figure 12

Scale and Content Variations for a Gas Utility

The relation between an engineering data base and a taxing authority data base is synergistic. Certainly neither would capture the others data for its own purposes. However, if the data is readily available, public utility service or potential service to an area may be verified by the taxing authority immediately. Correspondingly ownership information for a parcel may be identified by the public utility by just pointing at the parcel centroid. These are just two of many of the uses of a common data base.

Thematic and Statistical Systems

None of the systems that have been discussed have the analysis software provided by a state of the art Thematic and Statistical system such as the Geographic Information System offered by ESRI. Consider the analysis required to determine the location of a new multi million dollar shopping center. Factors such as:

> Accessibility
> Parking
> Zoning
> Population and Population Growth
> Affluence
> Competition Location
> Distance from Suppliers
> Land Cost and Projected Value Increase
> Availability
> Tax Assessment

must all be weighed to determine a successful location. This may be accomplished systematically be combining census tracts, major arteries, location of competitors, demographic information, etc. in the form of polygons or grid cell values.

For grid cells the area is divided into regular evenly spaced areas called cells. Attribute information is then stored about each cell. Values are determined for each variable considered for each grid cell which covers the geographic area. For example a value related to distance may be computed for each cell by examining its relative adjacency to other phenomena in surrounding cells (such as competitors, major arteries, suppliers). Grid cell overlay with weighting of selected classifications (such as population, deomographic information and zoning) results in a numeric surface of composite data expressing a systematic interpretation of the data. Finally all the data may be analyzed together using a search radius aggregation. This is a technique for moving a circle with a specified radius across a grid, cell by cell, searching outward from each cell. The routine then accumulates the number of events, activities or related phenomena which fall with the radius. The result is an enumerated value for each cell which can be related to site suitability. Plots of relative suitability are then made.

The results of such an analysis are often presented on small scale maps (1" to 2000' for example). This presentation is commesurate with the level of accuracy of the captured data. For taxing authority applications the data must be related to individual parcels. The point in polygon capability which determines which polygon the centroid of a lot is within does not tell the whole story. For example the evaluation of property partially within a flood plain, noise corridor or steep area is different from one next to it that is totally unencumbered. The real requirement is to zoom in to a much larger scale map that still has the results of the analysis displayed. In this manner the details of the property may be examined on both a macro and micro level. This is possible with the integration of Thematic and Statistical Systems with Cadastral Systems. This level integration is being developed now and the potential for imaginative users seem unlimited.

Figures 13, 14 and 15 provide example output produced by the INFORMAP system of data analyzed on the ESRI system. The data is in the El Torro township of Orange County, California. Figure 13 has urban hazards classified from which an airport can be clearly defined. Noise contours are presented in Figure 14. Figure 15 provides flood control acceptability categories. When Orange County purchases their Cadastral and Engineering System this data will be combined with detailed maps to provide integrated Land Records information.

The Suwannee River Project (Reference 4) combines Generalized Thematic and Statistical, Cartographic, Engineering and Cadastral mapping in a single project. Diverse systems are involved. However with complete vertical integration the work could be done on a single system.

Summary

Types of geographic systems have been categorized and related to the Land Records Information System requirements. A vertically integrated system is possible and the integration offers new horizons for synergistic data relations.

The INFORMAP system which offers Geographic Base File, Cartographic, Engineering, Cadastral and Property and Parcel system capability has been proven in numerous user sights. Forsyth County and Houston are examples of LRIS applications; the City of Bellevue and Burnaby are examples of joint use between public works and LRIS applications. The ESRI system has been proven in numerous planning and natural resource applications. Vertical integration will provide unique capability for the mapping and geographic data base community.

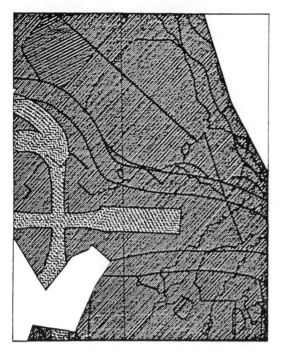

Figure 13
Urban Hazards - Crash Zone

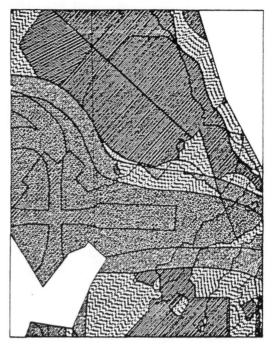

Figure 14
Noise Contours

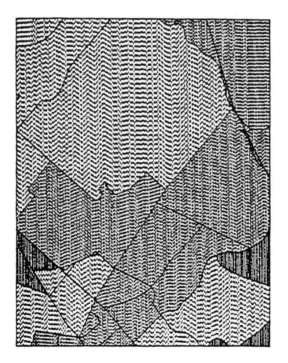

Figure 15
Flood Control

REFERENCES

1.   Dangermond, Jack, Software Components Commonly Used in Geographic Information Systems, 1982, Harvard Computer Graphics Week.

2.   Drinnan, Charles H., Implementation of Survey Commands in an Interactive Graphics Mapping System, 1981 ACSM Fall Technical Meeting. Pages 94 -105.

3.   Drinnan, Charles H., A Geographic Data Base Approach for the Utility Industry - Requirements, Solutions and Examples, 1982 Harvard Computer Graphics Week.

4.   Folchi, William L., Suwannee River Floodplain Mapping Project, 1982, Harvard Computer Graphics Week.

METROCOM: HOUSTON'S GEOGRAPHIC INFORMATION MUNICIPAL MANAGEMENT SYSTEM

Dr. Francis L. Hanigan, President
TCB Data Systems Inc.
P.O. Box 13089
Houston, Texas 77219

ABSTRACT

In September 1978, the Houston City Council approved the Department of Public Works Director's request that his department be authorized to embark upon the development of an interactive facility data management system, dubbed METROCOM (METROpolitan COMmon). In November 1978, the City Council further authorized the City's Tax Department to participate in this same project.

Since that time, the City's digital graphics/data base consultant has expended 316 man-years in developing the METROCOM Data Base. Of this effort, 121 man-years were devoted to digitizing the data and 55 man-years to system design and software optimization. The METROCOM system recently has been installed in the City Hall and now represents the largest municipal data base of its kind in the USA. This article presents a brief history of the project and discusses some of the many lessons learned.

INTRODUCTION

Houston's METROCOM system consists of an integrated collection of spatially-related municipal data accessible to all City departments. The METROCOM Data Base resides on mass storage devices which are available for instant inquiry by City offices. Output capabilities include a high-speed line printer, digital and electrostatic plotters, and CRT displays with associated hard copy devices. The key to the METROCOM system is geographic location. All data stored in the system are indexed to digital versions of the planimetric maps which the City has produced and continues to produce, using photogrammetric techniques. This continuous planimetric map covers some 1,760 square kilometers (680 square miles) and is overlayed with ownership and land-use information on some 554,000 parcels, as well as complete information on the water, sanitary, and storm sewer systems, the road network, and bridges servicing a population of approximately 1,7 million.

COMPUTER SYSTEM

The interactive graphics system which the City has chosen for its consultant's use in building and maintaining the METROCOM Data Base is Synercom Technology's INFORMAP system. This system uses the Digital Equipment Corporation's PDP 11/70 minicomputer with an RSX-11M operating system.

Figure 1 shows the configuration proposed by the consultant for installation in the City Hall. It consists of three PDP 11/70 minicomputers and 16 interactive graphic workstations. One PDP 11/70 will serve as host for the entire system. It will be dedicated to the task of managing the master METROCOM file and supplying data on request to the satellite systems through a DECNET™ data link. The host will also control the digital and electrostatic plotters as well as an array of other peripheral input/output devices. Present planning calls for the METROCOM system to have a disk storage capacity of 3.2 billion bytes. This will be achieved by using ten 300-megabyte disk drives and two 80-megabyte disc drives.

Eight interactive graphic workstations and one PDP 11/70 computer will be dedicated to the Department of Public Works. Three graphic workstations will be dedicated to the City's Planning Department, which will share the second satellite PDP 11/70 with the Department of Public Works. By using this distributive processing arrangement, the physical size of the specialized departmental files have been kept within reasonable limits. This concept also allows each department to access more rapidly that data which it frequently uses and to protect data of a restricted nature. Four graphic workstations will be retained in the METROCOM Service Center which will be responsible for hardware and software maintenance and master file updating.

OPERATIONAL INTEGRATION

The METROCOM system is not as yet a fully operational system. Tasks designed to make the system fully operational and scheduled for accomplishment in 1983 include the following.

- Transferring the systems' three PDP 11/70 CPU's, 16 Synercom Technology Graphic Workstations and other associated peripherals to an operating environment within City Hall.

- Networking this equipment and other City-owned computer systems.

- Training City personnel in the use of the system.

- Documenting operational procedures.

- Developing long-term system maintenance procedures.

- Planning for the expansion of the system to include data required by other City departments.

- Developing specific applications for the various City Departments.

As this article goes to press, the first of the two satellite CPUs and eight graphic workstations are being installed at City Hall. The host CPU and many of the

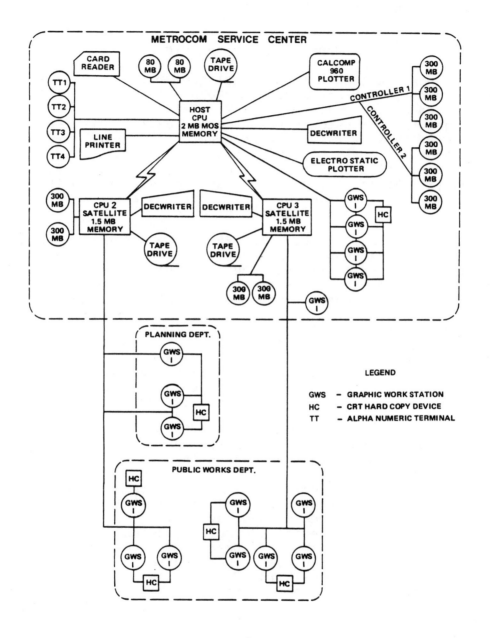

Figure 1

143

peripherals are scheduled for installation in January 1984. A date for the installation of the second satellite system has not been set.

COMPUTERIZED GEOGRAPHIC BASE

The foundation stones for the METROCOM Data Base are the more than 5,000 brass survey markers which have been implanted at intervals of approximately 610 m (2,000 ft.) throughout the Houston Metropolitan area.(1) The coordinates of these survey markers have been computed from Second Order, Class II Horizontal and Vertical Surveys, using National Geodetic Survey procedures and specifications.(2) Horizontal positions are adjusted using the NGS Traverse Program while vertical positions arre adjusted through use of the NGS VERT 02 program.

The 1:1,200 scale (1 in. = 100 ft.) maps which form the basis of the geographic base were originally compiled using 1:6,000 scale (1 in.= 500 ft.) color aerial photography acquired with a cartometric-quality camera. All maps were compiled to the National Horizontal Map Accuracy Standard of 1/40-inch at map scale for 90 percent of well-defined points.

The 1,530 map sheets produced before March 1979 were photogrammetrically revised using 1:12,000 scale (1 in. = 1,000 ft.) color photography prior to digitization on a graphic workstation. An additional 32,800 hectares (81,000 acres) were mapped at a scale of 1:2,400 (1 in. = 200 ft.), while orthophotos at a scale of 1:1,200 (1 in. = 100 ft.) were used to capture the planimetric detail of the large industrial sites along the Houston Ship Channel.

While most map revisions in the early stages of the project were performed using conventional (nondigital) mapping techniques, future updates and new mapping will be compiled directly from color aerial photography using a direct digital mapping system such as WILDMAP®. The next revision of the geographic base had been scheduled for 1983, at which time approximately one-third of that base was to have been updated; however, funding restrictions have delayed this revision until 1984 when the revision will most likely be accomplished in conjunction with the new mapping of recently annexed areas.

The complete METROCOM land base consists of some 2,700 map blocks, each covering a land area of approximately 65 hectares (161 acres). Each map block is edgematched to all adjoining and diagonal sheets. All map blocks are subdivided into 10 layers (annotation, roadways, railroads, drainage, sidewalks, fences, driveways, parking lots, buildings, major cultural features such as bridges, dams, power plants, etc., and miscellaneous cultural detail). The entire land base is stored as a continuous map on a single 300-megabyte disk pack. For more technical information on the mapping program, the reader is referred to Hanigan.(3)

COMPUTERIZED CADASTRAL SYSTEM

The second major subsystem of METROCOM is the real property ownership system. This system consists of a graphic inventory of all land parcels, both exempt and nonexempt, within the territorial limits of the City and the Houston Independent School District. (The boundaries of the two are not coterminous.) These digital files contain all the information needed to graphically represent approximately 554,000 land parcels, their deeded dimensions, and, where appropriate, their acreage.

The graphic segment of the real property ownership system was developed by fitting the best available data for each parcel to the computerized land base on a block-by-block basis rather than by using the sophisticated interactive geometric routines contained within the Synercom Technology's computer system to individually position each parcel. The graphic files of the real property ownership system also include the information needed to draw the city, county, and school district limit lines, subdivision boundaries, historic land survey lines and city, county, state, and federal rights-of-way. The nongraphic attributes stored in the computer for each land parcel are referenced to an apparent centroid which has been assigned to each parcel. These centroids can be graphically portrayed as the parcels are displayed on a CRT or plotted on a hard-copy map.

The nongraphic attribute information was obtained from two disparate sources. Current ownership data and other directly related information were obtained on magnetic tape for bulk loading into the interactive graphic system through a subcontract with a local title company. Property attributes relating to land use, the value of the land, and improvements thereon, as well as the nature and character of the improvements themselves, were obtained from a mass appraisal firm contracted for the reappraisal of all real property within the City and the Houston Independent School District. This information was bulk loaded into the METROCOM system as approved by the City and provided by the mass appraisal firm to the consultant. The reader is referred to Table 1 for a representative listing of the property characteristics available to the City's department through the METROCOM system.

TABLE 1 - REPRESENTATIVE PROPERTY CHARACTERISTICS

Land-use classification and neighborhood code
Tax account number
Names of all property owners and their percent interest
Service address of all parcels
Legal descriptions such as subdivision name, block, and lot
  number
Deeded and calculated acreage
Square footage of improvements
Date and price of last sale
Assessed value for land and improvements
Reference to appraisal review history

TABLE 1 (Cont'd)

Number and type of rooms (residential)
Style and type of residence
Existence of carports, garages, swimming pools, easements
Type of foundation and exterior walls
Type of heating and cooling system
Roofing material
Commercial structures: frontage, location, and parking
  availability codes
Year constructed and/or remodelled
Reference to outstanding building permits

COMPUTERIZED FACILITY MANAGEMENT SYSTEM

The third major subsystem of METROCOM is the facilities
management system. This system consists of a graphic
inventory of all City-owned utilities. The graphic segments of the facility management system were developed from
existing engineering plans and plats on file with the
Department of Public Works. During the data collection and
analysis phases of the project as many as 36 people were
involved in inventorying, collating, and reviewing over
40,000 sets of plans and 25,000 land development plats.
(No attempt was made to field verify these data.) From
these documents, those engineering drawings considered to
be the most current and accurate were selected for entry
into the facility management system. As each drawing was
digitized, the information contained on that drawing was
assigned a confidence factor indicating the level of confidence which the data evaluators believed could be placed
on the source documents.

As part of the data collection process, the inventory of
all plans and plats on file with the Department of Public
Works was computerized and geographically indexed to the
DIME (Dual Independent Map Encoding) file provided by the
City's Planning Department. This index will be maintained
as a permanent feature of the METROCOM system.

Included in the facilities management graphic files are
sufficient data to prepare 1:1,200 scale overlays of the
City's network of roads and bridges, the complete water,
storm sewer, and sanitary systems, as well as the distribution network for the City-owned Magnolia Park Gas Company.
In addition, the facilities management system contains
an assortment of nongraphic attributes related to these
utilities, as well as the City's streets and bridges.
The latter information was supplied in large measure by the
State Department of Highways and Public Transportation for
bulk loading into the METROCOM system. The reader is
referred to Table 2 for a representative list of the nongraphic attributes associated with the facilities management system.

TABLE 2 - REPRESENTATIVE FACILITY CHARACTERISTICS

Bridges: Deck material, design load, reference number,
            length, width, number of spans, span material,
            span type, and vertical clearance.

Roads: Classification, curb status, number of lanes,
            agency responsible for maintenance, width, and
            type of pavement.

Manholes: Various invert elevations, type construction,
            size of manhole cover, utility serviced.

Lines: Flow elevations at each end, type of material,
            shape, size, status (proposed, existing,
            abandoned).

Devices: Number, type, size.

All: Date digitized, source document number, evalua-
            tor's confidence level.

COMPUTERIZED LAND-USE SYSTEM

Although not designed as a land planning system, the METRO-
COM system can be used as such, as is evidenced by the map
shown at Figure 2. This map was prepared by combining the
graphic ownership mapping data with the land-use data
available in the system (see Table 3). These latter data
were acquired under separate contract as part of the City's
mass reappraisal program and bulk loaded into the system
late in 1982.

TABLE 3 - LAND-USE CODES

Principal Category	Subcategories
Residential, Private	11
Residential, Commercial	4
Commercial	60
Industrial	17
Institutional and Special Purpose Buildings	12
Communications	2

EFFORT

The magnitude of the conversion effort can be sensed from
the data given in Table 4. This table summarizes the
effort expended by the City's consultant between late 1978
and the end of 1982 in accomplishing the design and development tasks associated with the METROCOM system. The
table does not include the time expended on the preliminary
studies and the pilot project, nor does it include the
effort expended by the subcontract photogrammetric firms
and the title company responsible for researching and veri-
fying property ownership records and preparing same for
digitization.

Figure 2

148

TABLE 4 - METROCOM EFFORT (Man-Years)

	Organi-zation*	Collec-tion	Conver-sion	Data Processing**	Quality Control
Land Base	5.5	5.8	31.0	8.0	7.9
Real Property	11.8	14.2	52.1	27.4	3.5
Facili-ties	14.3	68.5	37.5	19.7	9.0
TOTALS	31.6	88.5	120.6	55.1	20.4

*Organization includes time spent on liaison with the City, project administration and supervision, and personnel recruiting and training.

**Data processing includes system design, programming, computer operations, and plotting.

An analysis of these data shows that the preponderance of the consultant's effort, or 47.1 percent (149 man-years), was devoted to the facility system. The ownership system accounted for 34.5 percent of the effort (109 man-years), while the land base required the other 18.4 percent (58.2 man-years). Organizations or persons contemplating such a project might note that the data collection effort for the facilities was almost twice the conversion effort. Were the hours expended by the various subcontractors on the collection of the real property data available, this same ratio would undoubtedly exist there, too.

LESSONS LEARNED

The over 300 man-years of effort expended by the consultant have taught everyone associated with the project innumerable lessons. Among the more cogent are the following.

- The availability of a staff skilled at organizing and managing a project of this magnitude is as critical, if not more critical, to its success than the availability of a staff skilled in interactive graphics technology.

- A well-designed and carefully executed pilot project is an absolute necessity.

- Heavy and continuous client involvement is a must throughout the project.

- A carefully phased project is to be preferred to a massive turnkey project.

- Innovative, project-oriented software can materially speed and simplify the conversion process; therefore, early emphasis should be given to the size and quality of the system's design and software development staff.

○ Even with a well-executed pilot, the data collection, evaluation, and preparation tasks will exceed even conservative estimates. Nonetheless, these tasks must be thoroughly accomplished if the completed system is to be credible.

○ Consistent work flow is critical. Data collection, evaluation, and preparation must stay at least 30 and preferably 45 days ahead of the conversion task.

○ The nature of much of the work involved is such that experienced engineering technicians are soon likely to lose interest in the data collection, analysis, and preparation tasks, yet their experience is vital to the timely and economic completion of the project. These tasks might better be done by the client whose data are being processed.

○ Early and continuous quality control by technically knowledgeable, well-trained, highly motivated personnel will save more than it will cost.

○ The quality control staff must be large enough to ensure rapid turn-around; otherwise, the entire project schedule will quickly deteriorate.

○ Subcontractor capabilities should be thoroughly investigated and tested during the pilot project. This is particularly true when the local situation requires the use of subcontractors whose prior knowledge of digital mapping, data processing, and Geobased Information System technology is limited.

SUMMARY

Houston's METROCOM system combines, on a heretofore unknown scale, high-order ground surveys, photogrammetrically developed planimetric maps, and modern techniques of digital cartography and data processing in a common interactive municipal management system.

When fully operational, the interactive nature of METROCOM will allow Houston's municipal planners and managers to explore alternative solutions to the pressing problems caused by the City's rapid expansion over the past decade. METROCOM will also reduce the cost of City operations by replacing currently inaccurate, inconsistent, and duplicative filing systems with a centrally maintained geographically oriented file of accurate and consistent data.

To the author's knowledge, the METROCOM system represents the largest application of computer graphics technology to date within a municipal environment. The work done on the METROCOM system has shown that this technology is capable of improving the manner in which municipalities manage their base, real property, and facility mapping systems.

The project has also shown that the implementation of a system such as METROCOM is not a trivial task. The proper application of this technology requires tight management, skilled personnel, a significant expenditure of resources, and a considerable period of time.

While not yet all things to all people, Houston's Metropolitan Common Digital Data Base should serve for years to come as an exemplary model of a comprehensive mapping and municipal data management system.

ACKNOWLEDGEMENTS

The author acknowledges the support of TCB Data Systems Inc. in providing time and administrative assistance in preparing this paper.

REFERENCES

(1) Gale, P. M., 1970, "Control Surveys for the City of Houston," Surveying and Mapping, xxx:1 (March 1970), pp. 95-101.

(2) Federal Geodetic Control Committee, Specifications to Support Classification Standards of Accuracy and General Specifications of Geodetic Control Surveys, U.S. Department of Commerce, National Ocean Survey, Rockville, Maryland, 1975, Reprinted 1977.

(3) Hanigan, F. L., 1979, "METROCOM: Houston's Metropolitan Common Digital Data Base - A Progress Report," Surveying and Mapping, xxxix:3 (September 1979), pp. 215-222.

PART 4
WATER RESOURCE MANAGEMENT

ANDREE YVONNE SMITH
RICHARD J. BLACKWELL
Jet Propulsion Laboratory
Pasadena, CA 91103

Development of an Information Data Base for Watershed Monitoring

Landsat multispectral scanner data, Defense Mapping Agency digital terrain data, conventional maps, and ground data were integrated to create a comprehensive information data base for Lake Tahoe and its environs.

THE LAKE TAHOE BASIN

THE GROWTH of human society impacts the environment in many ways. One region facing the problems associated with this

(492.1 km²) surface area receiving inflow from 63 tributaries with only one outlet at the Truckee River. The Tahoe Basin has traditionally attracted recreational activity due to its clear deep water and pine-forested

ABSTRACT: *Lake Tahoe is one of very few natural lakes in the United States which has remained relatively undamaged by the encroaching developments of man. The management of Tahoe's water quality has become a subject of intensive study in recent years in an effort to define and limit the effects of nonpoint source pollutants that are input from the contributing drainage basins.*

As an aid to the water quality management effort, the Image Processing Laboratory at Caltech's Jet Propulsion Laboratory has integrated Landsat multispectral scanner data, Defense Mapping Agency digital terrain data, conventional maps, and ground data to create a comprehensive information data base for Lake Tahoe and its environs. The project used the resources of the JPL-developed Image Based Information System (IBIS) programs to augment data gathered by the U.S. Forest Service for the Tahoe Regional Planning Agency.

The IBIS data base method allowed cross correlation of Landsat imagery and topographic data with a variety of environmental data relating to such parameters as surface runoff, drainage basin acreage, and terrain configuration. Parameters were evaluated and compared for each drainage basin defined by the Tahoe Regional Planning Agency (1977).

The methods used to construct and update the information data base will be described and evaluated. In addition, the utility of including Landsat imagery will be discussed.

growth is the Lake Tahoe Basin. The Tahoe Basin occupies over 500 square miles (1295 km²) situated in a graben straddling the boundary between California and Nevada. Lake Tahoe contains 126 million acre feet (155.4 km³) of water in a 190 square mile

shorelands coupled with its proximity to major metropolitan areas in Northern California. Since the 1950's the basin has experienced escalating demands for land development at the expense of the natural watershed. Discharge of sediments to the

PHOTOGRAMMETRIC ENGINEERING AND REMOTE SENSING,
Vol. 46, No. 8, August 1980, pp. 1027-1038.

lake has greatly increased due to accelerated human interference, and alterations to the natural drainage patterns are evident in some areas.

The problems which the Lake Tahoe Basin is presently confronting are certainly not unique to this area. These consequences of man's alteration of the natural environment are symptomatic of the pressures which can be brought to bear by the activities of an increasingly mobile populace.

The Jet Propulsion Laboratory's (JPL) Image Processing Laboratory (IPL) has for five years been concerned with developing techniques which can aid water quality management programs charged with the task of monitoring and assessment of the trophic status of these troubled lakes. In the past, the major emphasis has been placed upon assessing the viability of Landsat as a monitoring device and the development of a useable system for water quality management personnel (Blackwell and Boland, 1979; Smith and Addington, 1978). As a result, water quality in terms of the water body per se has been closely studied without major emphasis given to the contributing factors, such as nonpoint source pollutants from the surrounding land mass. In order to investigate the utility of a comprehensive system which takes into account the causes as well as the effects of lake eutrophication, the IPL has attempted to construct an integrated and workable data base, composed of currently available data sources, for the Lake Tahoe region. The purpose of such a data base is to integrate water quality related data from various sources into a comprehensive system which is capable of combining and cross-referencing such diverse elements as conventional maps, Landsat imagery, and tabular data obtained via in-situ methods. Such a system provides the environmental planner with the ability to visualize the integration of disparate data elements which formerly would have to be examined manually. The analyst can also combine data so as to generate new tabulations based upon cross-referencing of the various elements.

THE IMAGE BASED INFORMATION SYSTEM

To achieve the goal of a workable data base, IPL has relied upon the resources of its own information system, IBIS, which was designed specifically for this type of application. IBIS is a system composed of general purpose and specialized computer programs which can be grouped into logical steps to build an information data base (Bryant and Zobrist, 1978; Zobrist and Bryant, 1979).

Functionally, IBIS represents a selection of programs which operate under the Image Processing Laboratory's VICAR Image Processing System (Video Information Communication and Retrieval). The IBIS system is raster (image) based but allows integration of graphical and tabular data types as well. Image data sets can originate from a variety of sources such as Landsat or other imaging systems. Graphical data, such as from conventional maps, are electronically digitized and transformed into image format for integration into the system. Tabular data are linked to the system via an interface between the image data planes and entered through table-structured input which is keyed to a georeference plane (Yagi et al., 1978). The georeference plane is a map-based graphical representation of areas of interest, such as drainage basins. Data base storage, retrieval, and analysis operations are all performed using the VICAR digital image processing routines. Each image entered into the data base is geometrically corrected and registered to a planimetric base image creating a system of data planes, as illustrated in Figure 1. Each image plane is referenced to one or more georeference planes to which all tabular data are also keyed. The user is thus able to manipulate data from several sources which, despite their original disparity, are all referenced to a common base, usually geographic in origin.

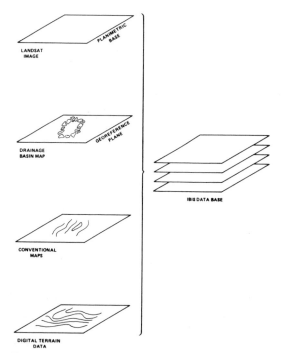

FIG. 1. Conceptual diagram depicting data planes in registration forming the IBIS data base.

CONSTRUCTION OF THE LAKE TAHOE BASIN DATA BASE

IPL chose to construct the Lake Tahoe Basin data base from data sources which were readily available and familiar to its analysts. Landsat imagery, which has formed the basis for the majority of IPL's water quality assessment efforts, was chosen as the planimetric base to which all other data would be registered. A subsection of a Landsat 2 scene from 27 August 1976 was extracted and transformed into a Lambert Conformal Conic projection to serve as the planimetric base. To this scene two other Landsat scenes were registered, one from Landsat 2, 21 July 1978, and one from Landsat 3, 30 July 1978. Registration was achieved through the aid of a piece-wise surface fitting algorithm. The surface transformation was defined through a series of tiepoints selected during the use of an interactive spatial pattern recognition routine in which the analyst selected common geographical features (Ruiz et al., 1977; Davis and Friedman, 1979). After all images were registered, color reconstruction was also performed on each scene using a color enhancement technique which approximates a Gaussian distribution using the principal components of each Landsat image separately (Madura and Soha, 1978). This type of enhancement was designed to attempt even distribution for all regions within an image to avoid the saturating effects of conventional linear enhancements.

Terrain data produced by the Defense Mapping Agency (DMA) were next integrated into the data base. Digital terrain tapes were acquired for the Lake Tahoe region from the National Cartographic Information Center. The tapes were prepared from U.S. Geological Survey (USGS) 1:25,000 scale topographic quadrangle map series. Map contour lines falling within a one degree block of latitude and longitude were digitized and a matrix of elevations generated with one elevation for every 0.01 inch on the map (200 feet/60.96 metres on the ground). The terrain data were reformatted for use in IPL's VICAR operating system. In reformatting, halfword integer elevation values were converted to single byte integers and scaled to the terrain variation within the area (Strahler et al., 1979).

After reformatting, the terrain data sets were rotated 90 degrees counterclockwise to orient the north at the top of the images. This compensated for data format as produced by the National Cartographic Institute. For the Lake Tahoe Basin, four separate terrain quadrangles were required to construct a complete image comprising the Tahoe Basin area. This necessitated the mosaicking of the four quadrangles before final registration to the data base. Figure 2 reproduces the four digital terrain quadrangles required to completely encompass the Lake Tahoe Basin and the final mosaic image. The relief-like effect portrayed in the images were produced by digitally shading between contour intervals.

Registration of the terrain mosaic with the Landsat planimetric base was achieved through the application of a resampling algorithm which applied a two-dimensional correction grid derived from selected control points (Yagi et al., 1978; Ruiz et al., 1977). Tiepoint selection was achieved interactively relating surface features through comparison of the Landsat planimetric base image and a relief-like version of the terrain image.

Graphical data for the Lake Tahoe Basin data base were acquired from conventional maps produced for the Tahoe Regional Planning Agency by the U.S. Forest Service. These maps presented hydrologic and climatologic data pertaining to the Lake Tahoe Basin. Before integration into the IBIS data base, graphical data must be transformed into image format. This was achieved by first digitizing the graphical data onto magnetic tape. Next, a least-squares affine transformation was applied which created a general surface fit. Corresponding line and sample coordinates, in terms of the Landsat planimetric base, were calculated from latitude and longitude coordinates taken from the graphical map data for geographical features common to both data sets. Registration was once again achieved through the selection of control points which linked geographical features from the graphical data file and the planimetric data base. In Figure 3, the transformation of a conventional map to a graphical data plane is illustrated. Step A represents the digitizing procedure in which the drainage basin map is converted to graphical image format.

A georeference image plane, which provided an interface between all data planes for the Lake Tahoe Basin data base, was created from the drainage basin map provided by the Tahoe Regional Planning Agency. As seen in Figure 3, this map identified the location and outlined the boundaries for the 63 drainage basins identified as providing watershed inflow to Lake Tahoe. The map was digitized and subsequently registered to the planimetric base. Each separate drainage basin was then assigned a

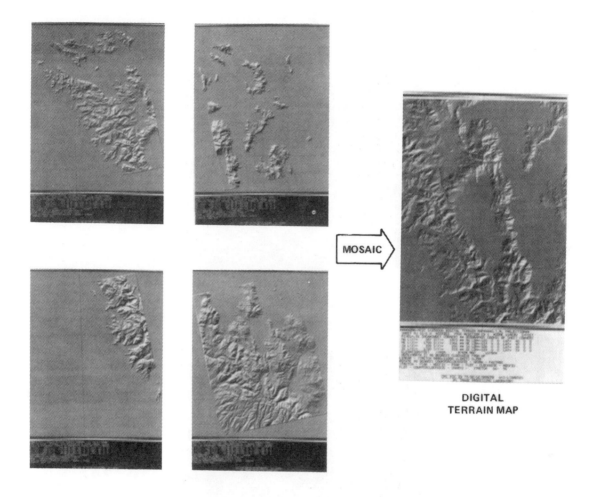

4—1 : 250,000
DIGITAL TERRAIN QUADS

FIG. 2. Four digital terrain quadrangles mosaicked to produce a terrain map which contains the Lake Tahoe Basin.

unique data number for all picture elements comprising that basin. Step B, in Figure 3, represents the process of encoding each drainage basin, a procedure known as "painting". The final product, the georeference plane, is reproduced at the right in Figure 3. All tabular data corresponding to the drainage basins were then entered into the data base by referencing the unique number assigned to each basin. The georeference plane is an integral element within the data base, for it is through this plane that all tabular and most image data are interfaced. Figure 4 is a conceptualization of the georeference plane for the Tahoe region. The magnified portion of the figure indicates the coded picture elements which comprise each basin. In a most basic application, the picture elements for each basin

are summed and transferred to an interface file in which the sums are stored according to basin code number. This interface file can be accessed at a later date to produce statistical output such as acreage estimates.

The Lake Tahoe Basin data base cannot be considered to be complete at this time. The elements presently contained within it are only a small sampling of data which can be integrated into such a system. The IBIS data base as IPL has designed it implies a dynamic system which can be constantly updated to reflect the most recent environmental and resource data available to the analyst. For instance, as new multispectral scanner data or other georeferenced information becomes available, it can be integrated into the system to replace or augment current imagery and related data.

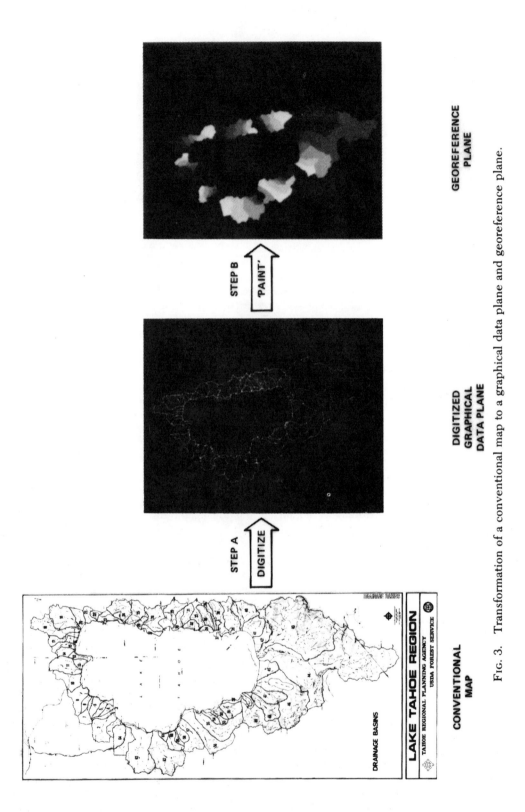

FIG. 3. Transformation of a conventional map to a graphical data plane and georeference plane.

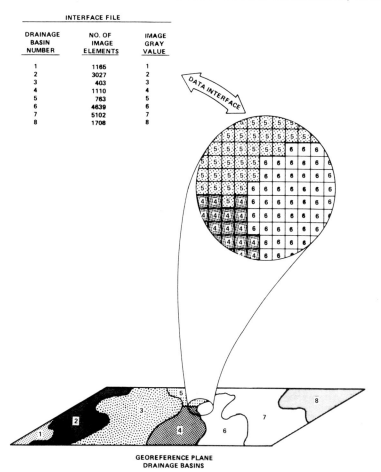

INTERFACE FILE

DRAINAGE BASIN NUMBER	NO. OF IMAGE ELEMENTS	IMAGE GRAY VALUE
1	1165	1
2	3027	2
3	403	3
4	1110	4
5	763	5
6	4639	6
7	5102	7
8	1708	8

GEOREFERENCE PLANE
DRAINAGE BASINS

FIG. 4. Conceptual drawing of the georeference plane and tabular interface file (after S. Z. Friedman, 1978).

APPLICATION OF THE IBIS CONCEPT TO THE LAKE TAHOE BASIN

Manipulation and integration of data planes comprising the data base are achieved through the implementation of VICAR and IBIS programs developed at IPL. Plate 1 illustrates standard hardcopy outputs produced from data plane integration. The Landsat scene for 27 August 1976 has been summed with three separate graphical data planes to produce overlay images. This type of imagery surpasses what is otherwise available to the analyst in the form of side-by-side comparison of conventional map versus Landsat image. The image to the left in Plate 1 defines the drainage basin boundaries. The middle image was produced by overlaying mean annual precipitation isohyets. The digital terrain mosaic was transformed to display 600-foot contour intervals and was overlayed on Landsat to produce the image to the right. Each of these images represents a visual tool, which provides the analyst with increased perspective, allowing more exact analysis of diverse data.

In future applications the mean annual precipitation isohyets will be converted to a continuous surface image. This process is similar to the conversion of USGS topographic maps to terrain relief images as depicted in Figure 2. The precipitation image will then be cross-tabulated with individual drainage basins to produce estimates of runoff coefficients.

Implementation of the georeference plane provides the opportunity for the analyst to examine more closely the unique characteristics of each drainage basin as revealed by the Landsat sensors. As depicted in Plate 2, each drainage basin mapped by the Forest Service can be extracted from the Landsat imagery and displayed separately. If a multispectral classification is then performed on the Landsat data, information can be similarly reproduced in a basin-by-basin format, thus permitting tabulations of land-cover type. The georeference plane can also be interfaced with tabular files to produce a statistical output such as that reproduced in Figure 5. In this table, the area of each uniquely coded drainage basin comprising the georeference plane was determined by summing picture elements and converting the sum to total acres. These data were then cross-referenced to tabular files which contain acreage estimates produced by the Tahoe Regional Planning Agency. All estimates were then ranked and output in the form of a computer listing, which provided a comparison of Landsat versus conventionally acquired acreage estimates.

The inclusion of digital terrain data provided an opportunity to integrate an important element of ancillary, ground-based information with remote sensing data. The Defense Mapping Agency developed these data by interpolating existing USGS 1:250,000 scale topographic maps to produce ultrafine

mesh digitized latitude, longitude, and elevation contour data. As described earlier, four DMA quadrangles were digitally mosaicked to form a unified, continuous surface elevation image. This image was then geometrically registered with the Landsat imagery to permit future cross-tabulations of elevation information with other information data planes within the data base. In an effort to quantify and process the digital terrain data, the elevation image was processed with VICAR software to produce a component representing slope magnitude. To develop this component it was necessary to compute the vector cross product between the horizontal (east-west) image elements and the vertical (north-south) picture ele-

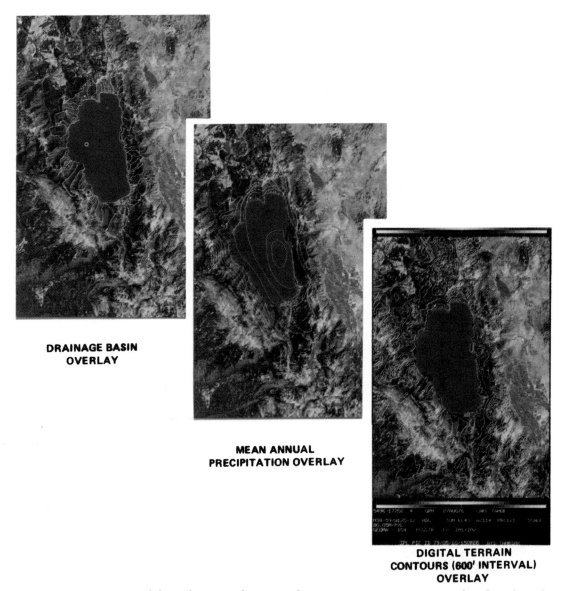

DRAINAGE BASIN OVERLAY

MEAN ANNUAL PRECIPITATION OVERLAY

DIGITAL TERRAIN CONTOURS (600' INTERVAL) OVERLAY

PLATE 1. Integration of data planes produces overlay images creating new visual and analytical tools which combine data from various sources.

160

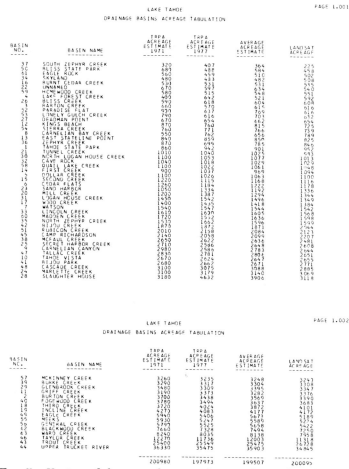

FIG. 5. Keying of the georeference plane with Landsat data and U.S. Forest Service data generated this tabular report.

ments. This vector product then provided an estimate of the slope between adjacent east-west and north-south elevation image elements. The outputs of this process were then coded to reflect slopes between 0° (no slope) and 90° (vertical slope). The angle in degrees was further coded for image output by rescaling 0° to be equal to 0 digital number (DN) or black and 90° to be equal to 255 DN or white. Figure 6 illustrates an application of these concepts. First, the slope magnitude was cross-tabulated with the drainage basin georeference plane. Each slope magnitude image element associated with the individual drainage basins was then extracted. Further cross-tabulation permitted the computation of a mean slope estimate by averaging the slope magnitude image elements for each drainage basin.

Figure 7 reproduces an IBIS table generated from the digital terrain imagery which lists azimuth and slope for each drainage basin. Eventually, it is hoped to develop a model for drainage basin terrain based on digital imagery. At present, however, difficulties have been encountered with discrepancies between individual DMA map quadrangles, especially along map edges, which have precluded the development of accurate models. Other types of digital terrain imagery are being investigated with the hope of integrating more reliable data into the Lake Tahoe Basin data base.

A data base, no matter how easily constructed and manipulated, is only as reliable as the elements which comprise it. Therefore, as new data are acquired, it is essential that these be tested and implemented into the data base to insure its viability. Although existing digital terrain imagery has proved difficult to adapt, Landsat data continues to work well within the data base concept. However, in terms of water quality, the subject lake and watershed area must be of such dimensions as to accommodate the resolution of the Landsat sensors.

Accurate correlation of changes in Lake Tahoe with various data elements contained

**GEOREFERENCE
PLANE**

INTEGRATE

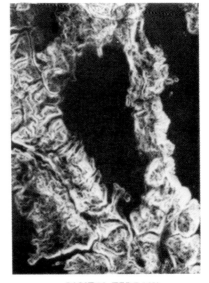

**DIGITAL TERRAIN
GRADIENT MAGNITUDE
IMAGE**

**DIGITAL
TERRAIN
INTERFACE**

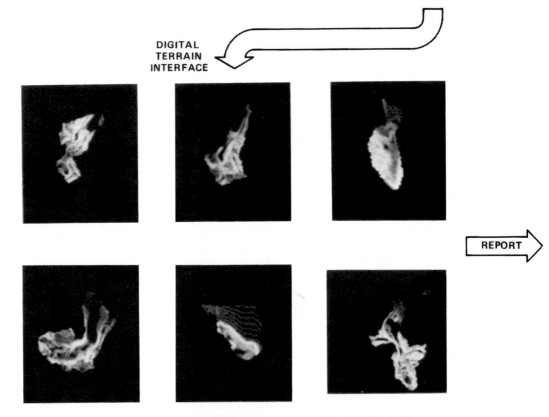

REPORT

GRADIENT MAGNITUDE IMAGES BY DRAINAGE BASIN

FIG. 6. The georeference plane has been interfaced with digital terrain data to extract individual basins and produce a statistical report (see Figure 7).

within the data base will require study over several years. It is hoped that the system over time can be used to monitor and evaluate causes for changes which effect the lacustrine environment. This will require the development of precipitation modelling, surface runoff models, and classification of drainage basin cover types. These elements must in turn be integrated and evaluated for accuracy before the system can be consid-

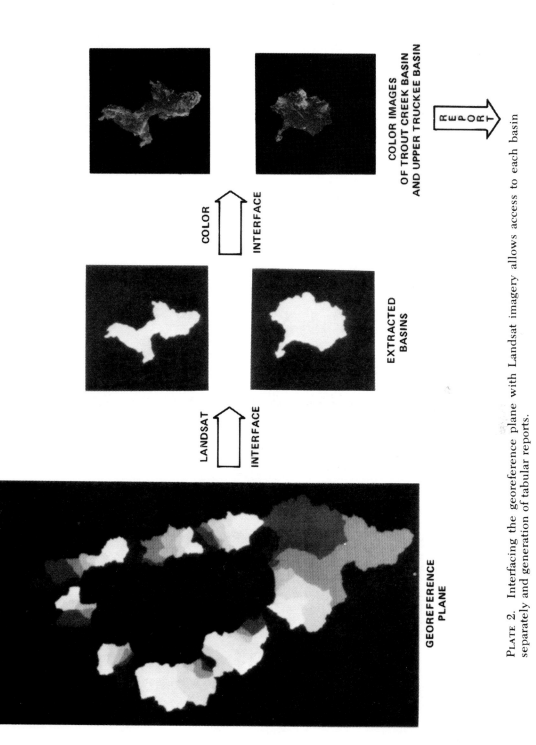

PLATE 2. Interfacing the georeference plane with Landsat imagery allows access to each basin separately and generation of tabular reports.

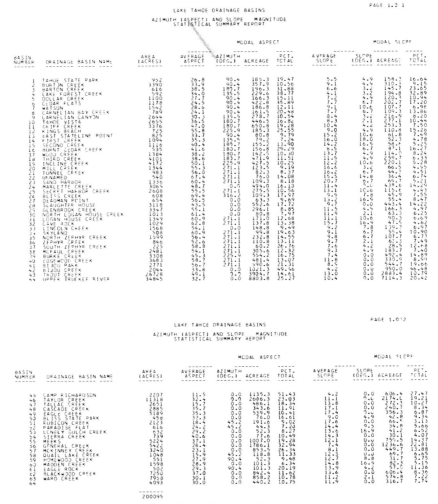

Fig. 7. The statistical report produced by interfacing the georeference plane with digital terrain data (see Figure 6).

ered usable. Such a system is feasible given the continued improvement of the remote sensing tools used to construct the data base and the data integrated into it. With the urban population continually encroaching upon delicate watershed systems such as Lake Tahoe, investigations into information systems for environmental monitoring will need to continue.

ACKNOWLEDGMENTS

The IBIS system was designed originally for land-use studies conducted at IPL. Two members of the Earth Resources Group were particularly instrumental in instructing the authors of this paper in the use of IBIS. The authors wish to extend their gratitude to David B. Wherry for his contributions to data storage and retrieval operations and Ron G. McLeod for his knowledge of DMA data and mosaicking techniques. The authors also wish to thank Terry Ann Bednarczyk for her efforts on this project.

This paper presents one phase of research conducted at the Jet Propulsion Laboratory, California Institute of Technology, under NAS7-100, sponsored by the National Aeronautics and Space Administration.

REFERENCES

Blackwell, J. R., and D. H. P. Boland, 1979. *Trophic Classification of Selected Colorado Lakes*, JPL Publication 78-100, January 1979.

Bryant, N. A., and A. L. Zobrist, 1978. *An Image Based Information System: Architecture for Correlating Satellite and Topological Data Bases*, 1st Users Conference on Computer Mapping, Harvard University, 1978.

Davis, J. B., and S. Z. Friedman, 1979. *Assessing Urbanized Area Expansion Through the Integration of Landsat and Conventional Data*, ASP Symposium, 1979.

Madura, D. P., and J. M. Soha, 1978. *Color*

Enhancement of Landsat Agricultural Imagery, JPL Publication 78-102, December 1978.

Ruiz, R. M., *et al.*, 1977. IPL Processing of the Viking Orbiter Images of Mars, *Journal of Geophysical Research*, Vol. 82, No. 28.

Smith, A. Y., and J. D. Addington, 1978. *Water Quality Monitoring of Lake Mead: A Practical Look at the Difficulties Encountered in the Application of Remotely Sensed Data to Analysis of Temporal Change*, 5th Canadian Symposium on Remote Sensing, Victoria B. C., August 1978.

Strahler, A. H., T. L. Logan, and C. E. Woodcock, 1979. *Forest Classification and Inventory System Using Landsat, Digital Terrain and Ground Sample Data*, 13th International Symposium on Remote Sensing of the Environment, Ann Arbor, Michigan, April 1979.

Tahoe Regional Planning Agency, 1977. *Lake Tahoe Basin Water Quality Management Plan*, Volume 1, Water Quality Problems and Management Program, July 1977.

Yagi, G. M., J. J. Lorre, and P. L. Jepsen, 1978. *Dynamic Feature Analysis for Voyager at the Image Processing Laboratory*, Conference on Atmospheric Environment of Aerospace Systems and Applied Meteorology, New York, N.Y.

Zobrist, A. L. and N. A. Bryant, 1979. *Elements of an Image Based Information System, International Journal on Policy Analysis and Information Systems*, Volume 1, No. 2, January 1979.

(Received 5 September 1979; revised and accepted 7 March 1980)

Thomas R. Loveland
Gary E. Johnson†
Technicolor Government Services, Inc.
EROS Data Center
Sioux Falls, SD 57198

The Role of Remotely Sensed and Other Spatial Data for Predictive Modeling: The Umatilla, Oregon, Example*

The spatial and remotely sensed data provided a unique look at past, present, and future characteristics of irrigation development.

Introduction

The U.S. Geological Survey's Earth Resources Observation Systems (EROS) Data Center (EDC) has the responsibility to assist land and water resources planning and management agencies in using remotely sensed data as a tool to help solve inventory and analysis problems. EDC assistance has matured from the early Landsat era

tion with other spatial data in models to provide answers to complex resource management questions. An example of this trend in resource analysis is a recently completed project involving EDC and the U.S. Army Corps of Engineers (COE), Portland District.

EDC and COE undertook a cooperative project in 1979–80 to explore the use of remotely sensed

ABSTRACT: *The U.S. Geological Survey's Earth Resources Observation Systems Data Center, in cooperation with the U.S. Army Corps of Engineers, Portland District, developed and tested techniques that used remotely sensed and other spatial data in predictive models to evaluate irrigation agriculture in the Umatilla River Basin of north-central Oregon. Landsat data and 1:24,000-scale aerial photographs were initially used to map the expansion of irrigation from 1973 to 1979 and to identify crops under irrigation in 1979. The crop data were then used with historical water requirement figures and digital topographic and hydrographic data to estimate water and power use for the 1979 irrigation season. The final project task involved production of a composite map of land suitability for irrigation development based on land cover (from Landsat), landownership, soil irrigability, slope gradient, and potential energy costs.*

The methods and data used in the study demonstrated the flexibility of remotely sensed and other spatial data as input for predictive models. When combined, they provided useful answers to complex questions facing resource managers.

when the desired end product was commonly a land-use or land-cover map. Today, satellite-derived inventory data are routinely incorporated into geographic data bases and used in conjunc-

*Presented at the Pecora VII symposium, Sioux Falls, South Dakota, 18–21 October 1981.

† Now with the Asian Regional Remote Sensing Training Center, Asian Institute of Technology, Bangkok, Thailand.

data in conjunction with other geographic data to assess several irrigation-related problems. The COE required information describing past and present measurements of irrigation and its impact in the Columbia River Basin and projections identifying lands suitable for future irrigation development. Specifically, the COE desired answers to the following questions:

Photogrammetric Engineering and Remote Sensing,
Vol. 49, No. 8, August 1983, pp. 1183-1192.

0099-1112/4908/-1183$02.25/0
© 1983 American Society of Photogrammetry

- What is the growth rate of center pivot irrigation?
- How many acres of land were irrigated in 1979?
- What types of crops were irrigated and what were their acreages in 1979?
- What were the irrigation crop-water requirements in 1979?
- What were the irrigation energy requirements in 1979?
- Which lands are likely to be irrigated in the future.

To fill these information needs, a geographic data base was developed, and analytical methods were selected that would provide answers to these questions.

The project included two distinct components: (1) data base development and (2) predictive modeling. The data base development aspects involved analyzing remotely sensed data to inventory land cover and crop types and collecting cartographic data relating to irrigation development to create a digital geographic data base. The predictive modeling involved manipulating the appropriate data in various spatial models to provide measurements of resource (water and power) consumption and to identify potentially irrigable lands. The project methods were tested in the Umatilla Basin, Oregon, to determine their suitability for investigating irrigation in other parts of the Columbia River Basin. Figure 1 summarizes the analytical flow of the study.

STUDY AREA DESCRIPTION

The Umatilla River Basin occupies 1.6 million acres in north-central Oregon (Figure 2). The northern edge of the basin is bounded by the Columbia River. The Blue Mountains occupy the east and southeast regions. The Umatilla River flows through the center of the basin. The proximity of the Columbia River has contributed heavily to the ongoing development of irrigation in the area. The basin's first large irrigation project was the Umatilla Project developed near Hermiston, Oregon, by the U.S. Bureau of Reclamation in the 1930's. In the 1970's, much center pivot irrigation was developed adjacent to the Columbia River. The mixture of both old and recent irrigation development contributes to the complexity of the region. A wide variety of field sizes, shapes, irrigation methods, and crop types increases the difficulty of inventorying the region's irrigated land resources.

DATA BASE INPUTS

Data describing land use, physical terrain characteristics, and administrative boundary restrictions were entered into a digital geographic data base in order to address project topics (Table 1).

All data were stored in cellular form with 63.6-m² (1-acre) grid cells registered to a Universal Transverse Mercator (UTM) map projection. Digital inputs to the data base were incorporated using

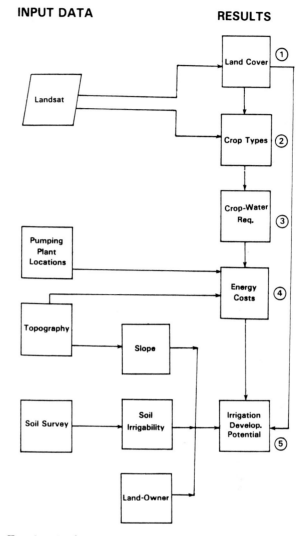

FIG. 1. Analysis sequence and spatial data used to assess Umatilla Basin irrigation.

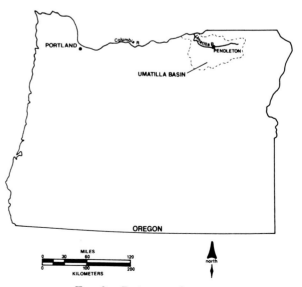

FIG. 2. Project study area.

167

TABLE 1. SUMMARY OF GEOGRAPHIC DATA BASE ELEMENTS.

Data Source	Input Form	Derivatives
Landsat	Photographic, digital	Land cover, crop type
Digital Terrain Tape	Digital	Elevation, slope
Soil survey	Map	Soil irrigability
Pumping plant location records	Tabular	Pumping distance
Administrative maps	Map	Land ownership

a least-squares registration process and resampling techniques appropriate to the characteristics of the data. Map inputs were digitized and georeferenced using an *x*-, *y*-coordinate digitizer.

REMOTE SENSING INPUTS

Landsat data, supplemented by 1:24,000-scale color-infrared aerial photographs, provided an inventory of 1973, 1975, 1977, and 1979 general land-cover patterns and 1979 irrigated crop type acreages. The remote sensing based inventory data were used for the measurement of the 1979 irrigation acreage, and the 1973–79 growth rate of center pivot irrigation, and the number of acres of specific types of crops irrigated in 1979. In addition, the inventories provided data elements needed to model current water and power consumption and to predict the areas of future irrigation development.

Land Cover Mapping. Maps of general land cover in the Umatilla Basin for 1973, 1975, 1977, and 1979 were produced by manual interpretation of Landsat false color composites (FCC's). A classification scheme comprised of the following Level I and III categories (Anderson *et al.*, 1976) was used:

  100 Urban and Built-Up Areas
    211 Dryland Agriculture
    212 Center Pivot Irrigation
    213 Other (noncenter pivot) Irrigation
  300 Rangeland
  400 Forestland
  500 Water Bodies
  600 Wetlands

Using Landsat images obtained late in the irrigation season (late July through August), the land-cover catagories were readily mapped on the basis of their image color, size, shape, and location. This resulted in generalized maps displaying the dominant cover types in the basin. The final maps were interpreted with a minimum mapping unit size of 40 acres. With this resolution, it was possible to examine irrigation expansion rates and to determine the types of lands which had been converted to irrigation agriculture. The precise estimation of the number of irrigated acres was not possible, however, because the coarse resolution of the Landsat imagery precluded the mapping of small nonirrigated areas found within intensively irrigated regions. Color-infrared aerial photographs (1:24,000 scale) for 1979 were thus used with the Landsat images in a stratified double-sampling ratio-estimation process in order to provide refined estimates of the 1979 center pivot and other irrigation acreages (67,835 ± 6,960 and 62,679 ± 7,083 acres, respectively).

Table 2 summarizes the land-cover acreages measured directly from Landsat FCC's in this phase of the project. By comparing the 1973–79 acreage estimates, an annual center-pivot irrigation growth rate of 7,758 acres was estimated, with almost 56 percent of this acreage converted from dryland agriculture. The 1979 land-cover map was digitized and entered into the geographic data base. In digital form, the land-cover map provided stratification data needed to increase the efficiency of the crop-type analysis. It also provided data describing current land-cover patterns and irrigation conversion trends needed to assess the basin's irrigation development potential.

Crop Type Analysis. 26 July 1979 digital Landsat data were analyzed to produce a map and the initial acreage estimates of crop types irrigated in the Umatilla Basin. The use of digital data was considered advantageous for this problem since

TABLE 2. LAND-COVER ACREAGE ESTIMATES DERIVED THROUGH THE DIRECT MEASUREMENT OF LANDSAT FCC's

Land-Cover Class*	Year			
	1973	1975	1977	1979
Center-pivot irrigation	24,587	59,304	74,995	85,867
Other irrigation	126,551	122,901	122,901	122,901
Dryland agriculture	553,187	527,681	518,646	512,423
Rangeland	419,577	413,197	403,452	402,711
Forestland	473,320	473,320	473,320	473,320

* Urban areas, wetlands, and water bodies were not shown because they occupied such a small percentage of the basin.

168

the spatial resolution of a 1.1-acre pixel provided more detailed information than is contained in a standard Landsat photographic image.

The crop classification process began with the geometric registration of the raw Landsat data into the data base. A least-squares registration using terrain control points and a nearest-neighbor resampling method were used to register the Landsat data to the 63.6-m^2 UTM referenced grid.

Because the irrigation in the Umatilla Basin was complex and variable, it was decided to stratify the Landsat data into analysis regions according to irrigation method. The digitized 1979 land-cover map was used as a mask to stratify two data sets—the center pivot stratum and other irrigation stratum—for classification. The use of a preexisting layer of the data base allowed narrowing and sorting of the Landsat data so that only the irrigable lands needed to be classified into crop-type categories, thus increasing analysis efficiency.

The crop types were classified independently for each of the two strata. A modified clustering technique (Fleming et al., 1975) was used to define spectral class statistics, and a maximum likelihood algorithm was used to classify the two stratified data sets. Each stratum classification resulted in 18 spectral categories. The spectral classes were compared to field-collected crop reference data, and initial attempts were made to assign the spectral classes to crop-type categories. However, because of similar spectral characteristics existing between different crop types (particularly between wheat, alfalfa, and potatoes), only 60 percent of the data could be placed into the desired single-crop categories. This problem necessitated the use of detailed spectral class descriptions to describe the exact crop-type composition of each spectral class.* In this approach, a sampling technique was used to develop statistics describing the individual crop-type acreages represented by each of the 36 spectral classes for the two strata. The spectral classification and category descriptions were then incorporated into the geographic data base for subsequent use in estimating other characteristics of the basin's irrigation.

Tables 3 and 4 show the crop acreage descriptions of the spectral classes derived using a double-sampling procedure with the Landsat classification and field-collected crop data. The associated variance and standard error statistics were also estimated for each class. Plate 1 is the irrigated crop-type classification and general land-cover map for the Umatilla Basin. As stated previously, 60 percent of the irrigated croplands fall into single-crop categories. The usefulness of the

remaining 40 percent that are in multicrop categories was salvaged by the development of detailed class descriptions that identify the specific characteristics of each spectral category.

OTHER SPATIAL INPUTS

Digital Terrain Tapes. Digital Terrain Tape (DTT) data acquired from the U.S. Geological Survey's National Cartographic Information Center (NCIC) provided elevation values that could be used to further evaluate regional irrigation conditions. The elevation data contained on DTT's are based on digitized contour lines from U.S. Geological Survey 1:250,000-scale topographic quadrangles, with an interpolation algorithm used to estimate a regular spaced grid of elevations (U.S. Geological Survey, 1980). The data were considered to be useful only for regional analysis and not site-specific investigations.

The DTT data, because of their digital form, were easily incorporated into the data base. The elevation data were registered to the 63.6-m^2 grid cells and resampled using bilinear interpolation. The registered elevation values were then used to calculate the average percent slope between each cell and the adjacent cells.

Soil Survey Data. The detailed county soil survey for Umatilla County (Soil Survey Staff, 1948) and the preliminary soil survey for Morrow County were digitized for inclusion into the data base. The soil polygons were converted to 63.6-m^2 grid cells and stored in 158 soil mapping unit classes. Based on the chemical and physical characteristics of each soil mapping unit, each of the 158 classes was assigned to one of five soil irrigability categories (excellent, good, fair, poor, or unsuitable).

Land-Ownership. U.S. Army Corps of Engineers (1979) maps provided government land-ownership data. Corps maps were digitized, and the ownership polygons converted to a grid cell format.

Pumping Plant Locations. COE documents (Swan et al., 1980) containing the locations of irrigation pumping plants along the Columbia River and its tributaries were used to establish a digital file of the locations of pumping plants. The pumping plant locations (identified by river-reach coordinates) were plotted on 7.5-minute quadrangle maps, digitized, and stored as point data.

PREDICTIVE MODELING

The inventory data derived through the interpretation of remotely sensed data, combined with the remaining spatial data contained in the digital data base, provided the elements needed to estimate 1979 irrigation water and power consumption and to identify the irrigation development potential of lands in the Umatilla Basin.

---

* This approach is being used with increasing frequency as a means to provide useful information in areas where "pure" classifications are not possible (Rohde and Miller, 1981).

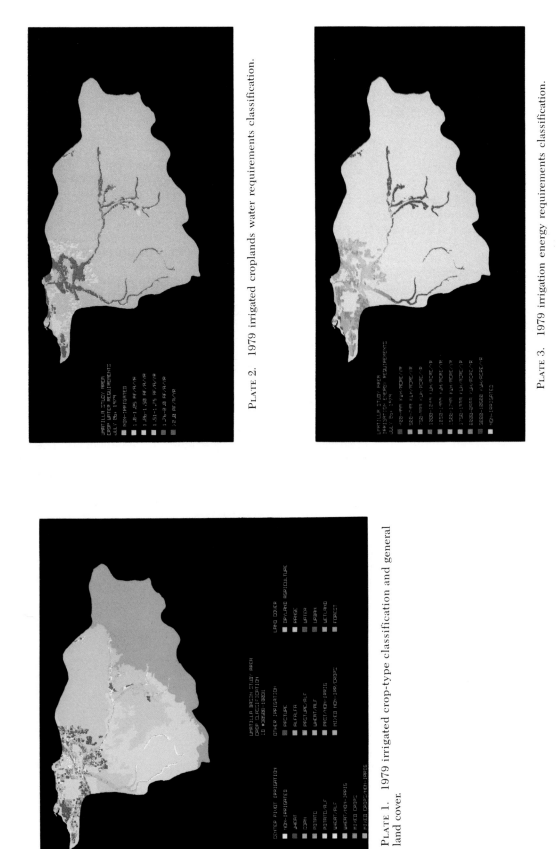

PLATE 2. 1979 irrigated croplands water requirements classification.

PLATE 3. 1979 irrigation energy requirements classification.

PLATE 1. 1979 irrigated crop-type classification and general land cover.

TABLE 3. CENTER-PIVOT STRATUM CROP ACREAGES BY SPECTRAL CLASS.

Spectral Class	Crop Type								
	Alfalfa	Corn	Potato	Wheat	Pasture	Peas	Beans	Mint	Onions
1	277	23	92	2,196	0	0	0	0	0
2	544	0	317	3,175	204	0	0	0	0
3	82	0	3,161	110	0	0	55	0	0
4	19	39	0	58	116	97	0	0	60
6	1,186	0	209	2,163	70	0	105	0	0
7	1,212	941	1,086	597	651	0	0	20	0
8	0	0	0	0	57	0	0	0	0
9	162	23	232	3,547	93	0	0	0	0
10	1,065	60	2,048	682	60	0	0	0	0
11	653	187	467	1,028	93	0	0	0	0
12	344	111	222	2,627	222	67	0	0	0
13	81	2,273	135	81	27	0	27	0	72
14	154	0	88	2,654	0	0	0	0	0
15	956	792	573	846	109	0	109	0	0
16	307	31	245	2,482	92	0	0	0	0
17	1,563	932	839	723	93	0	0	0	0
18	156	750	312	562	63	0	0	0	0
Total	8,761	6,162	10,026	23,531	1,950	164	296	20	132

PREDICTING CROP-WATER REQUIREMENTS

Estimates of 1979 irrigated crop-water requirements were calculated using the Landsat-based crop-type data and historic crop water requirement figures developed by the Soil Conservation Service (U.S. Soil Conservation Service, 1973) (Table 5). The specific water requirement per crop was multiplied by the area of each crop type found in each spectral class in order to determine the acre-feet of water required for each spectral class and overall. This simple translation of crop-type data to crop water requirements provided an estimate of 207,952 acre-feet of water needed for the basin's irrigation in 1979 and produced a file of spatial crop water requirements data. The 36-class crop water requirements classification is summarized in six categories and displayed in Plate 2.

It must be recognized that the analysis used water requirements figures based on an average of

TABLE 4. OTHER IRRIGATION STRATUM CROP ACREAGES BY SPECTRAL CLASS.

Spectral Class	Crop Type								
	Alfalfa	Corn	Potato	Wheat	Pasture	Beans	Mint	Straw-berries	Asparagus
1	101	0	0	505	0	0	0	0	0
2	1,335	0	157	0	786	157	0	0	157
3	1,496	0	0	0	45	0	0	0	0
4	0	0	0	0	0	0	0	0	0
5	1,270	0	868	68	1,804	201	467	0	668
6	412	0	0	0	0	0	0	0	0
7	697	0	620	154	3,756	0	39	0	77
8	1,856	0	53	212	4,030	0	106	53	0
9	133	0	927	88	2,207	0	88	0	221
10	697	131	218	0	6,135	0	87	0	0
11	0	0	0	0	51	0	0	0	0
12	3,923	0	531	200	2,261	67	665	133	67
13	1,021	0	34	0	5,205	0	0	0	0
14	1,422	414	432	1,483	494	0	0	0	0
15	1,518	0	34	0	2,624	0	207	0	0
16	716	0	0	1,312	119	0	0	0	0
17	1,814	209	35	0	391	0	107	0	0
18	2,058	0	105	53	2,553	0	369	0	0
Total	20,469	754	4,014	4,075	32,461	425	2,135	186	1,190

TABLE 5. HISTORIC CROP WATER REQUIREMENTS FOR UMATILLA BASIN CROPS (U.S. SOIL CONSERVATION SERVICE, 1973).

Crop Type	Water Requirements (acre-feet/acre/year)
Small grains	1.1
Legumes	1.1
Onions	0.9
Corn	1.2
Potatoes	1.6
Pasture	1.9
Mint	1.6
Strawberries	1.5
Alfalfa	2.2
Asparagus	1.6

many years and not just 1979 conditions. Because no attempts were made to model the effects of soil conditions or microclimates on water requirements, the estimates may vary from actual 1979 crop-water consumption rates. However, the method does provide a generalized estimate of the demand placed on regional water supplies and a perspective on regional patterns of water consumption.

MODELING IRRIGATION ENERGY REQUIREMENTS

The estimation of 1979 irrigation energy requirements was completed using a COE formula (Whittlesey and Buteau, 1980) in a physical model. It was developed as a tool to estimate power requirements in the Columbia River Basin but was never used in a spatial context. By using the equation with spatial data, it was possible not only to estimate total energy requirements but also to produce a map which spatially portrays power requirement patterns. The formula that estimates kilowatt-hours per acre (kWh/acre) of energy consumption is

$$kWh/acre = \frac{(1.024) \times (Diversion) \times (TDH)}{Epp}$$

where
   1.024 = a constant that standardizes the equation;
   Diversion = acre-feet of water diverted for irrigation;
   TBD = Total dynamic head based on the sum of the pump lift, friction loss during delivery, and the system operating pressure; and
   Epp = Pumping plant efficiency (which is estimated to be 85 percent in this area.)

The diversion requirements are based on the types of crops irrigated while TDH is primarily based on landscape position. The remainder of the equation uses nonspatial data and reduces to constant values for the study area.

Total Dynamic Head Analysis. The calculation

of TDH required analyzing the relationship between the water's source and the point of use. In this study, it was assumed that all water would be diverted from the Columbia River using existing pumping plants.

DTT elevation data were used to determine the required pump lift. By subtracting the pool elevation of the Columbia River from the basinwide elevation file, the pump lift distance in feet for each acre in the study area was calculated. Friction loss was estimated by calculating the minimum distance between the nearest pumping plant and all points (data-base cells) throughout the basin. The distances were determined using the pumping plant location file and a radius calculation procedure. The distances, calculated in one-half-mile intervals, were multiplied by 11.12 to estimate friction loss. The friction loss and pump lift data were then added to the system operating pressure (240 feet of static lift was the assumed operating pressure) to produce the TDH file.

Diversion. The diversion data needed in the energy equation were based on the 1979 crop-water requirements file created earlier. The water requirements estimates for each of the 36 categories were increased by 22 percent (to allow for water lost during transmission) in order to create the diversion file.

1979 Irrigation Energy Requirements. With the spatial elements of the energy equation assembled, they could then be linked in the model with the nonspatial terms to estimate 1979 energy requirements. The modeling was completed using simple multiplication and division as specified by the energy formula. The result was a spatial classification of lands based on their 1979 energy requirements (Plate 3). In addition, the analysis estimated that 292 million kWh of energy were used for the 1979 irrigation season. The combined results not only provided an estimate of the impact irrigation placed on regional power supplies but also indicated patterns of energy use that may aid in identifying future consumption trends.

Potential Irrigation Energy Requirements. The use of a hypothetical diversion figure permitted the classification of lands in the Umatilla Basin according to their potential energy requirements. This was accomplished using a diversion of 2.81 acre-feet (this is the diversion figure suggested by Whittlesey and Buteau for use in regional power analysis). The resulting potential energy requirements file was included in the geographic data base for later use in determining the irrigation development potential of the Umatilla Basin (Plate 4).

MODELING IRRIGATION DEVELOPMENT POTENTIAL

The final analysis task of the project involved the use of a cartographic overlay model to predict

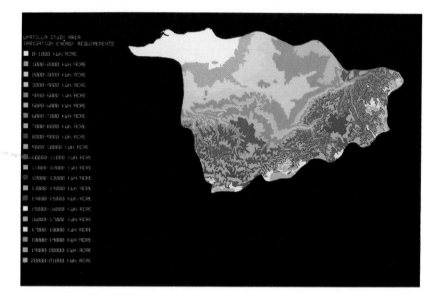

PLATE 4. Potential energy requirements for Umatilla Basin irrigation.

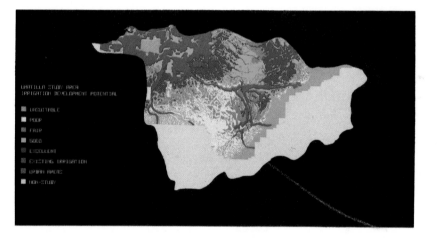

PLATE 5. Irrigation development potential composite map.

the irrigation development potential of the northern 1.1 million acres of the Umatilla Basin.* The model used data representing land cover, soil irrigability, percent slope, land-ownership, and potential energy requirements. Based on the characteristics of all the variables, the following assumptions were developed by EDC and COE scientists and used to determine variable weights for the overlay analysis. Table 6 contains the variable

* Soil survey data were only available for the northern 1.1 million acres of the 1.6-million-acre basin. For this reason, the irrigation potential modeling was restricted to the lands where complete soil survey existed. Most of the excluded lands are in National Forests.

weighting scheme developed from these assumptions.

- Energy costs have twice as much impact on potential irrigability as any other factor.
- Physical factors of soils, slope, and land cover are of equal importance.
- Land-ownership is important only for eliminating a land parcel from consideration for irrigation development.
- While both dryland agricultural land and rangeland are considered irrigable, the dryland agricultural lands are more favorable for irrigation development.
- Water bodies, wetlands, urban areas, and forestlands cannot be irrigated.

TABLE 6. WEIGHTING SCHEME DEVELOPED FOR COMPOSITE MAPPING. POTENTIAL ENERGY REQUIREMENTS (NOT SHOWN IN THIS TABLE) USED A 20 TO 1 NUMERIC RANGE WITH 20 REPRESENTING THE LOWEST COSTS (400 kWh/ACRE) AND 1 FOR HIGHEST COSTS (10,500 kWh/ACRE).

| Numeric Value (Weight) | Model Inputs | | | |
	Land Cover	Land Ownership	Soils	Slope (Percent)
10	Dryland agriculture	Private ownership	Excellent	0–3
9	—	—	—	—
8	—	—	—	—
7	Rangeland	—	—	4–7
6	—	—	Good	—
5				
4	—	—	—	8–12
3	—	—	Fair	—
2	—	—	—	—
1	—	—	Poor	—
not considered for irrigation suitability	Existing irrig. Water bodies Urban areas Wetlands	Wildlife area Military reservation State parks National forests	Unsurveyed	13+

- Lands with slopes from 0 to 3 percent are most favorable for irrigation, those with slopes from 4 to 7 percent are less favorable, and those with slopes from 8 to 12 percent are least favorable.
- Lands with slopes of 13 percent or greater are non-irrigable.
- All soils are irrigable although they range from excellent to poor.

An additive overlay process was used for the actual modeling. The scores assigned to the variables were summed cell by cell to produce a composite score of irrigation potential. At the same time, the presence of variable categories that rendered the area nonirrigable (wetlands, steep slopes, wildlife refuges, etc.) was tested in order to eliminate areas that have adverse conditions. The cumulative scores for the potentially irrigable land were evaluated and threshold scores were determined by EDC and COE scientists (Table 7). Using the thresholds, the raw scores were assigned to irrigation development potential categories and the final map was produced (Plate 5).

The irrigation development potential map illustrates the region's development potential based on a specific set of assumptions. As conditions change, the results would change. However, because of the flexibility afforded by the digital data base, it is possible to efficiently repeat the analysis procedure using new decision criteria. The modeling effort using physical variables (soils, slope, and land cover), administrative factors (land-ownership), and economic considerations (energy requirements) provided a unique opportunity to anticipate the impact of potential Umatilla Basin irrigation development in future years.

DISCUSSION

Remotely sensed data, used independently and with other spatial data, provided descriptive statistics and maps needed to understand irrigation characteristics in the Umatilla Basin. Basic land-cover and crop-type inventory data provided the COE with information describing the number of acres irrigated in 1979, the types of crops and their acreages irrigated in 1979, and the rate of center-pivot irrigation growth based on 1973–79 trends. The detailed digital crop classification results were combined with supporting data to estimate the impact of 1979 irrigation on regional water and power supplies. A simple translation from crop-type to water requirements provided an estimate of the acre-feet of water used for irrigation. A more complex physical model using pumping plant location and digital elevation data with Landsat-derived crop-water requirements produced an estimate of 1979 irrigation energy requirements. Finally, a land-cover interpretation and data describing soil irrigability, percent slope, land-ownership, and potential energy requirements were used in a cartographic model to determine the irrigation development potential for much of the Umatilla Basin.

Remotely sensed data were used in the study in several ways. Landsat data provided an efficient base from which area-wide inventories could be

TABLE 7. CATEGORIES, SCORES, AND FREQUENCY OF THE IRRIGATION DEVELOPMENT POTENTIAL COMPOSITE MAP.

Category	Composite Scores	Occurence (Percent)
Unsuitable	0–31	20
Poor	32–44	21
Fair	45–50	21
Good	51–53	19
Excellent	54–58	19

completed. The retrospective and repetitive nature of the data not only permitted a look at the basin in previous years so that trends could be detected, but also will allow updating baseline information. In coming years, new impacts can be measured, and development trends can be monitored. Landsat FCC's provided a comprehensive assessment of land-cover patterns without the need for complex analysis tools and techniques. In addition, the spatial generalization to the 40-acre minimum mapping unit was adequate for regional characterization of development suitability. Digital Landsat data were useful because the high resolution (1.1 acre) permitted the analysis of complex classes of information with the spatial resolution needed to produce more precise estimates. The digital data provided an effective sampling frame from which refined estimates could be made in order to mitigate misclassification.

Spatial data in a variety of forms were readily used with the remotely sensed data. No problems were identified that prevented the accomplishments of a project goal. The terrain data were relatively easy to reformat for the data base because they existed in a gridded UTM format. The remaining map-based data, such as soil, land-ownership, and pumping plant locations, required digitizing which increased analysis time and costs. Fortunately, these variables seldom change and thus will not need to be redigitized for future investigations. The spatial and remotely sensed data provided a unique look at past, present, and future characteristics of irrigation development in the Umatilla Basin. Increased benefits, at reduced costs, will be realized in the future, when modeling of irrigation potential is repeated using new scenarios or when new estimates of resource consumption are needed.

ACKNOWLEDGMENT

The work described in this paper was performed under U.S. Geological Survey Contract No. 14-08-0001-16439.

REFERENCES

Anderson, J. R., E. E. Hardy, J. T. Roach, and R. E. Witmer, 1976. *A land use and land cover classification system for use with remote sensor data:* U.S. Geological Survey Professional Paper 964, 28 p.

Fleming, M. D., S. S. Berkebile, and R. M. Hofer, 1975. *Computer aided analysis of LANDSAT-1 MSS data: a comparison of three approaches including a "modified clustering" approach:* Purdue University, Laboratory for Applications of Remote Sensing, LARS Information Note 072475, 9 p.

Rohde, W. G., and W. A. Miller, 1981. *Arizona vegetation resource inventory,* final report: U.S. Geological Survey EROS Data Center, Sioux Falls, So. Dak., 79 p.

Soil Survey Staff, 1948. *Soil survey—the Umatilla area, Oregon:* U.S. Department of Agriculture Series 1937-No. 21, Washington, D.C., 125 p.

Swan, G. A., T. G. Withrow, and D. L. Park, 1980. *Survey of fish protective facilities at water withdrawals on the Snake and Columbia River:* Seattle, Wash., Coastal Zone and Estuarine Studies Division, National Oceanic and Atmospheric Administration, 189 p.

U.S. Army Corps of Engineers, Portland District, 1979. *Columbia Basin water withdrawal environmental review;* Appendix A—Land Use: Portland, Ore., 26 p.

U.S. Geological Survey, 1980. *National Cartographic Information Center digital terrain tapes—a users guide:* U.S. Government Printing Office: 1980 0-311-344/184, 12 p.

U.S. Soil Conservation Service, 1973. *Oregon irrigation guide:* Portland, Ore., 91 p.

Whittlesey, N. K., and J. R. Buteau, 1980. *Pumping energy for irrigation development in the Pacific Northwest,* Portland, Ore.: U.S. Army Corps of Engineers, 13 p.

(Received 23 March 1982; accepted 4 November 1982; revised 3 March 1983)

CARTOGRAPHIC MODELING OF HYDROTHERMAL RESOURCE POTENTIAL

THOMAS W HODLER
University of Georgia / Athens / Georgia

ABSTRACT   The potential for utilizing hydrothermal (low temperature) resources can be evaluated by using a three-dimensional cartographic model. The four components of the model include temperature of the geothermal water, topographic slope, land elevation and supply economics. Each individual component is scaled ordinally and a perspective drawing of the site utility is computer generated. Temperature distribution is determined from an isothermal linear cartogram based on a constant temperature decay function. Slope and elevation are determined from 1:250,000 series topographic maps. Supply economics is defined by the pumping cost in transporting the water from hot spring to the point of utilization. The components are merged into a composite data set and the perspective representation of that data reflects the potential for hydrothermal utilization.

This paper describes a cartographic technique for a model analyzing the potential utilization of hydrothermal resources. Hydrothermal implies that the resource is a particular form of geothermal convective systems characterized by low temperatures ($< 150°C$) and surface manifestations, i.e., hot springs and/or geysers.[1] Such resources are found throughout the western United States and at other tectonically active regions of the world. As a consequence of lower temperatures, electrical power generation is not feasible. Thus, the utilization of the resource is restricted to a relatively small area surrounding the hydrothermal site.

Specific uses that are made of the resource are dependent upon the available water temperature (Table 1). For example, the utility of the water for space heating is dependent upon a minimal water temperature of 70° Celsius. Therefore, the hydrothermal resource utilization is site specific and temperature dependent. Traditionally, the water is used in close proximity to the source but in some locations, such as Iceland, the water is transported in insulated pipes over the land surface to points of application. The water temperature loss (temperature decay) is approximately 5°C for every 18 km transport (0.27°C/km).[2] Therefore, the potential utility of the resource decreases as transport distance increases.

To assist in resource analysis and evaluation of the potential for a given hydrothermal source, maps can easily be used. However, most mapping to-date focuses on the larger picture. Small scale maps exist identifying locations of known-geothermal-resource-areas or specific point locations of geothermal hot springs, geysers, or wells.[4] Geophysical and geomagnetic data are also included on maps designed for broad display of the resource distribution. The maps provide useful insight into national or regional data sets, however, they provide only limited input for large scale mapping associated with required environmental impact statements, etc. Such detailed analysis incorporates extensive evaluation of specific geothermal sites and their immediate surroundings.

A model and concomitant mapping technique which layers appropriate data in order to evaluate such resource utilization potential has been created. The model operates at either a planimetric level, utilizing the concepts of linear cartograms, or a three-dimensional level. Its implementation allows for either generalized or site specific analysis of the potential application and utilization of the resource.

THOMAS W HODLER is an Assistant Professor in the Department of Geography, University of Georgia, Athens, Georgia. The author wishes to thank Dr. Robert W. Schottman, Agricultural Engineering Department, University of Georgia, for his assistance in the economic portion of this paper. MS *submitted November 1982*

Table 1 POTENTIAL APPLICATIONS OF HYDROTHERMAL ENERGY[3]

Temperature (°C)	Applications
150	Refining sugar beets; barley malt, potato and tomato paste processing, salt evaporation.
140	Canning of food; blanching.
130	Evaporation of sugars.
120	Fresh water by distillation; refrigeration by medium temperatures; concentration of saline solutions.
110	Drying and curing of cement; onion and alfalfa drying.
100	Drying of organic materials; washing and drying of wool.
90	Space heating in older buildings with radiators; drying of stock fish; intense deicing operations.
80	Space heating in newer buildings; winter greenhouse space heating; direct heating.
70	Summer greenhouse space heating; pasteurization; refrigeration by low temperatures.
60	Greenhouses by combined space heating and hotbed heating.

(left margin brackets: ←saturated steam— spans 150 to 100; water spans 100 to 60)

THE MODEL BACKGROUND

The basic model contends that utilization of the geothermal water by existing or future inhabitants depends upon three factors. First, as described previously, the supply of water at a specific temperature is *the* most important factor in determining utilization potential. Application and temperature are so closely aligned that this component must be considered first. Second, locations with similar supply temperatures will not necessarily have the same utilization potential. Site characteristics of slope and elevation will determine the attractiveness of place for a given location. Lastly, the economics of transporting the water from its source to a point of utilization will determine the level of demand for that water.

The model operates under these basic assumptions. The first is that used by the U.S. Federal Government in its evaluation of geothermal resources, i.e., that we operate using existing or future technologies.[5] This implies that we can overcome the existing problems of extraction or undesirable chemical composition of the water. Secondly, it is assumed that the subsurface reservoir contains ample supply of geothermal water to warrant its extraction and that ample energy is available to extract it economically. Thirdly, the reservoir is recharged at a rate equivalent to the rate of extraction. This assumption is critical for continued use of the resource. Lastly, it is assumed that the need for the resource, either private or commercial, exists presently or will occur as the supply of other energy sources is reduced.

The framework for the model is based upon the planimetric mapping approach which uses concentric circle isotherms to delineate the area of supply of hydrothermal energy at specified temperatures (Figure 1).[6] The study area was a portion of southcentral Oregon and northeastern California. Utilizing the measured surface temperatures of four hot spring clusters and the temperature decay function, the 90°, 80° and 70°C transport locations were determined. The clusters served as centroids and the isotherms were constructed assuming a linear horizontal transport concept. Thus, the area within the 90°C isotherm could be supplied geothermal water with at least a 90°C temperature or hotter; within the 80°C

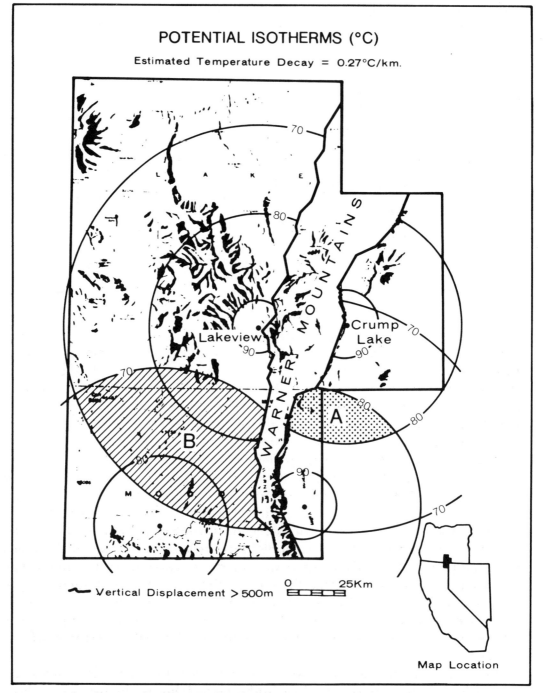

FIGURE 1 *Selected use-area locations defined by the isotherms. Location A can be supplied 80°C water from two sources and location B can be supplied water at 70°C from two different sources.*

isotherm, water temperature would be at least 80°C or hotter, etc. In essence, the linear cartogram defines not only areal locations of hydrothermal supply but locations where specific applications of the resource are possible.

Overlap of isotherms depicts locations where supply from two different source locations are possible (Figure 1). The east and west data sets are separated

178

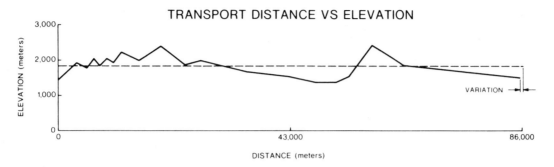

FIGURE 2. *Representative profile of surface transport distance (solid line) with a vertical exaggeration of 10× and a linear horizontal transport distance (dashed line) with no exaggeration.*

by a mountain range with extreme vertical displacement which effectively prohibits the supply of water across that barrier. This area of the northern basin and range physiographic region is characterized by highly faulted mountainous topography interspersed with broad flat valley floors.

MODEL COMPONENTS

Temperature

The use of the concentric circle isotherm approach for delineating areas of hydrothermal supply assumed only a linear horizontal transport function. This essentially eliminated the impact of rugged terrain upon the total length of pipe and thus the point at which specific water temperatures were located. In evaluating this assumption for use in this model the two Oregon hydrothermal sources were utilized (see Lakeview and Crump Lake, Figure 1).

Using 1:250,000 series topographic maps, azimuths were extended radially from each centroid at 15 degree intervals. Elevations and distance from the centroids were recorded for points of obvious change in slope as determined by the contours [C.I. = 200 ft. (61m)]. These data were collected along the azimuths until either the 70°C location, the edge of the study area boundary, or a major vertical displacement was encountered.

Two approaches were utilized to determine the variation in the position of the point where 70°C occurs. A computer program was generated to plot and calculate the length of pipe required to pump the water over the topographic surface to the 70°C location. A graphic enhancement of the plotted results is typified by Figure 2. The solid line represents the surface transport route plotted with a 10× vertical exaggeration. The program does not utilize the exaggeration in calculating the end point of the line. The dashed line depicts the linear horizontal function and its total length represents the distance required to reach 70°C. In both calculations, the same distance decay value was utilized. The variation between the two approaches is graphically displayed by the distance between the arrowheads.

The second approach of comparing the linear versus the topographic surface approach was to utilize the Pythagorean theorem. The topographic slope was considered uniform between sampling points and served as the hypotenuse of the

179

Table 2 PERCENTAGE VARIATION OF LINEAR HORIZONTAL PIPE LENGTH TO CALCULATED PIPE LENGTH

Azimuth	Horizontal Length (m)	Calculated Length (m)	% Variation
Lakeview Hydrothermal Source:			
15°	85,700	85,733.6	0.04
0/360°	85,700	85,808.8	0.13
345°	85,700	85,818.8	0.14
330°	85,700	85,940.3	0.28
315°	85,700	85,858.5	0.19
300°	85,700	85,856.7	0.18
285°	85.700	85,770.3	0.08
270°	85,700	85,751.7	0.06
255°	85,700	85,857.4	0.18
240°	85,700	85,820.2	0.14
225°	85,700	85,743.6	0.05
210°	85,700	85,733.4	0.04
195°	85,700	85,712.9	0.02
180°	85,700	85,732.4	0.03
165°	85,700	85,878.0	0.21
Crump Lake Hydrothermal Source:			
345°	22,100	22,269.9	0.77
0/360°	89,300	89,342.9	0.05
15°	89,300	89,392.4	0.10
30°	89,300	89,423.0	0.14
45°	74,500	74,515.4	0.02
60°	59,400	59,410.4	0.02
75°	59,400	59,416.2	0.03
90	52,000	52,036.4	0.07
105°	54,200	54,212.5	0.02
120°	59,700	59,751.3	0.03
135°	67,800	67,830.5	0.07
150°	89,300	89,341.6	0.05
65°	89,300	89,369.6	0.08
180°	89,300	89,387.2	0.10
195°	89,300	89,358.5	0.07
210°	55,700	55,859.0	0.29
		$\bar{X} =$	0.12

triangle. The summation of all segments along the azimuth served as the comparative figure of transport distance. It was found that the 70°C isotherm location varied between the two techniques by an average of 0.12 percent or approximately 100 meters over an 86km transport distance (Table 2).* Therefore, the length of pipe required to transport the geothermal water to the resultant temperature is insignificantly different between the surface and linear methods. Positioning of the isotherms is so similar that the use of the concentric circle isotherm analysis technique is recommended for the construction of the linear cartogram for defining the areas of potential application of the resource.

A 93 by 82 cell grid was overlaid on the topographic maps containing the two hydrothermal sites referred to above. At the topographic map scale, each grid cell represented a four square kilometer area. A suitability rating for resource utiliza-

---

* The seemingly small value of variation is contributed to the scale of the map base utilized. The use of larger scale maps would undoubtedly produce a slightly greater level of variation. However, the model concentrates on using the largest map scale available everywhere in the u.s., i.e., 1:250,000.

tion was assigned by converting the temperature data to a five class ordinal data set. The temperature distribution defined by the isotherm technique was encoded by grid cell using the temperature-class relationship cited below.

Suitability Value	Temperature Range (°C)
5	90–94
4	85–89
3	80–84
2	75–79
1	70–74

The higher temperature class was assigned the higher suitability value so that the area of greater potential hydrothermal utilization is graphically emphasized when produced using a three-dimensional plotting program. Figure 3 was produced using THREE-D CALCOMP perspective drawing software in conjunction with the Versatic electrostatic plotter at the University of Georgia. The hotter hydrothermal temperatures are exemplified by the higher surface planes in the center of the figure. The temperature decay function is defined by the descending step-like terraces radiating outward from the centroids. Note, the mountainous area between the two data sets (refer to Figure 1) was encoded into the matrix as a zero value (as well as all locations outside of the 70°C isotherm) and appears as a statistical valley between the two areas of hydrothermal supply.

Attractiveness of Place
Slope. There are two components that are used to describe the attractiveness of a specific location for hydrothermal resources utilization, slope and elevation. Using the topographic contours, the slope of the land was determined for each grid cell and encoded according to the following relationship:

Suitability Value	Percent Slope	Slope Characteristics[7]
3	0–5	flat to gentle
2	6–20	moderate
1	>20	steep

The flatter the slope, the more potential a location has for site development. Excluding engineering modifications, steeper slopes are less desirable and therefore assigned a lower suitability value.

The upper surface of Figure 4 represents the land area where the attractiveness of a place for hydrothermal utilization is ideal when based upon slope characteristics. Where the suitability surface appears pitted, the slopes are greater. It can be noted that in the southeast portion of the study area the land surface is dominated by undulating and steeply sloping topography. Large lakes and reservoirs were removed from the model by a zero slope classification.

Elevation. The second component of attractiveness of place, mean elevation, was determined from the maximum and minimum elevations within each grid cell. Potential development and thus resource utilization will most likely occur on a flatter slope at a lower elevation prior to such development of a flatter slope at a higher elevation. The mean cell elevation was evaluated relative to a datum

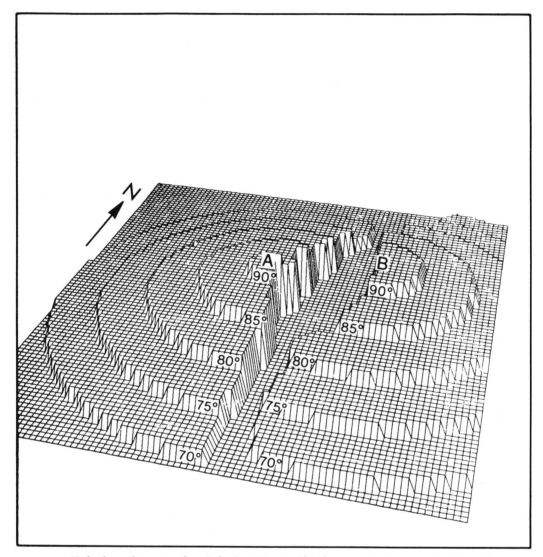

FIGURE 3. *Hydrothermal transport from Lakeview (A) centroid with a source temperature of 93°C and Crump Lake (B) centroid with a temperature of 94°C. A temperature decay value of 0.27°C/km was utilized.*

representing the hot spring elevation. Variations between cell and hot spring elevations of 100 meters or less were assigned higher site suitability values. The following suitability definitions were used to generate the map depicting area of preferred site utility based upon elevation (Figure 5). The raised suitability surface emphasizes the locations where elevation most closely approximates the hot spring elevation and is not referenced to sea level.

Suitability value	Approximate meters above hot spring elevation	Mean cell elevation (meters above sea level)
3	<100	<1500
2	101–450	1501–1850
1	>450	>1850

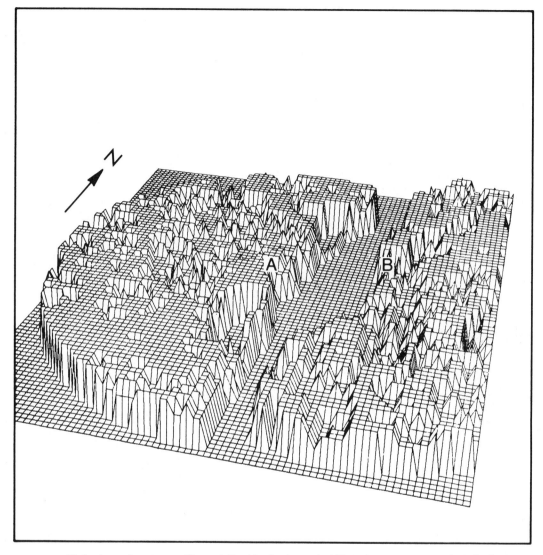

FIGURE 4. *Hydrothermal resource utility as defined by the slope suitability values. A represents the Lakeview centroid and B represents the Crump Lake centroid.*

As a result of the horst-and-graben topography of this region, slope or elevation alone does not easily define the attractiveness of a given location. In several instances, a relatively flat slope occurs at the top of a fault block mountain. Therefore, in order to adequately describe the attractiveness of a place for resource utilization both components should be included.

Economics

The fourth component of the model is designed to evaluate the apparent cost of transporting the geothermal water from the hot spring source to the point of application. Engineering formulae for determining the cost of pumping irrigation water were utilized to determine the cost of pumping one acre-foot (1,233.3m³ or 325,851 U.S. gallons) of water from the hot spring to the site. The

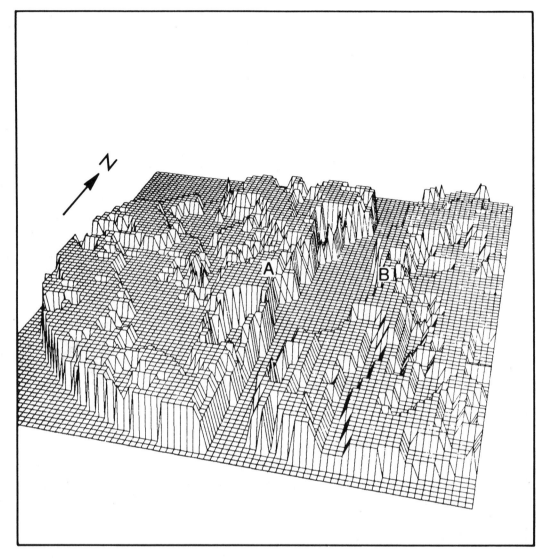

FIGURE 5. *Hydrothermal resource utility as defined by the elevation suitability values. A represents the Lakeview centroid and B represents the Crump Lake centroid.*

formulae utilized were:[8]

$$\text{total head} = (\Delta E + C_1)D$$

Where: ΔE = change in elevation (in feet) from hot spring to application site
C_1 = a constant friction of 0.0115
D = linear distance of transport from source
and, $\text{Kw-hr/A-ft} = (\text{total head})C_2$
Where: total head = as previously defined
C_2 = a constant of 3.373

These formulae determine the number of kilowatt-hours required to pump the acre-foot of water. The pumping efficiency and charge per kilowatt-hour must also be considered in order to convert the cost into dollars.

The pumping cost data, utilized to generate Figure 6, were collected along the 15° azimuths at a temperature decay distance of every two degrees Celsius. The calculated cost was assigned to the land area centered about the two degree increment and 7½ degrees either side of the azimuth. Pumping costs were found to range between 10 and 576 dollars utilizing a 50 percent pumping efficiency and a $0.10 per kilowatt-hour energy cost. For comparison, an acre-foot of water can be supplied in an urban location for approximately 175 dollars (based on water rates in Athens, GA 1982).

Figure 7 graphically displays the pumping cost per acre-foot based upon the following assigned suitability values.

Suitability Value	Cost per A-ft	Suitability Classification
3	<$100	high
2	$101–250	medium
1	>$250	low

As expected, the figure indicates that pumping costs are lowest closest to the hot springs. Cost is also low along the valley extending northward from the hydro-thermal sources. Locations in the southeastern portion, where elevation and slope are least suitable, produce higher pumping cost. Otherwise, as distance from the source increases, site utility based on pumping cost decreases.

COMPOSITE ANALYSIS

The four model components are merged into a final data set which depicts the potential for resource utilization. The four suitability factors per grid cell are multiplied to generate a new cell value. Multiplication is utilized to emphasize the five class temperature distribution. The importance of temperature supplied to a site is the most critical component in determining application possibilities. The new data set yields 136 possible values per grid cell (0 to 135).

Of the 7,626 grid cells, 5,084 fall within the area of 70°C or hotter. Less than one percent of those observations fall in the highest possible class, i.e., the locations with the greatest potential for resource utilization. The mid-point value of each data set (i.e., a value of 3 for temperature and 2 for the other data sets) would produce a composite value of 48. Slightly less than ten percent of the grid cells generated a composite value of 48 or higher. Thus, the model suggests that less than ten percent of the area which could receive 70°C hydrothermal water is suitable for utilization of that water.

Figure 7 depicts the distribution of the potential for utilizing hydrothermal energy. It is apparent that the areas in closest proximity to the hot springs possess the greatest resource potential. The impact of temperature is greatest in that same area which produces the higher composite cell values. Beyond the 85°C isotherm (approximately), the variation in potential site suitability is less pronounced. Yet, certain areas outside of the 'central core of higher values' can be identified for further investigation. Two of those areas include the central area west of the Lakeview geothermal source and north-northeast of the Crump Lake source.

185

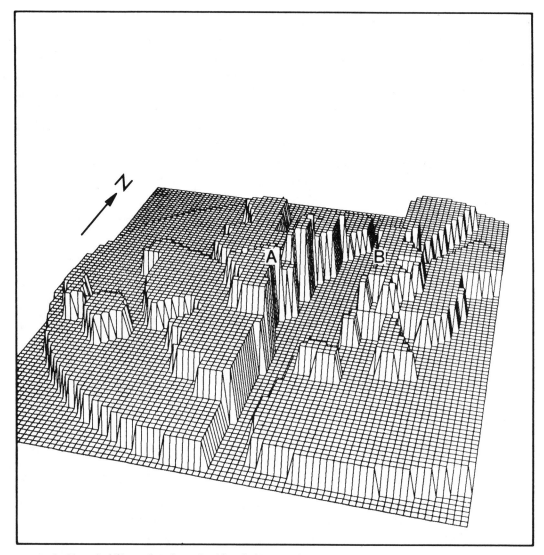

FIGURE 6. *Site suitability surface determined by relative pumping costs per acre-foot of water. A represents the Lakeview centroid and B represents the Crump Lake centroid.*

EVALUATION OF THE MODEL

The model allows for the evaluation of both generalized and site specific locations for using the resource. The three-dimensional technique provides for the graphic perspective display of the resource utilization potential. As was expected, the areas relatively close to the geothermal source will provide greater opportunity for utilization than will locations several kilometers away. This model is not designed to predict utilization location but to assist in the location of areas with potential for utilization.

The identification of the concentric circle isotherms' position is essential to the application of the model. Placement of the isotherms according to calculations based on a constant temperature decay function and a linear horizontal pumping system best defines the areas of thermal supply.

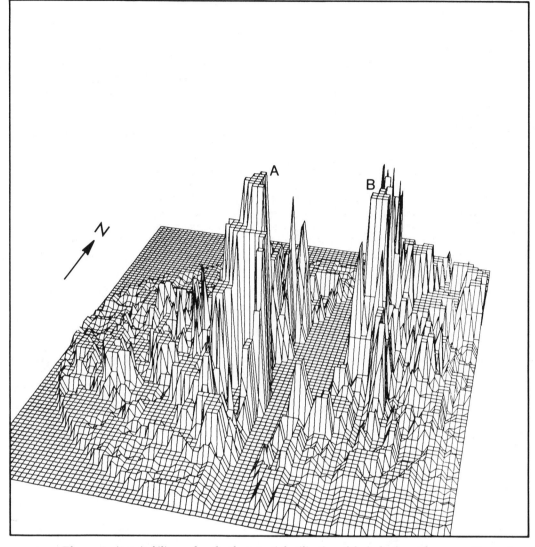

FIGURE 7. *The composite suitability surface for the potential utilization of the hydrothermal resource. The surface is created by the product of the four model components. A represents the Lakeview centroid and B represents the Crump Lake centroid.*

The model is flexible in that individual needs or values can easily be incorporated. The scaling of each component can be adjusted according to the priorities of the investigation. The number of classes can be either enlarged or reduced. In any case, it will assist the analyst by cartographically displaying the areas of preferred and of lesser preferred potential. The perspective view can then be related back to existing topographic or planimetric maps for further evaluation.

As the need for resource development increases, early evaluation abilities become necessary. This model can easily be applied when attempting to determine the locations where hydrothermal water can be utilized. Specific application is temperature dependent; however, utilization must also include site and economic parameters. This can serve as an effective tool for analyzing the resource's potential for utilization.

NOTES

[1] For extended discussions of hydrothermal resources see, for example, Muffler, L.J.P. (ed), *Assessment of geothermal resources of the United States, 1978*. U.S. Geological Survey Circular 790, 163 p.; White, D.E. and D.L. Williams (eds), 1975. *Assessment of geothermal resources of the United States, 1975*. U.S. Geological Survey Circular 726, 155 p.; Sammel, Edward A. and Robert W. Craig, 1981. *The geothermal hydrology of Warner Valley, Oregon: a reconnaissance study*. U.S. Geological Survey Professional Paper 1044-I, USGPO, 147 p.; and, Kruger, Paul and Carel Otte (eds), 1973. *Geothermal energy: resources, production, stimulation*. Stanford: Stanford University Press.

[2] Kruger and Otte, *op. cit.* (ftnt 1).

[3] Framework for this table assembled from Lindal, Baldur, 1974. "Geothermal energy for process use" in Lienau, Paul J. and John W. Lund (eds), *Multipurpose use of geothermal energy*. Geo-Heat Utilization Center, Oregon Institute of Technology, pp. 16–42; and, Raschen, Rory and William S. Cook, 1976. *Exploration and development of geothermal resources*. Menlo Park, USGS, Conservation Division, (An Internal Working Document).

[4] For example, Muffler, *op. cit.* (ftnt 1); Bowen, R.G. and N.V. Peterson, 1970. *Thermal springs and wells in Oregon*. Portland: Dept. of Geology and Mineral Industries. Miscellaneous Paper 14; Waring, Gerald A., 1965. *Thermal springs of the United States and other countries of the world: a summary*. U.S. Geological Survey Professional Paper 492; and, Stearns, Norah D., Harold T. Stearns, and Gerald A. Waring, 1937. *Thermal springs in the United States*. U.S. Geological Survey Water Supply Paper 679-B:55-206.

[5] Muffler, *op. cit.* (ftnt 1).

[6] The model was first suggested by Hodler, Thomas W., 1982. A mapping strategy for hydrogeothermal evaluation, *Journal of Geological Education*, v. 30, pp. 168–171.

[7] Slope categories patterned after Nichols, D.R. and J.R. Edmundson, 1975. Text to slope map of part of west-central King County, Washington: U.S. Geological Survey Miscellaneous Geological Inventory map I-852-E.

[8] Derived from Hansen, Vaughn E., Orson W. Israelsen, and Glen E. Stringham, 1980. *Irrigation principles and practices*, 4th ed. New York, John Wiley and Sons, pp. 272–293.

PART 5
SOIL RESOURCE MANAGEMENT

Geographic information systems for natural resource management

By Stephen J. Walsh

SECTION 208 of the Federal Water Pollution Control Act Amendments requires planning agencies to formulate means by which agricultural nonpoint pollution sources can be identified and to set forth methods for the control of such sources. Devising best management practices for nonpoint pollution control necessitates evaluation of alternative strategies—determining the location, spatial distribution, and acreage affected by nonpoint pollution problems (3).

With the use of a geographic information system, planners can correlate landcover and topographic data with a variety of environmental parameters relating to such indicators as surface runoff, drainage basin acreage, and terrain configuration. This approach permits water quality data from various sources to be integrated into a comprehensive system capable of combining and cross referencing such diverse data elements as conventional maps, Landsat imagery, and tabular data obtained "on the ground." Environmental planners thus can visualize disparate data elements that otherwise must examined manually.

Computer-based information can also be used to refine such models as the universal soil loss equation. The result is reasonable predictions of agricultural pollutant loads and the potential transport of non-

Stephen J. Walsh is an associate professor of geography and director of the Center for Applications of Remote Sensing, Oklahoma State University, Stillwater, Oklahoma 74078.

point-source pollutants based on watershed parameters, including soil, slope, vegetative cover, acreage, and spatial proximities.

Potential nonpoint-source analysis

To identify and prioritize pollution problem areas within a watershed, planners must define the contributing areas, or hydrologic units, within the watershed or subwatershed. A control strategy can then be designed using a listing of best management practices that will have maximum impact on water quality improvement. Such a strategy should include these practices that will control the pollutants involved most effectively, considering soils, topography, land use, and rainfall. Specific actions to improve water quality can be developed for each targeted area. For example, if a pesticide pollution problem originates from a certain field, planners can examine crop records and pesticide usage to identify specific locations and activities for control emphasis.

As one example, the Oklahoma Conservation Commission asked Oklahoma State University's Center for Applications of Remote Sensing (CARS) to produce a detailed surface hydrology map of selected watersheds for sampling water quality and for depicting surface pollutant movement arteries. CARS personnel interpreted U.S. Geological Survey topograhic maps and aerial photographs to identify intermittent

streams and seasonally filled ponds. A locational, township-range-section grid and a Universal Transverse Mercator grid also were provided for the watershed.

Using soil erodibility data supplied by Soil Conservation Service (SCS) personnel, CARS developed a map overlay of the watershed showing highly and very highly erodible soils. That map was produced through the digitization and extraction of soils data in 10-acre cells. Initially, the information was obtained from soil survey map sheets and processed on the CARS' computer system.

Map overlays also were produced delineating areas with slopes exceeding 5 percent. Slope data came from soil survey information and from National Cartographic Information Center (NCIC) digital terrain data. Finally, CARS personnel used a series of Landsat digital tapes to map the areal distribution of row crops, mainly peanuts and cotton, and cropland, primarily wheat and alfalfa. Agricultural land use was discriminated on the basis of local crop calenders and spectral patterns. All maps were overlayed and a composite map made showing areas with high potential for soil erosion and subsequent deposition (7).

Water samples were collected and intensive field surveys were conducted within the watersheds where the map overlays indicated the greatest pollution potential. Land use modification strategies are being developed to reduce potential soil erosion within the watershed, with special empha-

sis on the critical zones identified by the mapping process.

The composite map was produced by manually overlaying five thematic maps representing hydrology, landcover, topography, and high and very high soil erosion. This process is fairly efficient for a small number of overlays and for observing limited interrelationships.

Many variables with complex patterns and data extraction requirements, however, warrant an automated overlay and analysis approach. Systems that integrate layers of spatially oriented information, whether manually or automatically, are considered geographic information systems (4).

Geographic information systems

Development and use of an automated geographic information system can expedite data integration problems and the time-consuming process of synthesizing tremendous amounts of information for the spatial examination of nonpoint pollution. An automated information system, through which geographically referenced data can be input, manipulated, and analyzed, can improve the decision-making process of an organization (10).

Technologies of the 1980s, including remote sensing and geographic information systems, are attractive because of their capabilities for analyzing large and small areas, integrating numerous variables into the evaluation process, and easily updating

data bases (6). Data elements entered into the system, data resolution, analytical models, data output products, and the particular system applications depend upon the investigator and the desired informational requirements (8).

The OGIRS system

Responding to a need for a computer-based, geographically referenced natural resource information system in Oklahoma, CARS developed OGIRS, the Oklahoma Geographic Information Retrieval System. OGIRS is a highly interactive, grid-oriented data collection, storage, and manipulation software package (2). Through OGIRS, CARS can collect and store data from disparate sources in a common format; edit that data; and produce a variety of outputs ranging from tabular data to color images to selective single or composite maps (2).

Data entry

Geographic information systems are characterized by the spatial character of input data. Points, lines, or polygons that represent desired geographical facts must be converted to a format that can be stored, manipulated, and displayed by a computer-based information system. Spatial information represented on a map first must be transformed into an array of X, Y, and Z coordinates that describe the data.

The X and Y values relate to the geographic position of the grid point or cell, while the Z value indicates the cell's value, such as soil type or percent slope. This transformation usually is performed with a graphic digitizer. The computer-compatible array of X, Y, and Z coordinates is then stored in the computer and arranged in an ordered data structure. Points may be digitized and thereby geographically located in a township-range-section grid, the Universal Transverse Mercator grid system, or a user-defined coordinate system. Actual boundaries of geographic units are not stored in a grid information system. Such systems are developed by converting the geographic units into an array of grid cells. All data are summarized by the grid cell. Cells thus become the key to all data storage, manipulation, and display.

The array of data cells in a grid information system is constructed by overlaying the desired map with a grid. Each grid cell is then examined to determine which geographic unit occupies the majority of the cell's area. The value of the geographic unit is then applied to the entire grid cell. The degree of detail in the resulting map depends upon the size of the grid cell—the concept of resolution. Smaller grid cells portray greater spatial variation.

Data storage

Within OGIRS, data are stored in a consistent format in thematic libraries. The thematic library is more a convenience to the user than a system necessity. Thematic libraries allow the user to organize the input data into major categories and then subdivide the major headings into smaller units representing individual data sets. A soils library, for example, may contain the soil series name, depth to bedrock, depths of the individual horizons, porosity, soil moisture, shrink/swell potential, and other soil-related characteristics as separate levels or channels within the data library. Editing capabilities exist at the level of data entry and as a function of library update (2).

Soils. Soil type, permeability, available water-holding capacity, and hydrologic groups can be digitized from SCS county soils survey sheets and used in hydrologic models.

Permeability is based on known relationships among the soil characteristics observed in the field—soil structure, porosity, and texture—that influence the downward movement of water in soil. Available water-holding capacity is based on soil characteristics that influence the ability of the soil to hold water and how available it is to plants. Important characteristics are organic matter content, soil texture, and soil structure. The hydrologic groups are used to estimate runoff after rainfall. Soil properties that influence the minimum infiltration rate into the bare soil after prolonged wetting are depth to layers of slowly or very slowly permeable soil.

Landcover. Remotely sensed data can be used to extract landcover information. Classification techniques are used to characterize landcover through statistical measures. Statistical and spatial analyses combined with field checking provide the mechanism for efficient linking of statistical classes into meaningful landcover types. Landcover data from the Landsat satellite series is effective due to its synoptic view, repetitive nature, and computer compatibility (9).

A simplified scenario for obtaining a digital classification of landcover from unprocessed Landsat data would include: (a) preprocessing of Landsat data to remove unnecessary banding or striping and reformating the data into a more efficient format; (b) selecting the specific study area coordinates to minimize the overall processing time; (c) developing a set of spectral categories using a search routine to establish the foundation for the maximum likely classifier; (d) running the classification program for the study area, deriving a completed thematic classification of the landcover present and recognized by Landsat; (e) fine tunning the classification

by combining spectral classes as necessary; and (f) geographically referencing the Landsat thematic classification to the Universal Transverse Mercator coordinate system, producing a geometrically and spatially accurate map.

Terrain. Slope angle as well as aspect and elevation data can be secured from the NCIC digital terrain tapes.

Climate. Precipitation data can be obtained from a network of rain gauges within or surrounding a watershed. These point measures of precipitation can be interpolated into per cell measures of monthly precipitation with the Thiessen polygon approach. Precipitation polygons are identified for precipitation, and each cell within a particular polygon is assigned the precipitation total from the rain gauge within the polygon.

A similar approach can be used for mean monthly temperatures. Grid cells within each of the polygons are digitized to contain the mean monthly temperature of the appropriate recording station.

Data display

OGIRS output products include electrostatic printer/plotter maps and color images. In addition to the two-dimensional

map representations, the cell values (Z cooridnates) for a data set can be listed in tabular format. Digital image data sets are generated from either the individual channels of a data library or from the digitizer directly.

Data manipulation

Manipulation of stored data is by a standard set of arithmetic functions, five value selection modes, and three mapping functions. The arithmetic functions include add, subtract, multiple, divide, square root, exponentiation, trigometric functions, and logarithms. The value selection modes are equal value, greater than a value, less than a value, greater than and less than a value, and less than a value or greater than a value. The three mapping functions allow any number of individual data sets from any number of data libraries to be combined as the union, intersection, or exclusion.

The five relation modes each generate a new data set that is a subset of the original. Each of the three mapping functions, union, intersection, and exclusion, first allow the user to place conditional restrictions on the data sets with one of the five relation modes. Finally, the selected mapping function is applied to the resulting restricted data sets to produce a new output data set or map.

Data transformation from original map to grid base (top). Data overlaying for composite mapping (bottom).

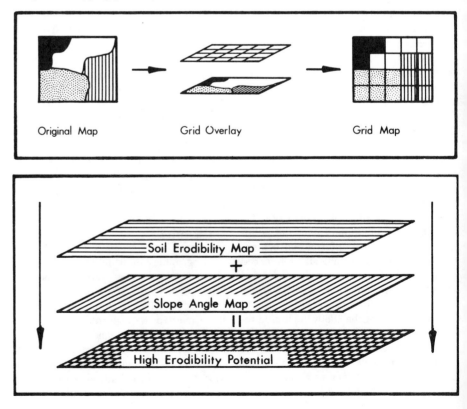

Original Map   Grid Overlay   Grid Map

Soil Erodibility Map

+

Slope Angle Map

=

High Erodibility Potential

The interactive capabilities of OGIRS allow users to selectively isolate specific natural resource features and to examine the consequences of interaction between disparate data sources. Location of a single soil type may be extracted from the soil series map and the areas displayed as a separate map. Other relation modes may be used to generate, for example, a map showing all elevations over 500 feet or all slope angles less than 8 degrees. Features may also be bracketed to display slope aspects between north and northwest or areas that receive less than 10 inches or more than 40 inches of rain per year.

In another application the mapping functions can be used to compare the acreage planted in wheat on one soil type to that of wheat planted on another soil type. The user extracts the areas with wheat from the general landcover data set and then intersects these data with the specific soil types to obtain a new map showing the locations of wheat on the specific soil types.

Many user-defined models in OGIRS are of the suitability analysis type. A basic suitability analysis model locates and displays from the data base only those areas that contain all of the characteristics described by the user. By describing the requirements of variables, such as soil type, slope angle, elevation, or depth to bedrock, the intersection mapping function isolates only those characteristics.

Model development

As well as having the capability to discover and display information gained through the testing of interactions between phenomena, geographic information systems can be used to organize and appraise variable coefficients for predicative models. Numerous models have been developed for evaluating nonpoint-source pollution and implementing best management practices to limit pollution to an acceptable level. Because of the spatial variation of climate, soils, and agricultural practices, however, no single group of control measures can be used in every region. Physically based mathematical models are a logical step to regional analyses. Various types of models have been developed to evaluate components of nonpoint pollution (1, 5). Although hydrology is only one component of the total system, water is the principal element; it causes erosion, carries chemicals, and is an uncontrolled natural input. Each climatic region and physiographic area has its own characteristics that affect the response of the system. Watershed-type models require updates of landcover, pre-

cipitation, and other data and an assessment of the dynamic spatial patterns affecting runoff, infiltration, and potential nonpoint pollution. Remotely sensed data organized and manipulated in a geographic information system can greatly benefit these analyses.

Conclusions

The ability to turn analog information into digital data and then to edit, store, and display the data as maps or color images has a variety of applications in resource management and regional planning. Land use planners, engineers, agriculturalists, and many others use maps or images displaying soil type, geology, land use, hydrology, and other natural resource data on a daily basis. The advantages of archiving these data in a geographic information system include ease of retrieval, variety of output products to fit almost any need, and the ability to discover and display information gained by testing the interactions between natural resources phenomena and to organize and appraise variable coefficients for predicative models.

A need exists to estimate runoff as a function of soil, vegetation, and antecedent moisture and to simulate physically based elements of the natural system on a temporal and spatial basis. Organizing data in a geographic information system permits the appraisal of nonpoint pollution over extensive areas and on a repetitive schedule. Such a system also provides a framework for identifying and managing areas that contribute in relative amounts to the runoff problem.

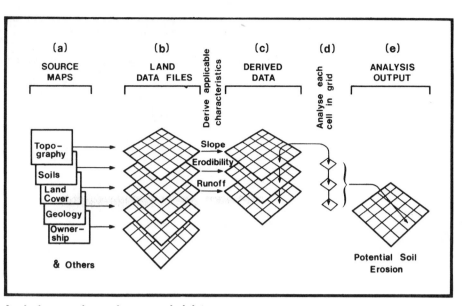

Analysis procedure using geocoded data.

REFERENCES CITED
1. Beasley, D. B. 1977. *Answers: A Mathematical Model for Simulating the Effects of Land Use and Management on Water Quality.* Ph.D thesis. Purdue Univ., West Lafayette, Ind. 266 pp.
2. Blanchard, W. A. 1982. *The Oklahoma geographic information retrieval system.* In Proc., Nat. Conf. Energy Resource Mgt. Conf. Publ. 2261. Nat. Aeronautics and Space Admin., Washington, D.C. pp. 239-249.
3. Campbell, W. J. 1979. *An application of Landsat and computer technology to potential water pollution from soil erosion.* In *Satellite Hydrology.* Am. Water Resources Assoc., Minneapolis, Minn. pp. 616-621.
4. Cicone, R. C. 1977. *Remote sensing and geographically based information systems.* In Proc., 11th Int. Symp. Remote Sensing Environ., Environ. Res. Inst., Univ. Mich., Ann Arbor. pp. 1,127-1,137.
5. Crawford, N. H., and R. K. Linsley. 1966. *Digital simulation in hydrology: Stanford watershed model IV.* Tech. Rpt. 39. Dept. Civil Eng., Stanford Univ., Stanford, Calif.
6. Estes, J. E. 1982. *Remote sensing and geographic information systems coming of age in the eighties.* In *Remote Sensing: An Input to Geographic Information Systems in the 1980s.* Am. Soc. Photogram., Falls Church, Va. pp. 23-40.
7. Ripple, B. J., and S. B. Miller. 1982. *Remote sensing and computer modeling for water quality planning in South Dakota.* In Chris J. Johannsen and James L. Sanders [eds.] *Remote Sensing for Resource Management.* Soil Cons. Soc. Am., Ankeny, Iowa. pp. 309-316.
8. Smith, A. Y., and R. J. Blackwell. 1980. *Development of an information data base for watershed monitoring.* Photogram. Eng. and Remote Sensing (46): 1,027-1,038.
9. Wilson, C. L., and F. J. Thomson. 1981. *Integration and manipulation of remotely sensed and other data in geographic information systems.* In *Remote Sensing: An Input to Geographic Information Systems in the 1980s.* Am. Soc. Photogram., Falls Church, Va. pp. 303-317.
10. Wilson, C. L., and P. J. Thomas. 1977. *A general design scheme for an operational geographic information system.* Infor. Systems Tech. Lab., Fed. Rocky Mountain States, Fort Collins, Colo. pp. 1-44. □

Erosion control potential with conservation tillage in the Lake Erie Basin: Estimates using the universal soil loss equation and the Land Resource Information System (LRIS)

T. J. Logan, D. R. Urban, J. R. Adams, and S. M. Yaksich

ABSTRACT: The universal soil loss equation (USLE) was used with the U.S. Army Corps of Engineers' Land Resources Information System (LRIS) to estimate soil loss in the U.S. portion of the Lake Erie drainage basin. Annual soil loss averaged 2.2 tons per acre (4.9 metric tons/hectare) in the basin, from 0.4 to 4.4 tons per acre (0.8-9.8 metric tons/hectare) for individual watersheds. The Raisin (Michigan); Maumee, Sandusky, Huron, Vermilion (Ohio); and Cattaraugus (New York) River Basins had the highest soil losses. Reducing soil loss on all soils to the tolerance value (T) would reduce soil loss in the basin by 40 percent. Reductions of 46 percent and 69 percent would be possible by using conservation tillage and maximum conservation tillage (includes no-till on some soils), respectively.

SINCE its inception in 1974, the Lake Erie Wastewater Management Study (LEWMS), directed by the Buffalo District of the U.S. Army Corps of Engineers, has focused on the input of various sources of phosphorus to Lake Erie. P is thought to be the nutrient most limiting the growth of algae in the lake (6) and the element most amenable to control by management.

According to LEWMS, 45 to 55 percent of the total P load in Lake Erie is from diffuse sources, and rural sources account for 84 percent of the diffuse load. P response models for Lake Erie (2) indicate that elimination of most point-source P loads to Lake Erie (primarily wastewater treatment plant effluents) will not adequately improve the lake's water quality. Significant reductions in the lake's diffuse P load are therefore necessary. If concentrations of total P in effluents of wastewater treatment plants with a flow greater than 1 million gallons per day are reduced to 1.0 micrograms per milliliter, diffuse-source loads would have to be reduced an estimated 50 percent to meet the target loads for lake water quality improvement.

A Land Resource Information System (LRIS), developed by LEWMS (1), shows that 57 percent of the Lake Erie drainage basin on the U.S. side is in cropland. Furthermore, LRIS shows this land use contributes significantly to the rural, diffuse P load in the lake. Soil erosion is the primary cause of P losses from cropland, although some loss of soluble P is the result of P fertilization (6).

We investigated the feasibility of reducing soil loss in the Lake Erie Basin by soil and crop management, including conservation tillage. We examined several scenarios with the universal soil loss equation (USLE). We combined the extensive LRIS data base on land use, topography, and soils with the USLE to determine soil loss.

The LRIS data base

LRIS is a variable cell-size [9.8-90 acres (4-36 hectares)], computer-based system that contains data, at each cell, on land use, soil characteristics, and political and hydrologic boundaries (1). Land use was determined by photointerpretation of high-altitude aerial photography. Soil data came from soil surveys. Additional variables describing slope length, slope percentage, and suitability for conservation tillage were added to LRIS for each soil mapping unit. In our analysis with the USLE, land use at each individual cell was used to calculate soil loss. Soil properties were given for each soil mapping unit.

Determining the USLE variables

U.S. Department of Agricultural scientists developed the USLE to predict long-term annual soil loss. The equation, in its simplest form, is a linear function that re-

T. J. Logan is an associate professor in the Department of Agronomy, Ohio State University, Columbus, 43210. D. R. Urban is with the Soil Conservation Service, U.S. Department of Agriculture, Washington, D.C. 20250. J. R. Adams and S. M. Yaksich are on the staff of the Lake Erie Management Study, Buffalo District, U.S. Army Corps of Engineers, 1776 Niagara Street, Buffalo, New York 14207.

lates soil loss to climatic, soil, and vegetative conditions:

$$A = RKLSCP$$

where A is annual soil loss, R is the rainfall erosion index, K is inherent soil erodibility, LS is the combination of slope percentage and slope length, C is the cover and management factor, and P is the conservation practice factor.

In using the USLE with LRIS to give distributed estimates of annual soil losses in the Lake Erie Basin, we derived some factors from the soils data file (LRIS); others we computed from regional information. We solved the equation initially on the assumption that only conventional tillage practices were used. We also solved it for several scenarios in which some form of conservation tillage or other means of reducing soil loss were used.

Developing the data for each of the USLE factors involved the following consideration:

Rainfall erosion index. Rainfall erosion index data is in the form of annual isoerodent lines (*12*). We used this data directly in our analysis, but interpolated to obtain a single value for each county in the basin. These varied from 75 in the northern reaches of the basin to 150 in the southwest corner. Most R values in the basin are between 100 and 138.

Soil erodibility. Soil erodibility values were developed by USDA scientists for U.S. soils. LRIS includes these for each mapping unit.

Conservation practices. Conservation practices, such as contour strip cropping and terraces, are not common in the Lake Erie drainage basin. Statistical information on the distribution of these practices is not readily available. Although these conservation practices are used to a limited extent throughout the basin, we assumed such use to be negligible and thus assigned the P factor a value of 1.

Slope and slope length. Use of the USLE with LRIS required a degree of slope and a slope length to be assigned by soil mapping unit. Given the level of detail of our study, this means that the same soil series and slope phase would be assigned the same slope percentage and slope length wherever it occurred in the basin. Because the USLE is unreliable when used at very high slope percentages and because some surveys only indicate slopes greater than 35 percent instead of giving the actual slope, we held the degree of slope for any slope phase in excess of 35 percent at 35 percent.

Slope length is an important factor in the USLE. We made an extra effort to assign a realistic slope length to each soil

series and slope phase (*1*). Our selection was based on several sources, including slope lengths measured in the basin for the National Erosion Survey conducted by the Soil Conservation Service, local experiences of SCS personnel in the basin, and the Maumee River Basin Level-B Study (*11*), which estimated soil loss with the USLE in that basin. We then determined the LS factor with the equation provided in Agriculture Handbook 537 (*12*).

Cover and management factors. We determined crop management factors for each county in the basin according to the distribution of crop acreage in the county as provided by the U.S. Crop Reporting Service in Indiana, Michigan, and Ohio. These were then combined to give three crop types: R, row crop (primarily corn and soybeans); Sg, small grains (wheat, oats, barley); and M, meadow (hay and pasture).

We chose eight rotations to represent cropping conditions in the basin (Table 1), then calculated C factors using Handbook

537 and the "Ohio Erosion Control Guide" (*3*) for various management practices (Table 1). Estimated percentages of fall and spring plowing in each county ranged from 10 to 70 percent. Using the factors in table 1, we next assigned or fitted crop acreages in each county to the eight rotations. This was an iterative process, which for a high grain-producing county would be: (a) assign all meadow acreage (M) to RRSgM, (b) assign remaining small grain (Sg) acreage to RRSg, and (c) assign remaining row crop acreage to continuous R. Where meadow (M) acreage was high, the sequence might be RRSgMMM, followed by RSgM or RRSg. After assigning all cropland acreage to the rotations, we determined an average factor for the county by weighting the individual factors for the rotation by the acreage each represents.

We assigned all of the meadow acreage (10,000 acres) to the first rotation, the remaining small grain acres (42,900 minus 10,000) to the second rotation, and the remaining row crop acres [134,600 −

Table 1. Assigned crop management (C) factors for various crop rotations and tillage practices in the Lake Erie Basin.

Crop Rotation*	Spring Plow Residue Left†	Fall Plow Residue Left†	Winter Cover‡	Conservation Tillage§	No-Till‖
R Sg Sg M	0.070	0.080	0.075	0.075	0.025
R R Sg M	0.120	0.140	0.105	0.085	0.030
R Sg M	0.055	0.070	0.065	0.065	0.030
R Sg M M M	0.035	0.045	0.040	0.040	0.020
R R Sg	0.250	0.270	0.210	0.110	0.040
R Sg Sg M M	0.055	0.060	0.060	0.060	0.030
Continuous R	0.380	0.430	0.320	0.130	0.030
Continuous M	0.005	0.005	0.005	0.005	0.005

*R = row crop (primarily corn and soybeans); Sg = small grain (wheat, oats, barley); M = meadow (hay and pasture).
†Moldboard plow. Residue left after harvest.
‡Winter cover seeded after row crop and plowed under in spring.
§Plow first year after meadow. Use chisel plow, disk, etc., with row crop to give 1,700-2,200 kilograms of residue per hectare.
‖Assumes 3,400-4,500 kilograms of residue per hectare.

Table 2. Weighting process for determining the C factor, Allen County, Ohio.

Crop	Acres
Corn	66,900
Soybeans	67,900
Row crop (R)	134,600
Wheat	35,900
Oats	7,000
Small grains (Sg)	42,900
Hay	10,000

Area-weighted C factor for spring plowing

Rotation	Acres for Rotation Sequence	Years in Rotation	Acres in Rotation	Percentage of Area in Each Rotation	Rotation C Factor from Table 1	Area Weighted C Factor
RRSgM	10,000	4	40,000	21.3	0.12	0.025
RRSg	32,900	3	98,700	52.7	0.25	0.131
Continuous R	48,800	1	48,800	26.0	0.38	0.098
			Area weighted C factor for spring plowing in Allen County, Ohio			0.254

2(10,000) – 2(32,900)] to the third rotation. We then calculated an area-weighted C factor from the C factors in table 1. These were summed to give an area-weighted C factor for the entire county (Table 2). Similar calculations were made for each management practice in each county in the Lake Erie Basin.

The C factors calculated for "present conditions" were based on the estimated percentages of fall- and spring-plowed acreage in each county and the C factors for fall and spring plowing in that county. These percentages were based on surveys of SCS district conservationists and extension agents in the counties.

The "present conditions" factor also assumed that moldboard plowing was used whenever there were row crops (R) in the rotation. This is not entirely true, and it overestimated soil loss for those counties with significant amounts of conservation tillage or no tillage. However, it was not possible to assess accurately the extent of these practices in the basin. A survey of

Lake Erie Basin farmers conducted subsequent to this study showed that an average of 12.2 percent of their 1979 acreage was in minimum tillage and only 1.7 percent in no-till (4). In using these county C factors in the USLE analysis, we assumed that crops are produced randomly on all soils in the county.

We used C values for grassland (idle and pasture) and woodlands assigned in the basin during SCS's 1-percent National Erosion Survey. A wide range of values were obtained, but we used only the median value in our analysis. This approach seemed adequate because cropland management for erosion control was the major objective. C values assigned to vineyards primarily found in New York, were based on discussions with an SCS district conservationist and an extension service grape specialist in New York as well as information from the SCS "Technical Guide" for Chautauqua County, New York. Following are the C values used throughout the basin: Grassland, 0.003; woodland, 0.005; vine-

yards (soil management group 1), 0.020; and vineyards (soil management groups 2-5), 0.090.

Soil management groups

Soils in the Lake Erie Basin were placed in soil management groups (9) to indicate the land's general suitability for conservation tillage (1).

Soil loss reduction scenarios

We evaluated seven management scenarios for reducing soil loss on Lake Erie Basin cropland. Table 3 gives the scenarios and management practices selected for each soil management group. Scenario 2 reduced soil loss to the soil loss tolerance (T) level on soils with a present soil loss greater than T. Scenario 2 did not change the percentage of soil loss on soils with losses already less than or equal to T. Scenario 6 used conservation tillage on all soil management groups except 3, 5, 7, and 9, which were moldboard plowed. Scenario 7 applied no-till to those soil management

Table 3. Soil loss for various management scenarios in the Lake Erie drainage basin.

Scenario	Soil Management Group									
	1	2	3	4	5	6	7	8	9	10
1. Present conditions	PC*	PC	PC							
2. Reduce soil loss to T	T=PC† T=T‡	T=PC T=T	T=PC T=T	T=PC T=T	T=PC T=T	T=PC T=T	T=PC T=T	T=PC T=T	T=PC T=T	T=PC T=T
3. Spring plow only	SP§	SP	SP	SP	SP	SP	SP	SP	SP	SP
4. Fall plowing only	FP‖	FP	FP	FP	FP	FP	FP	FP	FP	FP
5. Winter cover crop	WC#	WC	WC	WC	WC	WC	WC	WC	WC	WC
6. Conservation tillage	CT**	CT	PC	CT	PC	CT	PC	CT	PC	CT
7. Max. conservation tillage	NT††	NT	PC	CT	PC	NT	PC	CT	PC	NT

*Present conditions.
†If the existing soil loss calculated for a cell is less than the soil loss tolerance, T, soil loss remains at the present condition.
‡If the existing soil loss calculated for a cell is greater than the soil loss tolerance, T, for the soil, soil loss is set equal to T.
§Spring moldboard plow only.
‖Fall moldboard plow only. This will usually increase soil loss over present condition.
#Winter cover crop planted in the residue of the previous crop. Spring tillage precedes the next crop.
**Conservation tillage (chisel plow, disk, etc.) in spring or fall as substitute for moldboard plow.
††No-till.

Table 4. Soil loss in the western Lake Erie drainage basin for various management scenarios.

River Basin	Present Conditions	Reduce Soil Loss to T	Spring Plow Only	Fall Plow Only	Winter Cover Crop	Conservation Tillage	Maximum Conservation Tillage	Cropland Grassland and Forest	Other Land Use
			tons					acres	
Belle	62,964	55,286	62,101	71,833	62,958	50,690	21,761	86,226	10,223
Black	86,272	84,766	85,004	96,985	86,272	75,967	36,684	267,705	30,948
Clinton	263,196	253,588	259,784	299,111	260,429	168,697	85,835	426,135	61,600
Mill Creek	59,331	57,718	58,502	67,914	59,331	47,707	20,584	94,608	-
Rouge	230,038	162,549	227,075	255,286	227,822	131,532	63,924	262,456	44,190
Raisin	2,462,627	1,024,031	2,356,470	2,616,390	2,378,991	1,106,973	425,544	566,571	71,224
Maumee	11,081,131	6,179,702	10,498,421	11,631,026	10,554,470	5,418,653	2,969,085	3,634,019	323,877
Portage	547,152	369,691	515,085	568,659	517,252	232,256	125,705	245,180	10,247
Huron	552,648	466,785	543,857	625,109	540,403	319,453	138,881	376,676	104,207
Direct drainage	1,213,329	998,107	1,156,006	1,279,440	1,171,429	771,766	631,201	1,380,428	252,812
Total soil loss	16,558,689	9,652,224	15,762,305	17,511,751	15,859,357	8,323,693	4,519,203	7,340,003	909,368
Soil loss rate (tons/acre/year)	2.3	1.3	2.1	2.4	2.2	1.1	0.6		
Percentage soil loss reduction		42	5	(–6)*	4	50	73		

*Soil loss increased.

Table 5. Soil loss in the Central Lake Erie drainage basin for various management scenarios.

River Basin	Present Conditions	Reduce Soil Loss to T	Spring Plow Only	Fall Plow Only	Winter Cover Crop	Conservation Tillage	Maximum Conservation Tillage	Cropland Grassland and Forest	Other Land Use
				—tons—				—acres—	
Sandusky	2,789,267	1,655,505	2,674,484	2,980,948	2,734,208	1,302,277	577,262	758,725	35,074
Huron	737,802	492,884	704,523	777,761	710,753	387,982	220,097	220,400	15,886
Vermilion	412,190	284,902	397,056	445,271	400,587	274,473	180,474	132,172	8,580
Black	486,112	343,435	463,419	525,169	476,969	363,695	277,301	248,453	12,690
Rocky	285,306	159,127	271,541	306,830	278,496	191,949	131,668	149,423	20,529
Cuyahoga	159,619	127,148	157,739	170,096	159,097	127,115	99,218	397,520	47,557
Euclid Creek	6,315	5,891	6,283	6,660	6,210	6,055	5,912	11,711	3,031
Big Creek	8,004	8,004	8,004	8,004	8,004	8,004	8,004	21,907	2,113
Chagrin	85,686	55,176	85,086	91,946	83,984	64,759	52,176	126,795	29,600
Grand	298,073	243,013	294,408	304,822	293,775	228,254	144,643	433,430	18,646
Ashtabula	49,610	45,119	48,816	58,377	49,542	44,685	27,694	80,345	6,592
Conneaut Creek	158,062	115,110	154,712	180,884	157,416	120,765	79,508	108,958	10,205
Direct drainage	1,254,533	861,683	1,202,262	1,337,898	1,210,477	772,362	564,500	716,748	
Total soil loss	6,730,579	4,396,989	6,468,333	7,230,668	6,569,516	3,892,375	2,368,456	3,406,608	313,209
Soil loss rate (tons/acre/year)	2.0	1.3	1.9	2.1	1.9	1.1	0.7		
Percentage soil loss reduction	0	35	4	(−7)*	2	42	65		

*Soil loss increased.

Table 6. Soil loss in the eastern Lake Erie drainage basin for various management scenarios.

River Basin	Present Conditions	Reduce Soil Loss to T	Spring Plow Only	Fall Plow Only	Winter Cover Crop	Conservation Tillage	Maximum Conservation Tillage	Cropland Grassland and Forest	Other Land Use
				—tons—				—acres—	
Mill Creek	6,417	3,826	6,318	7,060	6,367	5,086	3,927	3,736	2,135
Raccoon Creek	1,751	1,675	1,719	1,957	1,735	1,132	593	1,661	158
Cattaraugus Creek	785,562	373,050	774,967	882,073	785,562	507,902	361,637	248,926	23,952
Delaware Creek	5,960	4,932	5,828	6,754	5,960	5,235	3,720	4,670	623
Eighteen-mile Creek	42,589	27,120	41,747	47,640	42,589	35,681	21,241	21,105	2,357
Direct drainage	966,673	688,193	950,633	1,088,044	963,690	720,919	535,157	530,440	99,990
Total soil loss	1,808,952	1,098,794	1,781,211	2,033,528	1,805,904	1,275,956	926,274	810,538	129,216
Soil loss rate (tons/acre/year)	2.2	1.4	2.2	2.5	2.2	1.6	1.1		
Percentage soil loss reduction	0	39	2	(−12)*	0	29	49		

*Soil loss increased.

groups (1, 2, 6 and 10) for which there would be no crop yield reduction with no-till versus conventional tillage. On more restrictive soils (groups 4 and 8), conservation tillage was used instead of no-till. Moldboard plowing was used on soil management groups 3, 5, 7, and 9. We then applied the USLE for each cell in LRIS using the factors as developed in the previous discussion. Soil loss was summarized by various subunits, for example, county, sub-basin, and basin.

Results

Tables 4, 5, and 6 give soil loss by river basin in the western, central, and eastern basins of the Lake Erie drainage basins (Figure 1). The western basin (Table 4), which receives drainage from Michigan and western Ohio, had a total soil loss under "present conditions" of 16,558,689 tons (15,022,042 metric tons). The per acre

rate was 2.3 tons (5.16 metric tons/hectare). The Maumee River Basin, the largest river entering the Great Lakes, contributed most of the soil loss. This region has a high percentage of cropland (69%) in intensively cultivated crops, primarily corn, soybeans, and wheat. Areas that were not included in one of the major subbasins were aggregated under the heading "direct drainage."

The scenarios show that meeting the soil tolerance objective (T) would reduce soil loss by 42 percent. Conservation tillage would reduce erosion 50 percent, while maximum conservation tillage would reduce soil loss by 73 percent. Winter cover crop and spring plowing had little effect on soil loss. Fall plowing would increase soil loss by 6 percent.

The distribution of rainfall erosivity (R) given in Agriculture Handbook 537 (12) for this part of the United States predicts

that most erosion will occur in June, July, and August. Soil cover during this period is affected very little by tillage or cover management treatments the previous fall, winter, and spring. At present, the R value neither adequately describes snowmelt and runoff during the late winter and early spring thaw, nor does it account for sediment transport by runoff during this period. Most of the sediment load in the Maumee River Basin occurs in February, March, and April (5). Significant soil loss also occurs in the basin during this period (5).

The Sandusky River accounts for the greatest soil loss in the central basin (Table 5). Unit area soil losses are less than in the western basin because of the lower percentage of cropland (48%) rather than soil differences. Percentage reductions in soil loss possible under the conservation tillage scenarios were similar but somewhat lower

January-February 1982

than in the western basin.

The eastern basin (Table 6) is small compared with the other two, and low in cropland acreage (38%). However, soils are much steeper than in the western and central basins, and unit area soil losses are similar to the others. Percentage reductions in soil loss with conservation tillage were lower in this area because of the limited acreage in row crops, which reduced the opportunity for conservation tillage.

Table 7 summarizes soil loss in the Lake Erie Basin. Annual soil loss under "present conditions" is 25 million tons (22.7 metric tons). The annual rate of soil loss is 2.2 tons per acre (4.9 metric tons/hectare). A 40 percent soil loss reduction could be achieved by meeting the T value, which is similar to the 46 percent soil loss reduction with conservation tillage. The maximum potential for soil loss reduction is 69 percent, using a combination of no-till and conservation tillage. Other scenarios, such as winter cover crops and shifts in fall or spring plowing, had minimal effects on soil loss.

Highest unit area soil losses are in the Raisin River Basin of southeastern Michigan; the Sandusky, Maumee, Huron, and Vermilion River Basins in Ohio; and the Cattaraugus Creek Basin in New York (Table 8). The Michigan and Ohio watersheds are almost contiguous (Figure 1). These watersheds have similar soils and land use. The glacial-till-derived soils are primarily medium to fine textured. They support a high percentage of cultivated crops. New York's Cattaraugus Creek watershed, on the other hand, has little cultivated cropland, but its slopes are steeper than in the Ohio and Michigan watersheds.

We also summarized our USLE analysis

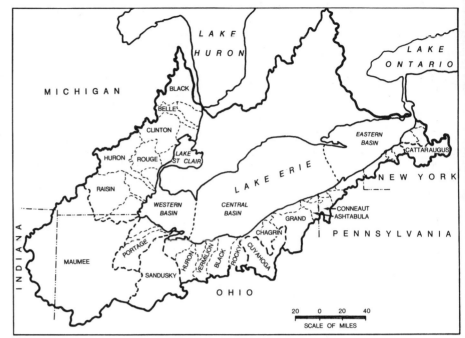

Figure 1. The Lake Erie drainage showing the western, central, and eastern basin segments and the major watersheds draining into them on the U.S. side.

by county. This information is detailed in the county management packages being produced by the Corps of Engineers for counties in the Lake Erie Basin (1).

Discussion

Our analysis with the USLE showed that mean soil loss in the Lake Erie Drainage Basin is about 2.2 tons per acre per year (4.9 metric tons/hectare/year), with a range from 0.4 to 4.4 tons per acre per year (.8-9.8 metric tons/hectare/year) for individual watersheds. Although this rate is not high from the standpoint of maintain-

ing crop productivity, individual soils experience soil losses exceeding T. The scenarios showed that soil loss could be reduced 40 percent in the Lake Erie Basin by bringing erosion on all soils down to T.

This reduction, however is inadequate to meet the 50-percent reduction in diffuse-source P loads in Lake Erie that is required to meet target loads for improved water quality (8). A reduction in P load of this magnitude may require about a 75 percent reduction in mean soil loss in the basin (10). Table 7 shows that even maximum use of conservation tillage, which would

Table 7. USLE—estimated soil losses in the Lake Erie drainage basin (U.S. side) and potential for soil loss reductions with alternative land management strategies.

	Present Conditions	Reduce Soil Loss to T	Spring Plow Only	Fall Plow Only	Winter Cover	Conservation Tillage	Maximum Conservation Tillage
	Western Basin						
Tons	16,558,689	9,652,224	15,762,305	17,511,751	15,859,357	8,323,693	4,519,203
Tons/acre/year	2.3	1.3	2.1	2.4	2.2	1.1	0.6
Percentage reduction*	-	42	5	−6†	4	50	73
	Central Basin						
Tons	6,730,579	4,396,989	6,468,333	7,230,668	6,569,516	3,892,375	2,368,456
Tons/acre/year	2.0	1.3	1.9	2.1	1.9	1.1	0.7
Percentage reduction	-	35	4	−7	2	42	65
	Eastern Basin						
Tons	1,808,952	1,098,794	1,781,211	2,033,528	1,805,904	1,275,956	926,274
Tons/acre/year	2.2	1.4	2.2	2.5	2.2	1.6	1.1
Percentage reduction	-	39	2	−12	0	29	49
	Total Lake Erie Basin (U.S. side)						
Tons	25,048,220	15,148,007	24,011,849	26,775,947	24,234,777	13,492,024	7,813,933
Tons/acre/year	2.2	1.3	2.1	2.3	2.1	1.2	0.7
Percentage reduction	-	40	4	−7	3	46	69

*Reduction as percentage of present conditions.
†Soil loss increases with fall plowing.

include no-till on some soils, would only provide a 69-percent soil loss reduction. This means that other measures to reduce P inputs to the lake must be used. These might include reducing P concentrations in some sewage treatment plant effluents below the 1.0 microgram per milliliter now required, or it might require more judicious use of P fertilizer on cropland where soil test levels have been built up to adequate or even excessive levels in previous years.

Our USLE analysis also identified those areas in the Lake Erie Basin where soil loss is greatest (Table 8). These occur primarily in the western and central basins. Because the eutrophication problem in Lake Erie is in the central and western basins, erosion control in watersheds draining into these lake segments should receive the greatest attention. The Pollution from Land Use Activities Reference Group, in its final report to the International Joint Commission for the Great Lakes (7), identified watershed areas draining into the western and central Lake Erie Basins as having the highest soil loss and sediment loads in the Great lakes and recommended erosion control on cropland in this area as a high priority in the effort to control pollution of the Great Lakes.

REFERENCES CITED
1. Adams, J. R., T. J. Logan, T. H. Cahill, D. R. Urban, and S. M. Yaksich. 1980. *A land resource information system (LRIS) for water quality management in the Lake Erie Basin.* J. Soil Water Cons. 37(1): 45-50.
2. Bierman, V. J. 1980. *A comparison of models developed for phosphorus management on the Great Lakes.* In R. C. Loehr, C. S Martin, and W. Rast [eds.] *Phosphorus Management Strategies for Lakes.* Ann Arbor Science, Ann Arbor, Mich. pp. 235-259.
3. Bone, S. W., R. Christman, L. M. Feusner, B. H. Nolte, B. L. Schmidt, and J. Shupert. 1975. *Ohio erosion control and sediment pollution abatement guide for agricultural land.* Bull. No. 594. Ohio Coop. Ex. Serv., Columbus. 20 pp.
4. Forster, D. L. 1981. *Adoption of reduced tillage and other conservation practices in the Lake Erie Basin.* Tech. Rpt. Series. Lake Erie Wastewater Manage. Study. U.S. Army Corps Eng., Buffalo, N.Y. 26 pp.
5. Logan, T. J., and R. C. Stiefel. 1979. *The Maumee River Basin pilot watershed study. Volume 1. Watershed characteristics and pollutant loadings.* EPA-905/9-79-005-A. Great Lakes Nat. Program Office, Environ. Protection Agency, Chicago, Ill.
6. Oloya, T. O., and T. J. Logan. 1980. *Phosphate desorption from soils and sediments with varying levels of extractable phosphate.* J. Environ. Qual. 9: 526-531.
7. Pollution from Land Use Activities Reference Group (PLUARG). 1978. *Final report.* Int. Joint Comm., Windsor, Ont.
8. Thomas, N. A., A. Robertson, and W. C. Sonzogni. 1980. *Review of control objectives: New target loads and input controls.* In R. C. Loehr, C. S. Martin, and W. Rast [eds.] *Phosphorus Management Strategies for Lakes.* Ann Arbor Science, Ann Arbor, Mich. pp. 61-91.
9. Triplett, G. B., D. M. Van Doren, and S. W. Bone. 1973. *An evaluation of Ohio soils in relation to no-tillage corn production.* Res. Bull. 1068. Ohio Agr. Res. and Dev. Center, Wooster.
10. U.S. Army Corps of Engineers. 1979. *Lake Erie management study.* Methodology Rpt. Buffalo, N.Y. 146 pp.
11. U.S. Department of Agriculture. 1975. *An estimation of soil loss and sediment yield for the Maumee River Basin using the Universal Soil Loss Equation and linear program models.* Great Lakes Basin Comm., Ann Arbor, Mich. 128 pp.
12. Wischmeier, W. H., and D. D. Smith. 1978. *Predicting rainfall losses—A guide to conservation planning.* Agr. Handbook No. 537. U.S. Dept. Agr., Washington, D.C. □

Table 8. Unit area annual soil loss in Lake Erie Basin watersheds.

Watershed	Soil Loss (t/a/yr)
Western Basin	
Raisin	4.35
Maumee	3.05
Portage	2.23
Huron (Michigan)	1.47
Rouge	0.88
Belle	0.73
Mill Creek	0.63
Clinton	0.62
Black (Michigan)	0.32
Central Basin	
Sandusky	3.68
Huron (Ohio)	3.35
Vermilion	3.12
Black (Ohio)	1.96
Rocky	1.91
Conneaut	1.45
Grand	0.69
Chagrin	0.68
Ashtabula	0.62
Euclid Creek	0.54
Cuyahoga	0.40
Big Creek	0.37
Eastern Basin	
Cattaruagus Creek	3.16
Eighteen-mile Creek	2.02
Raccoon Creek	1.55
Delaware Creek	1.28
Mill Creek	1.26

GIS FOR SOILS AND RANGELAND MANAGEMENT

R.G. Best and F.C. Westin*

*Remote Sensing Institute, South Dakota State University, Brookings, SD

ABSTRACT

Large areas of rangeland that are being cultivated may be susceptible to soil erosion. These
areas require extensive management in order to preserve the soil resource. Conventional field survey
techniques for identifying and monitoring problem
areas are not adequate or cost efficient. This investigation demonstrates how a computerized geographic information system (GIS) can be used in an
operational soils and rangeland management program.
A composite map of landuse, (interpreted from Landsat imagery), soil capability class and slope was
produced with the Area Resource Analysis System, a
cellular GIS. Cultivated areas on capability
classes V-VIII and/or slopes greater than 6% were
identified as problem areas. In this pilot project area, more than 60% of the cultivated land
(3922 hectares) was on soils not suited for cultivation. The GIS was also used to produce maps
useful for formulating a management plan.

INTRODUCTION

Substantial areas of rangeland are being cultivated because of inflated land values and Governmental farm programs. Some of those areas are
not suitable for cultivation and may be susceptible to soil erosion. These critical areas require extensive management in order to preserve
the soil resources.

If "Sodbuster" legislation that limits agricultural program incentives to farmers and ranch-
ers who cultivate rangeland on soils not suited
for cultivation is enacted, methodology will have
to be developed to identify and monitor areas of
non-compliance. Both soils and landuse data must
be analyzed in order to accurately identify problem areas. In this country, the area of potentially arable rangeland is large and conventional
field survey techniques would probably not be
adequate. Conventional methods would require extensive manpower to complete in a reasonable
length of time and would not be cost efficient.

The use of a computerized Geographic Information System (GIS) may be a viable alternative.
Geographic information systems differ from other
computerized managerial data systems in their
ability to retain the spatial integrity of the
input data. Most systems have the ability to
input, store, analyze and output map data. The
objective of this project was to demonstrate the
operational use of a GIS for the identification
and management of cultivated rangeland sites susceptible to soil erosion. In the project, Landsat
derived landuse and soil survey data published by
the Soil Conservation Service, were analyzed in
the Area REsource Analysis System developed at the
Remote Sensing Institute, South Dakota State University.

Area Resource Analysis System

AREAS is an interactive cellular GIS that
operates on a Prime 400* mini-computer. Two options are available for spatial data entry (coordinate digitizing) in the system. Data can be
entered in a vector format as segments and nodes
with a 40" x 60" Summagraphics coordinate digitizing table interfaced to the computer or manually
entered with a terminal from a cellular grid overlay. The vector data that result from the digitizing table are converted to a cellular array and
data from both methods are stored in a compressed
change-point format to minimize storage requirements. Output from the system can be tabular summaries, maps, or color and black-and-white displays. Maps can be plotted at any scale on the 3-
pen Houston Instruments plotter or generated as
symbol maps on a dot matrix printer. Spatial output data can also be down-loaded onto 8" floppy
diskets and color and black-and-white displays
generated on the microprocessor based Remote Infor-
mations Processing System (RIPS).

The associated software comprises the rest of
the system. The primary functions available on the
system are summarized in Table 1. In addition to
the functions listed in Table 1, there are numerous utility programs for error checking, data set
correction and general bookkeeping applications.
The most recent user executable version of AREAS
and more detailed documentation is available from
the Remote Sensing Institute, South Dakota State
University, Brookings, SD, 57007.

---

*The inclusion of registered trade names or trade-
marks are provided solely for documentation and
do not constitute an endorsement by the authors
or the Remote Sensing Institute.

Table 1. Summary of Primary Functions in AREAS

COORDSIN - Convert table digitizer vector output
into cellular array.
COMPOSIT - Overlay 2 to 4 data sets into a composite data set.
DIFFER - Composite and interpret 2 data sets into
single map with 2 codes "Same" and
"Differ".
GRID - Generate cellular grids, any dimension with
any cell size.
IMAGEIN - Convert digital image to AREAS format.
IMAGEOUT - Convert an AREAS data set to digital
image format.
INTERPRT - Generalize AREAS data set by combining
map units with similar characteristics.
MERGER - Join AREAS data sets at sides or top and
bottom.
PLOTTER/PLOT - Create file and plot AREAS data set
on Houston Instruments plotter.
POLYNUM - Convert AREAS data set to polygons.
PRINTMAP - Pattern maps on dot matrix printer.
SHRINK - Increase cell size and reduce matrix dimensions by aggregated cells.
TABULATE - Areal summaries of AREAS data set.
WINDOW - Create subsets from larger data arrays.

METHODS

A township (T6N, R28E) in central South Dakota was selected for analysis in this project (figure 1). During the last decade more than half of the rangeland in the township has been cultivated. Soils data for the project area were derived from the "Soil Survey of Stanley County, South Dakota".[1] The 1980 level I landuse[2] for the township was interpreted on an enlargement print from a Landsat false color composite (E-22038-16485). The Landsat landuse interpretation was enlarged and registered to the 1:24,000 soils map on a reflecting projector.

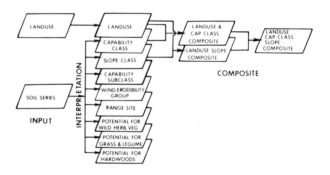

Figure 1. Location of study site.

Both maps were digitized on a Houston Instruments coordinate digitizing table. The resulting vector data sets were converted to cellular arrays with 0.8 mm square cells (0.1 acres @ 1:24,000). Figure 2 illustrates the GIS approach used to identify potential problem areas. A soil capability class map, slope class map, capability subclass map, wind erodibility group map, range site map and maps that indicate the potential for establishing and maintaining wild herbaceous vegetation, grasses and legumes and hardwood trees were produced from the soils data set with the INTERPRT function in AREAS. The capability class map, slope class map, capability subclass map and wind erodibility group maps were composited with the landuse map to identify cultivated areas with potential problems. In order to simplify the landuse/capability class and landuse/slope class composite maps, they were interpreted to show only cultivated land on soils with severe limitations (classes V-VIII) and on strongly undulating soils (slopes >6%), respectively. The resulting composite maps were composited to produce a single map of cultivated land on capability classes V-VIII and/or slopes greater than 6%. Plotted maps were produced at scales of 1:60,000 and 1:24,000. In addition to the plotted maps, symbol maps, black-and-white and color displays were produced for selected maps. Areal tabulations were run for each of the maps.

Figure 2. Flow chart of GIS analysis used to identify problem areas.

RESULTS AND DISCUSSION

There were 23 different soil mapping units in the project area (figure 3). Each of the soil mapping units has a unique combination of properties that distinguish it from the others. Most of the soil mapping units have at least some properties in common with the others. The properties of the soils significant to this analysis are summarized in Table 2. The detailed soil series map is useful only if one is familiar with the properties of each of the soil mapping units. It is easier to identify areas where problems may occur if soils are cultivated, on maps interpreted from the original soil series data. Figures 4 and 5 are soil capability class and soil slope maps interpreted from the soil series map. The capability class map has been further simplified from eight classes to three where: capability classes I-III have been combined, capability class IV is separate

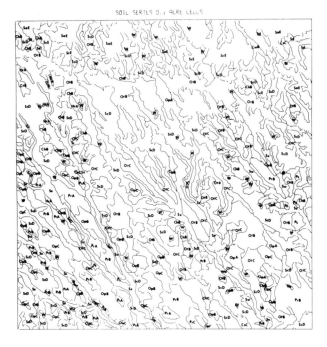

Figure 3. Original soil survey data plotted from AREAS data set.

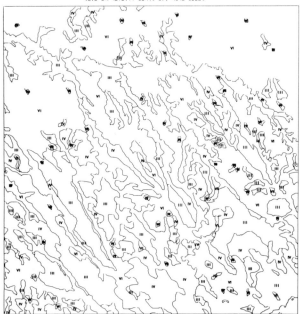

Figure 4. Soil capability class map interpreted from soil series map with AREAS.

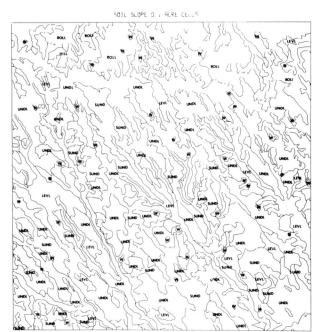

Figure 5. Soil slope class map interpreted from soil series map with AREAS.

and capability classes V-VIII have been combined. The resulting map shows the spatial distribution of regions that have minor limitations to cultivation, are marginal for cultivation and that have severe limitations to cultivation. Tabulations of this data set show that almost 54% of the project area has severe limitations to cultivation. In addition, 42% of the landscape has slopes greater than 6% and may be susceptible to erosion if a close growing vegetation is not present. Slope is considered in assigning capability class but not the only factor. Maps based on soil properties alone only indicate areas that need to be monitored. Rangeland with close growing grasses is the best landuse for these areas. Extensive management is necessary only when critical areas are cultivated.

The next objective in an operational soil conservation program would be to identify which of these areas are currently being cultivated in order to concentrate the management effort. Generally, it is not necessary to identify which crops are being grown, only that the land is being cultivated, which removes the year-long close growing vegetation. The repetitive, synoptic coverage of the Landsat system makes it a good source of data for interpreting and monitoring changes in general landuse for large areas. Its small scale and course spatial resolution would not limit its use in this application.

Figure 6 is a landuse map of the project area interpreted from Landsat imagery. Sixty-three percent of the township was cultivated in 1980. Obviously, some of the cultivated land is on soils that are not suited for cultivation. Landuse com-

Table 2. Properties of soils that occur in T6N, R28E, Stanley County, South Dakota.

SOIL TYPE	ABREV.	CAPABILITY CLASS*	CAPABILITY SUB-CLASS**	SLOPE CLASS†	WIND ERODI-BILITY GROUP‡‡	RANGE SITEΨ	POTENTIAL FOR ESTABLISHING AND MAINTAININGΨΨ — WILD HERBACEOUS VEG.	GRASS/LEGUME	HARD-WOOD TREES
Chantier clay	ChB	VI	s	UNDL	4	IV	2	1	2
Chantier-Sansarc clay	CSC	VI	s	UNDL	4	IV	2	1	2
Chantier-Swanboy clay	CwB	VI	s	UNDL	4	IV	2	1	2
Hurley silt loam	HrA	VI	s	LEVL	6	VI	2	1	2
Hurley-Slickspots complex	Hs	VI	s	LEVL	6	VI	2	1	2
Opal clay	OpA	III	s	LEVL	4	III	4	3	3
Opal clay	OpB	III	e	UNDL	4	III	4	3	3
Opal clay	OpC	IV	e	SUND	4	III	4	3	3
Opal-Chantier clay	OtB	III	e	UNDL	4	III	4	3	3
Opal-Chantier clay	OtC	IV	e	SUND	4	III	4	3	3
Opal-Promise clay	OxB	III	e	UNDL	4	III	4	3	3
Promise clay	PrA	III	s	LEVL	4	III	4	3	3
Promise clay	PrB	III	e	UNDL	4	III	4	3	3
Promise clay	PrC	IV	e	SUND	4	III	4	3	3
Promise-Hurley complex	Ps	III	s	LEVL	4	III	4	3	3
Sansarc clay	SaF	VII	e	ROLL	4	IV	3	1	2
Sansarc-Opal clay	ScD	VI	e	ROLL	4	IV	3	1	2
Sansarc-Opal clay	ScE	VII	e	HILL	4	IV	3	1	2
Sansarc-Rock outcrop complex	Sd	VII	e	LEVL	4	IV	3	1	2
Swanboy clay	Sw	VI	s	LEVL	4	IV	2	1	2
Swanboy-SLickspot complex	Sx	VI	s	LEVL	4	IV	2	1	2
Wendte clay	Wd	VI	w	LEVL	4	II	3	3	2

* Classes I-III minor limitation to cultivation, IV marginal for cultivation, V-VIII severe limitations to cultivation
** s-soil limitations, e-erosion limitations, w-water limitations.
† LEVL-level (0-3%), UNDL-undulating (2-6%), SUND-strongly undulating (6-9%), ROLL-rolling (6-15%), HILL-hilly-steep (15-40%).
‡‡ 4-more than 35% clay, moderately erodible; 6- 18-35% clay, very slightly erodible.
Ψ I-VI decreasing productivity.
ΨΨ 1-very poor, 2-poor, 3-fair, 4-good.

posited with capability class or slope class produce two different maps, neither of which shows the total extent of the problem area. A map of cultivated land on capability classes V-VIII and/or slopes greater than 6% can be produced by compositing landuse with both capability class and slope (figure 7). The extent and distribution of all problem areas can be identified on this single map. In 1980, more than 5900 hectares were cultivated in the twonship. Results of this project indicate that more than 60% of this cultivated land (3922 hectares) is on soils not suitable for cultivation.

The GIS can also be used to produce information useful for formulating management plans. Maps of management related data can also be interpreted from soils data. A range site map that can be used to estimate the potential productivity of rangeland based on soil properties was interpreted from the soils data. Stocking rates can be calculated from this data and it can be used to justify the conversion of cultivated areas back to rangeland. In this project, maps that showed the potential for establishing and maintaining wild herbaceous plants, hardwood trees and grasses and legumes were also produced from the soils data. Other maps related to soil properties can also be

generated from the soils data[3]. The maps that can be produced are determined by the user's needs and available data.

The computer cost for completing this project was $775.00 at a rate of $.02/CPU sec. Almost 75% of the cost was for data input and the rest for analysis and output. The costs for digitizing are proportional to the complexity and size of the map, because bigger more complex maps have more line segments and nodes. More than 80% of the digitizing costs were for entering the soil series map. Eight maps were generated from the soils map and if the input costs are distributed across all eight, the per map input cost would be less significant. Automatic digitization might reduce the input cost and would reduce the input time[4]. The data analysis costs are related to the size of the data array, which is determined by the size of the area, scale and cell size. Optimal cell size should be determined from the objectives and the desired map accuracy[5]. The cost per unit area would be less for maps at smaller scales with comparable cell sizes, which would probably be used in a large area operational soil conservation program.

LANDUSE 1980 0.1 ACRE CELLS

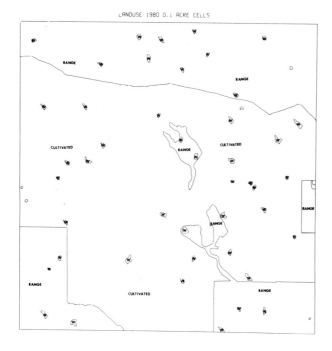

Figure 6. Level 1 landuse data interpreted from Landsat imagery and plotted from AREAS data set.

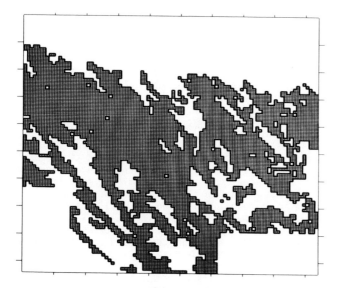

Figure 7. AREAS composite symbol print map of cultivated land on capability classes V-VIII and/or slopes greater than 6% (1.0 hectare cells).

SUMMARY AND CONCLUSIONS

This project demonstrates how a GIS can be used in an operational, soils and rangeland management program. Areas of rangeland that were cultivated on soils not suited for cultivation were identified by compositing a landuse map derived from Landsat imagery and soil capability and slope maps produced from a soil survey map. The GIS composite map shows the location and distribution of problem areas. Quantitative areal summaries were also generated in the GIS to determine the magnitude of the problem. The GIS produced a range site map that indicates the potential productivity of rangeland on the soils in the project area. Maps showing the potential for establishing and maintaining wild herbaceous vegetation, hardwood trees and grasses and legumes were generated to help formulate a management plan. The technique would be more time and cost efficient than conventional field survey techniques. Cost of using a cellular GIS will vary with the CPU rate, the size and complexity of the map and the cell size.

REFERENCES

[1] G.A. Borchers, "Soil survey of Stanley County, South Dakota," U.S. Dept. of Ag/Soil Conservation Service Rep., 135 pp., 1980.

[2] J.R. Anderson, E.E. Hardy, J.T. Roach and R.E. Witmer, "A landuse and landcover classification system for use with remote sensor data," USGS Prof. Paper 964, U.S. Gov't. Print. Office, Washington, D.C., 28 pp., 1976.

[3] M.E. Wehde, K.J. Dalsted and B.K. Worcester, "Resource applications of computerized data processing: the AREAS example," J. Soil and Water Conservation, Vol. 35, No. 1, pp. 36-40, 1980.

[4] D.F. Marble and D.J. Peuquet, "Geographic Information Systems and remote sensing," *Manual of Remote Sensing 2nd ed.*, Vol. I, R.N. Colwel, J.E. Estes and G.A. Thorley, eds., American Society Photogrammetry, Falls Church, VA., pp. 939-940, 1983.

[5] M. Wehde, "Grid cell size in relation to errors in maps and inventories produced by computerized map processing," Photogrammetric Engineering and Remote Sensing, Vol. 48, No. 8, pp. 1289-1298, 1982.

PART 6

VEGETATION RESOURCE MANAGEMENT

A SPATIAL ANALYSIS OF TIMBER SUPPLY[1]

Joseph K. Berry and John K. Sailor[2]

A computer-assisted map analysis system is discussed that characterizes timber supply by harvesting accessibility and haul distance to a proposed mill site. Factors governing accessibility of timber include effective skidding distance to haul roads as modified by barriers and slopes. Haul distance is expressed in units that take into account the relative ease of travel along various road types. The timber supply is grouped into spatial units, termed "timbersheds", of common access to particular haul road segments. The analysis techniques are demonstrated on a cartographic database covering an area surrounding Petersham, Massachusetts.

INTRODUCTION

The demand for stumpage throughout the United States has been increasing in recent years. This increased demand has stimulated new technologies such as whole tree chipping and chip recovery operations during milling. Significant growth in the pulp and particleboard industries has occurred and the potential use of chips to generate base load electricity may cause the demand to accelerate over the next decade.

The growing demand for wood chips necessitates the development of new methods for assessing roundwood supply. In the past, wood chip supplies have been dependent on mill-oriented activities or self-sufficient pulping operations on large ownerships. Whole-tree chipping on diverse ownerships is the most likely source of additional supplies. Under these new demand conditions, it is important to identify sources of roundwood according to their spatial distribution throughout the supply area.

[1] Paper presented at the In-Place Resource Inventory: Principles and Practices Workshop, August 9-14, University of Maine, Orono.

[2] Dr. Joseph K. Berry is Assistant Professor of Forestry at Yale School of Forestry and Environmental Studies. John K. Sailor is a recent graduate of the master's degree program at Yale School of Forestry and is presently with New England River Basins Commission.

This is particularly important when considering supply relative to a proposed forest products or energy facility, because of the costs of procuring harvesting and transporting the wood may determine the economic viability of such a facility.

One approach for assessing timber supply, identified by Berry and Tomlin (1980), assesses forest cover, harvesting acessiblilty and ownership characteristics so that timber inventory, harvesting costs, and owner's willingness to sell stumpage are intergrated into an overall supply schedule. This paper extends the original approach by describing the structure of a computer oriented model for assessing roundwood supply relative to a proposed (hypothetical) mill site. The model characterizes supply by:

haul distance: considering road type, and by
accessibility: considering harvesting factors.

Based on these factors, the model determines spatial units of common access ("Timbersheds"). For this paper a constrained version of the model is demonstrated on a 17.7 sq. km. area surrounding Petersham, Massachusetts.

Social factors that may influence the availability of timber have been considered in an earlier study (Berry and Mansbach 1981). Further studies are planned that will combine accessibility information with timber availability based on social factors.

FUNDAMENTAL CONSIDERATIONS

Information on the extent and locaton of timber resources serves as primary input. This information can be extracted from several sources, such as existing forest cover type maps, or maps generated from aerial imagery. Inventory information alone, however, may not yield useful estimates of actual supply as affected by physical and economic considerations. Some stands may be far removed from existing roads or may be found on highly erodable soils or steep slopes that would make certain harvesting techniques inappropriate (Dykstra and Riggs, 1977). Given the existing roads and terrain features, some timber resources may be effectively too remote for economic harvest and transport to a processing facility. In such instances, the forest inventory will overestimate actual supply.

In order to determine if the supply of timber is adequate for a proposed facility, the forested areas must be characterized by the accessibility of the timber for harvesting, and the transport distance required to haul it to the mill. Areas of accessible timber within hauling range can be grouped into units of common accessibility to a haul road segment. These units, termed "Timbersheds", indicate all of the timber areas that are economically accessible to a particular haul road segment.

SPATIAL DATA BASE

Data encoding, analysis, and display capabilities for this study were provided through the use of software developed at Yale University as part of the Map Analysis Package (Tomlin, in preparation). Information on the biological, physical, and cultural features of a given geographic area is encoded to correspond with a grid cell data structure. Each grid cell is assigned a value which represents one member of a set of mutually exclusive categories (e.g. dry land, stream, pond, lake) (Tomlin and Berry, 1979).

Information used in this study is part of a general purpose data base being developed for the Harvard Forest vicinity (Petersham, Massachusetts). Each map represents 1770 hectares or 17.7 sq. km. (1/4 hectare per cell, 7080 cells in all). The maps used for this study include vegetation cover types, elevation, roads, and water features.

CARTOGRAPHIC MODEL

Figure 1 is a flow chart of the spatial analysis model. It presents a logical sequence of

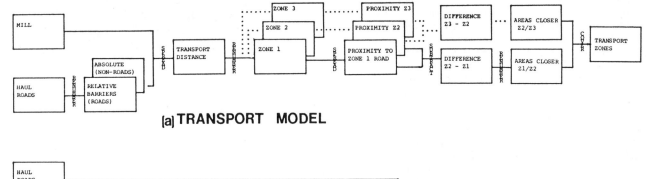

(a) TRANSPORT MODEL

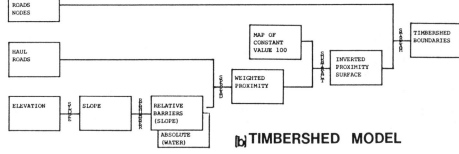

(b) TIMBERSHED MODEL

Figure 1. Flowchart of spatial analysis model.
   a) Transport model measures haul distance from mill along roads, weighted by relative ease of travel along road types. Haul distances are divided into 3 zones. Accessibility of timber to roads in these zones is then measured considering slopes and barriers, yielding a map of transport cost zones for the study area.
   b) "Timbershed" model measures accessibility of timber considering slopes and barriers. The resulting weighted proximity map is used to generate boundaries of zones of common access to particular haul road segments ("timbersheds").

207

Map Analysis Package operations represented as arrows which transform maps that are represented as boxes.

The model consists of two submodels: the transport distance model, and the "timbershed" model. The transport distance model (fig. 1a) uses a map showing a proposed mill site and measures haul distance and harvesting accessibility of timber. Haul distance is measured along the existing road network. Constraint maps are used so that distance is measured only along roads and is weighted by road type. For the purposes of this demonstration, timber hauling along secondary roads was assumed to take 50% longer as that along primary roads; hauling on tertiary roads was assumed to take 100% longer than on primary roads; and hauling on unimproved roads was assumed to take 200% longer than on primary roads; From the resulting map of weighted haul distance, a map of haul distance zones was created. A map of accessibility of timber, or effective yarding (skidding) distance to roads in these haul zones, was then generated taking into consideration steep slopes and water areas that must be circumvented. Slopes were used to weight the distance such that moderately limiting slopes of 11-15% required 100% more time to traverse (and thus for cost purposes are twice as "far" away), severely limiting slopes of 16-20% required 200% more time to traverse, and slopes greater than 20% were avoided. The map resulting from this measurement was converted to show zones of weighted harvesting distance (or cost). The resulting zones are then combined with the vegetation map to produce tables of transport cost versus timber supply.

The second submodel (fig. 1a) is concerned with finding minimum transport distances to haul roads. A map of weighted distance from haul roads is made using the same characteristics as in the transport distance submodel. Using a road intersection map, a map of boundaries of equal haul distance to two or more roads is produced. This map is created by generating lines from road intersections that are constrained to being on the weighted distance margin between two roads. These boundaries are areas that are equally accessible to two or more roads, using the weighted accessibility measure. On either side of these boundaries, timber is more accessable to a particular haul road, and therefore will be cost least to harvest if it is yarded to that road.

DEMONSTRATION RESULTS

The thirty-four vegetation types occurring the the Petersham area were collapsed into nine classes of merchantable forests. For display purposes, these are grouped into five categories (Figure 2). Forested areas comprise 83.6% of the study area. However, these areas have different accessibility and transport costs which must be considered in determining potential supply.

Figure 3 shows important intermediate maps associated with the transport submodel. For display and tabulation the roadnet was divided into

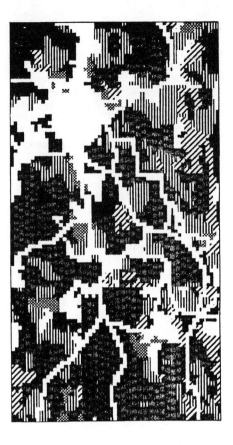

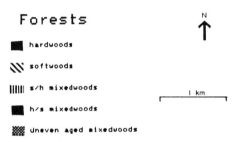

Figure 2. Forest Type map for the study area. The nine forest classes are grouped into five categories for display.

3 zones of haul distance, and the timber areas were divided into 2 zones of accessibility. Because the distances measured are a function of the roads or terrain traversed, the maps can be considered travel time maps or transport cost surfaces. The units can be expressed in time, cost, or in distance equivalents. In this case, "distances" are expressed relative to kilometers travelled on primary roads for haul distance and kilometers skidded on level ground for accessibility. Table 1 shows the forest types as a function of transport costs zones.

208

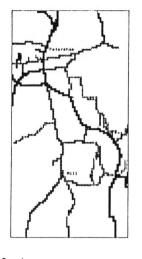
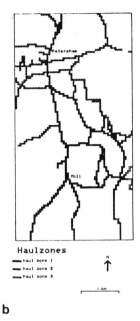

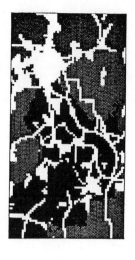

Roads
- ━━ Primary road
- ▰▰▰ Secondary Road
- ━━ Tertiary Road
- ╍╍ Dirt Road

N ↑

1 km

a

Haulzones
- ━━ haul zone 1
- ━━ haul zone 2
- ▰▰▰ haul zone 3

N ↑

1 km

b

Zone 3 Skidding
- ■ less than .5 km skid
- ■ .5 to 1 km skid
- not zone 3

N ↑

1 km

c

Transport Zones
- ■ closest to zone 1
- ■ closest to zone 2
- ▨ closest to zone 3
- non-forest

N ↑

1 km

d

Figure 3. Intermediate and final maps from the transport
model. (a) The existing roads;(b) weighted haul distance
along these roads ;(c) harvest (skidding) accessibility to
haul zone 3 roads , and (d) the final transport zones ;
are shown as represented by a Hewlett-Packard 2648A
graphics terminal.

Table 1.--Tabulation of forest classes by transport
distance zones.

Forest[2] Class	Total Area	Transport Zone[1] (Haul/Access) (ha)					
		1/1	1/2	2/1	2/2	3/1	3/2
1	26.25	0	0	3.5	0	21.75	1.25
2	22.75	1.75	0	10.5	0	9.25	1.25
3	47.0	2.0	0	10.75	0.75	11.5	22.0
4	76.0	7.75	0	12.25	0	44.25	11.75
5	73.0	11.0	0.75	23.75	0	21.25	16.25
6	235.75	21.25	1.5	48.5	0	132.0	32.5
7	477.25	45.5	5.25	142.0	1.5	194.75	78.75
8	53.75	104.0	4.25	83.0	3.75	177.75	52.0
9	53.75	9.0	0.5	14.0	2.0	12.75	17.0

[1]Transport Zones (for example 1/2 is Haul Zone 1,Access Zone 2.)

Haul Zones
Haul Zone 1 - less than 2 km. haul
Haul Zone 2 - 2 km. to 4 km. haul
Haul Zone 3 - greater than 4 km. haul
  Haul distances weighted by road type and
  expressed relative to hauling on primary road.

Access Zones
Access Zone 1 - less than .5 km skid
Access Zone 2 - .5 km skid or greater
Skid distances weighted by slope and expressed
relative to skidding on flat surface

[2]Forest Classes
Class 1 - hardwoods; 41-60 ft.; 81-100% closure
Class 2 - hardwoods; 61-80 ft.; 81-100% closure
Class 3 - softwoods; 21-60 ft.; 30-80% closure
Class 4 - softwoods; 61-80 ft.; 81-100% closure
Class 5 - mixedwoods(S/H);21-60 ft.;30-80% closure
Class 6 - mixedwoods(S/H);61-80 ft.;30-80% closure
Class 7 - mixedwoods(H/S);21-60 ft.;30-80% closure
Class 8 - mixedwoods(H/S);61-80 ft.;30-80% closure
Class 9 - mixedwoods(uneven aged)

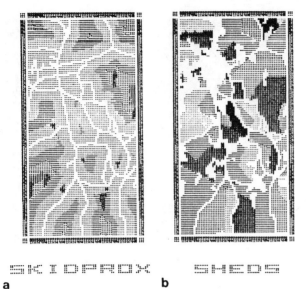

SKIDPROX   SHEDS

a   b

Figure 4. Intermediate and final maps from the "timbersheds" model. (a) Accessibility of timber (weighted skidding proximity); and (b) areas of common access to haul road segments ("timbersheds"); are shown in line printer representation.

Figure 4 shows intermediate and final maps from the "timbersheds" model. Two example areas of common access to haul road segments are compared with transport costs in Table 2. These "timbersheds" are compared with forests types in Table 3. This information would be useful in analyzing the fixed costs of harvesting, as it provides an indication of the minimum number of landings required to economically harvest the supply.

Table 2. - Forest classes in two example "timbershed" areas.

	Total Area	Forest Classes[1] (ha)				
Timbershed		4	5	6	7	8
6	32.25	6.5	2.25	6.75	5.5	11.25
28	23.25	3.0	0.0	4.0	9.75	6.5

[1] Forest Classes are listed in Table 1.

Table 3. Transport Zones in two example "timbershed" areas.

	Total Area	Transport Zones[1]			
Timbershed		1/1	1/2	3/1	3/2
6	32.25			24.5	7.75
28	23.25	23.0	0.25		

[1] Transport Zones are listed in Table 1.

Table 4.--Transport zones combined with forest areas when relative cost of skidding on slopes is increased by 50%.

Forest[2] Class	Transport Zone[1] (Haul/Access) (ha)					
	1/1	1/2	2/1	2/2	3/1	3/2
1	0	0	3.5	0	21.75	1.25
2	1.75	0	10.25	0.25	9.25	1.5
3	2.0	0	10.75	1.0	10.75	22.75
4	7.5	2.5	12.25	0	43.5	12.5
5	11.0	0.75	23.25	0.5	21.25	29.25
6	21.25	1.5	48.0	0.5	130.0	34.5
7	45.5	5.25	141.0	2.5	192.75	81.0
8	104.0	4.25	73.75	10.0	168.25	61.0
9	9.0	0.5	14.0	2.0	12.75	17.0

[1] Forest classes are shown in Table 1.

[2] Transport zones are shown in Table 1.

The results of the cartographic model allow the strategic planner to characterize the timber supply surrounding a mill site in terms of harvesting costs. This information can then be used to determine if the supply, within economic transport and access "reach", is sufficient to sustain the proposed facility.

An advantage of computer-assisted map analysis is that once a model is developed and the appropriate data are encoded, repeated simulation of the model using different constants can be performed rapidly. For example, the model can be run again with a different cost factors associated with skidding on slopes. In this way, sensitivity analyses can be performed to investigate the importance of the accuracy of cost factors to the model results. Increasing the cost of skidding on slopes changes the access costs as shown in Table 4. Similarly, the model could be evaluated for several mill sites.

CONCLUSION

This model can serve as a strategic planning tool to spatially characterize the harvest and transport costs for the timber supply surrounding an existing or proposed forest products facility. The analysis, however, is not intended to provide output to the harvesting crew. Rather it is intended to extend the utility of forest cover type maps for characterizing timber supply in the vicinity of a forest products facility.

LITERATURE CITED

Berry, J.K. and A. Mansbach. 1981. Extending the Utility of Forest Cover Maps. Proceedings Eastern Regional Remote Sensing Applications Conference, Danvers, MA.

Berry, J.K. and C.D. Tomlin. 1981. A Cartographic Model for Assessing Roundwood Availability. in WOODPOWER: New Perspective in Forest Usage, Pergamon Press, 1981.

Dykstra, D.P. and J. L. Riggs. 1977. An application of Facilities Location Theory fo the Design of Forest Harvesting Areas. Transactions of the American Institute of Industrial Engineers 9(3): 270-277

Tomlin, C.D. (in preparation). Digital Cartographic Modeling Techniques in Environmental Planning. Doctoral Dissertation, Yale School of Forestry, New Haven, CT.

Tomlin, C.D. and J.K. Berry. 1979. A Mathematical Structure for Cartographic Modeling in Environmental Analysis. Proceedings of the American Congress on Surveying and Mapping, 39th Meeting, ppp 269-283.

The Use of Computer Graphics in Deer Habitat Evaluation

by C. Dana Tomlin, Stephen H. Berwick, and Sandra M. Tomlin

PROBLEM

In 1977, the Yale School of Forestry and Environmental Studies initiated a study of the white-tailed deer *(Odocoileus virginianus)* population at the Yale Forest in northeastern Connecticut. The Yale Forest property comprises some 7,800 acres located in the towns of Union, Ashford, Eastford, and Woodstock, Connecticut, about 30 miles northeast of Hartford. The area is one of second growth forest, low hill topography, and distinctively rural New England character.

As one part of the deer population study, standard pellet group counts[1] were conducted during the spring of 1977 over an area of approximately one square mile. Within that area, five vegetative cover types were sampled to develop estimates of deer utilization. Among these cover types were red pine *(Pinus resinosa),* white pine *(Pinus strobus),* hemlock *(Tsuga canadensis),* mixed hardwoods (predominantly *Acer saccharum* and *Betula allegheniensis*), and open field. Sampling sites were located along transects within each cover type. These were grouped at five sites per transect and five transects per cover type for a total of 125 sites. The first site of each transect was located randomly. Subsequent sites were then established at uniform distances along each transect line. At each sampling site, pellet groups were counted over a circular area of .01 acre or approximately 24 feet in diameter.

Analysis of the 1977 pellet group counts found statistically significant differences between cover types and indicated the following habitat preference ratings, expressed as estimates of the number of deer per square mile of each habitat type:

hemlock	124
white pine	71
mixed hardwoods	54
red pine	23
open field	1

Not yet considered at this point, however, (and often overlooked in studies of this sort) were effects due to the spatial nature of data being examined. Among the most important of these were, first, the inherent lack of statistical independence among observations made at neighboring locations, and secondly, the influence of specific spatial relationships between sampling sites and other locational features such as roads, water bodies, or areas of a particular vegetation type. It is this second group of spatial effects which has most recently been examined as part of the Yale Forest study and which is discussed below in terms of the computer graphic techniques employed.

The stated purpose of this phase of the deer habitat study was to examine, in a quick, flexible, and largely intuitive manner, the relationship between pellet group counts and certain factors associated wit the spatial context of each sampling site. Its more general purpose was, in fact, to demonstrate the speed, flexibility, and intuitive nature of one particular approach to that problem in the form of computer mapping. This effort was to be directed toward a local audience of student foresters generally unfamiliar with digital technology, uncomfortable with involved mathematics, and suspicious of anything not advertised in the L. L. Bean spring catalogue.

METHODS

In keeping with these objectives, a decidedly small set of data was examined. Only two maps of the mile-square study area were encoded. The first was simply a collection of point values indicating pellet group counts for each of the 125 sampling sites. The second was a map of areas characterized by land cover types and site features. These included roads, structures, hemlock stands, red and white pine stands, hardwood areas, areas of mixed hardwood and softwood growth, open fields, and wetlands. This map is shown in Figure 1.

Both of these digital maps were derived from conventional maps compiled at a scale of 1:15,840 from field observations and aerial photographs. Pellet group counts were digitized as z coordinate values associated with x,y corrdinate locations. Land cover types and site features were defined by boundary lines digitized as sequences of x,y coordinate pairs.

These digitized coordinates were then used to create grid maps for subsequent analysis. A grid cell size of approximately 30 feet by 30 feet was chosen as one which would provide the resolution necessary to distinguish between sampling sites without requiring excessively large amounts of storage space. Each grid map contained 28,900 cells, dimensioned at 170 rows by 170 columns, and each cell was represented as a 16-bit integer. In the case of the pellet count map, each cell con-

---

[1] Pellet group counting is a method of estimating wildlife population density and forage use based on measurement of the periodic accumulation of fecal droppings at sampling sites within a given area. In this case, a defecation rate of 12.7 pellet groups per deer per day was assumed.

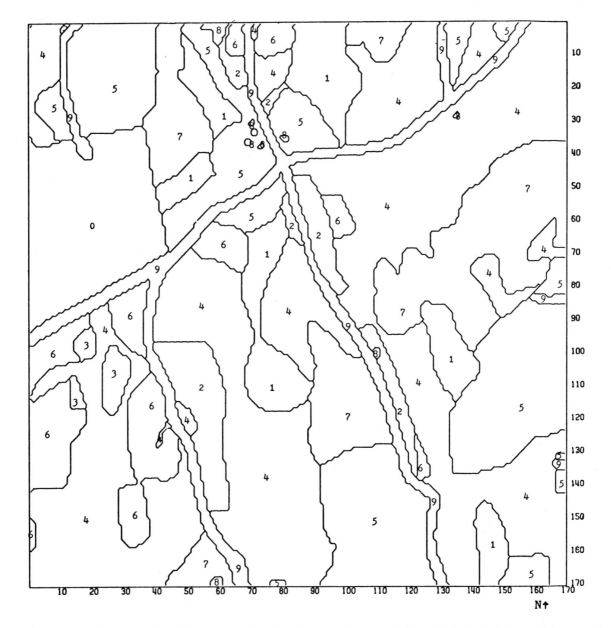

Figure 1. Land Cover Types and Site Features: 0. Non-Yale land; 1. Wetland; 2. Open field; 3. Red pine; 4. Hardwoods; 5. Mixed hardwoods and softwoods; 6. White pine; 7. Hemlock; 8. Structure; 9. Road.

213

taining the center of a sampling site (no cells contained more than one) was assigned the pellet count (x100) recorded at that site. All other cells were set to a value of zero. In the case of the land cover and site features map, values were assigned as follows:

0 land outside of the Yale Forest property
1 wet land
2 open field
3 red pine
4 hardwoods
5 mixed hardwoods and softwoods
6 white pine
7 hemlock
8 structure
9 road

Cells containing boundary lines, and therefore including more than one of the above categories, were assigned to whichever of those categories had the greatest value. The above values were initially assigned with this in mind such that each cell would be characterized according to its most "important" attribute.

Analyses of the spatial effect of land cover types and site features on deer pellet group counts generally followed a common procedure involving transformation of the cover type map to generate a new map of some particular spatial phenomenon; and, examination of pellet group counts, by cover type, within each region of this new map.

In one case, the cover type map was used to generate a map of "effective proximity" to roads and structures. This map is shown in Figure 2. Note here that "effective" proximity is clearly not measured in terms of Euclidean distance. Rather, it is defined in terms of units which may be consumed at rates which vary among cover types according to the degree to which those types either support or impede the movement of deer, i.e. units analogous to minutes of travel time or dollars of travel cost. In creating this map, it was assumed that areas of greater vegetation density (e.g. hemlock stands, hardwood stands, and areas of mixed hardwood and softwood growth), which generally offered better protection, were for that reason more conducive to deer movement. Open areas (e.g. wet lands and fields), on the other hand, were regarded as zones of higher "friction" relative to the movement of deer. It is for this reason that the proximity contours shown in Figure 2 are closer together in areas crossing wet lands and fields.

An alternative interpretation of "effective proximity" to roads and structures was also considered. Here, it was argued that it is not the movement of deer but the dissipation of noise, activity, and influence of man which should be measured and mapped. In this case, higher rather than lower "friction factors" were applied to areas of greater vegetation density.

Typical of the other factors examined in this manner were spatial effects such as the "local diversity" of cover types, the "narrowness" of forested areas, and the "shape complexity" of individual forest stands. A map of "local cover type diversity" is shown in Figure 3. Here, the value assigned to each grid cell indicates the number of different cover types occurring within a radius of approximately 25 meters around that cell. In creating the "forest narrowness" map, a value was assigned to each forest cell according to the shortest total distance from that cell to two diametrically opposing forest edges. "Shape

complexity" values were computed by isolating individual forest stands (i.e. contiguous areas of a single cover type) then characterizing each according to the ratio of its perimeter to the square root of its area.

Once mapped, these and other spatial factors such as stand size, proximity to hemlock stands, and visual exposure to roads and structures, were examined both individually and in weighted combination. In each case, categories of the spatial factor map, the land cover map, and the pellet count map were then compared on an overlay basis to generate cross-tabular statistics.

All of these analyses were performed in an informal and exploratory manner using packaged software in an interactive mode. The user-oriented nature and modular structure of that software proved to be particularly accommodating in this regard. Consider, for example, the following man-machine dialogue. **See page 218 for missing text.**

This is the command-response sequence generated in creating the effective proximity map shown in Figure 2. Here, "renumber," "spread," and "contour" are processing operations; "covermap," "development," "friction," and "proximitymap" are user-defined map file names; and "for," "assigning," "to," "thru," and "and" are parts of command-modifying phrases. "Relumber" is a mistake.

RESOURCES

All software, hardware, manpower, and funding in support of this project were made available through the Yale School of Forestry and Environmental Studies. Data encoding, analysis, and display capabilities were provided through the use of software being developed as part of the Map Analysis Package (C.D. Tomlin, 1979). The Map Analysis Package is written in FORTRAN IV and may be used either interactively or in batch mode. At present, the package employs a grid-cell data structure for all analytic operations. Input data, however, may exist in the form of digitized points, lines, or polygons and output may be produced in the form of line-plotter graphics.

For this study, spatial coordinate data were encoded using a Talos RP648a digitizer and a GT44 graphics terminal linked to a PDP 11/40 processor. These coordinate files were then stored on disk packs accessible to Yale's IBM 370/158 mainframe computer. The 370, operating under IBM's Time-Sharing Option (TSO) and accessed through a Hewlett Packard 2648A graphics terminal, was used for all subsequent analyses. Preliminary graphics were generated by high-speed line printer while final maps were produced using a Calcomp 763 drum plotter.

Throughout this phase of the deer habitat study, slave labor was employed wherever possible. Digitizing was done by a first-year graduate student[2] of requisite steady hand and weak mind but without previous experience. The authors called upon similar traits in performing the analyses and preparing graphics. The analyses were performed at a total cost of about five CPU minutes.

---

[2]P.V. Brylski

214

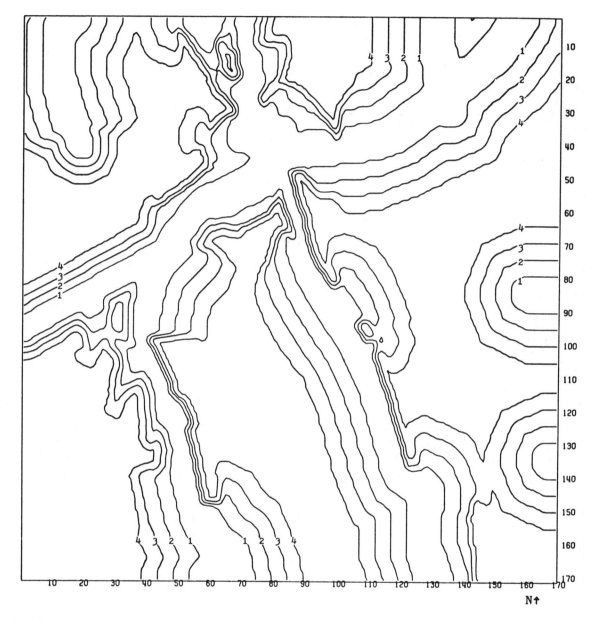

Figure 2. Effective Proximity to Roads and Structures: 1. One unit away; 2. Two units away; 3. Three units away; 4. Four units away.

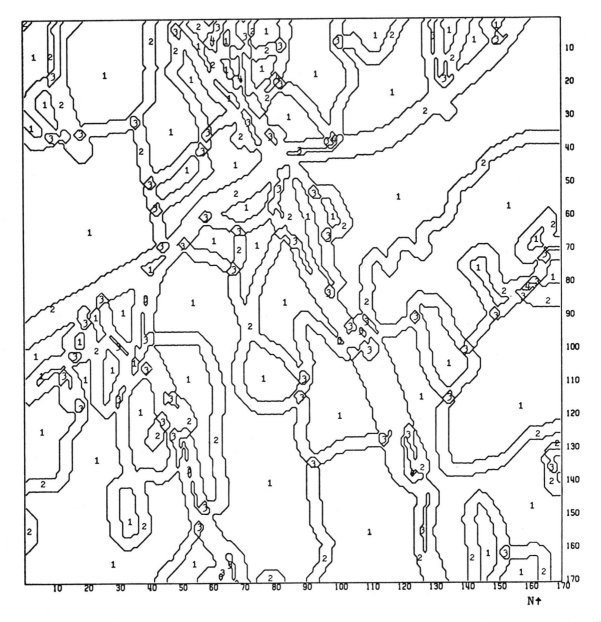

Figure 3. Local Cover Type Diversity: 1. One cover type nearby; 2. Two cover types nearby; 3. Three cover types nearby; 4. Four cover types nearby.

PROBLEMS

Even in light of the limited scope and informal nature of this project, surprisingly few problems were encountered. By far the most costly and time-consuming aspects of the work were those associated with data encoding. This was largely due to specific problems in dealing with recently installed hardware.

Also somewhat frustrating was the fact that sampling sites were not uniformly spaced over the study area. As mentioned earlier, the pellet count data used in this study were not collected with spatial analysis in mind. While this presented no serious problems in terms of demonstrating the potential for applying computer mapping techniques, it did limit the ability to draw specific conclusions about the deer population in this particular area. The same was true in terms of the small number of independent variables considered. Most conspicuous by its absence was any consideration of topographic relief. Elevation data are now being encoded in hopes of incorporating factors such as slope, aspect, and cool air drainage.

DOCUMENTATION

Two reports (Lyons, 1978 and 1979) are available describing the initial phases of the Yale Forest deer population study. Documentation on the more recent phases of that work involving computer graphics is forthcoming.

The Map Analysis Package is being developed as part of a doctoral dissertation on cartographic modeling techniques in environmental planning. A user's guide to the package is presently available in draft form. The Map Analysis Package approach to cartographic modeling is also supported by a graduate level course (Berry and Tomlin, 1979) and a series of user-oriented workshops sponsored by the Yale School of Forestry and Environmental Studies under the direction of Joseph K. Berry.

GRAPHICS

The intuitive nature of graphics in general and the speed and flexibility of computer graphics in particular were found to be of critical importance to this project. This was true not only in the presentation of results but also in the formulation of the analyses themselves.

Examples of graphics produced are presented in Figures 1 through 3. All are grid maps reproduced in line-plotter form. Column/row reference numbers are indicated at the bottom and to the right of each map. Figure 2 was created by tracing contour lines over the three dimensional surface represented by effective proximity values. Figures 1 and 3 were created by tracing boundaries between cells of different values then rounding off right angles with diagonal lines. Labeling in each case was done by hand.

RESULTS

Among the clearest findings of this project regarding the deer population in the Yale Forest study area were a mild direct relationship between pellet group count and local diversity of land cover types; and a mild inverse relationship between pellet group count and visual exposure to roads and structures.

In light of the lack of spatial distribution among sampling sites and the small number of factors considered, however, the significance of these results was felt to be questionable at best.

Despite this, the potential for applying computer graphic techniques in this area was clearly demonstrated. To the extent that this exercise might be characterized as a tool in search of an application, it did succeed in finding one.

REACTION

Local reaction to the substantive content of this phase of the deer population study has focused with appropriate skepticism on the informality of the analyses performed. Reaction to the general approach demonstrated by this work, however, has been very enthusiastic. Particularly encouraging is the variety of projects making use of the Map Analysis Package at the University of Connecticut and at Yale. Among these are studies of bowhead whale populations in the north Pacific (Friedlander, 1979); airborne lead transport in Idaho (Von Lindern, 1979); spruce budworm infestation in Maine (Thorpe, 1979); the effect of microtopography on vegetation in New Hampshire (Gemborys, 1979); and watershed management (S.M. Tomlin, 1979), highway corridor siting (Kennard, Lefor, and Civco, 1979), and land use (Berry et al., 1979) in Connecticut. Developmental versions of this particular software are also now being used by several other universities and environmental consulting firms. Favorable reaction to the Map Analysis Package has generally focused on its intuitive command language, its range of modeling capabilities, and its organization as a series of independent "building block" operations.

HINDSIGHT

In retrospect, several points have become increasingly clear. Had the initial phases of the deer habitat study anticipated the present concern for spatial effects, an alternative sampling strategy would almost certainly have been chosen. A larger set of data would also have been considered.

In terms of its ulterior motives in demonstrating the wonders of digital cartography, however, the simplicity of this study has proved to be one of its major assets. This has been the case not only in terms of software accessibility and analytic informality but also in terms of the conceptual simplicity of the grid-cell data structure.

TRANSFERABILITY

Most of the data-processing capabilities used in this study can be found in many of the increasingly large number of packaged geographic information systems currently available. Experience at Yale and elsewhere suggests that such capabilities, if packaged in a sufficiently generalized manner, can accommodate a surprisingly wide range of applications.

Versions of the Map Analysis Package itself have so far been implemented on IBM, DEC, and HP computers. A CDC version may also be released. The package will soon be made generally available through Yale University.

BIBLIOGRAPHY

Berry, J.K. and C.D. Tomlin, 1979. "An Academic Approach to Cartographic Modeling in Natural Resource Management." Second International User's Conference on Computer Mapping Hardware, Software, and Data-Bases.

Berry, J.K.; C.D. Tomlin; A. Friedlander; A. Howard; P. Korotky; P. Lewis; T. Niemann; J. Sailor; P. Scanlon; J. Tam; J. Thorpe; and I. Von Lindern, 1979. "Guilford Project Report: Cartographic Analysis of Residential Development Plans and Identification of Potential Landfill Sites." Yale School of Forestry and Environmental Studies.

Friedlander, A.I., 1979. "Analysis of Spatial Distribution of the Western Arctic Bowhead Whale (*Balaena mysticetus*) Population: Computer Mapping." Yale School of Forestry and Environmental Studies.

Gemborys, S.R. 1979. "Tree Distribution in Relation to Microtopography: The Bowl, Wonalancet, New Hampshire." Harvard Forest, Harvard University.

Kennard, W.C.; M.W. Lefor; and D.L. Civco, 1979. "Computer-Assisted Analysis of Earth Resources Information." University of Connecticut Department of Natural Resources Conservation.

Lyons, J.R., 1978. "The Status of Research on the Yale Forest, Tolland and Windham Counties, Connecticut." Yale School of Forestry and Environmental Studies.

Lyons, J.R., 1979. "Union Forest Deer Study: Report 2." Yale School of Forestry and Environmental Studies.

Thorpe, J.P., 1979. "Site Factors Affecting Spruce Budworm Infestation in Maine." Doctoral dissertation in preparation, Yale School of Forestry and Environmental Studies.

Tomlin, C.D. 1979. "Cartographic Modeling Techniques in Environmental Planning." Doctoral dissertation in preparation, Yale School of Forestry and Environmental Studies.

Tomlin, S.M. 1979. "A Digital Cartographic Data-Base for the Holcomb Farm Watershed." University of Connecticut Department of Natural Resources Conservation.

Von Lindern, I.H. 1979. "An Empirical Air Quality Model for Heavy Metal Contamination Surrounding a Primary Lead Smelter." Doctoral dissertation in preparation, Yale School of Forestry and Environmental Studies.

```
OK, MAP ANALYSIS PACKAGE READY

.
relumber covermap for development /
NO, THERE IS NO OPERATION NAMED "RELUMBER";
        FOR HELP, TYPE "EXPLAIN"

.
renumber covermap for development assigning 0 to 1 /
                thru 7 and 1 to 8 thru 9
OK, "DEVELOPMENT" CREATED BY RENUMBERING "COVERMAP"
        ASSIGNING        0 TO 1    = WET LAND
                         THRU 7    = HEMLOCK
        ASSIGNING        1 TO 8    = STRUCTURE
                         THRU 9    = ROAD

.
renumber covermap assigning 0% to 4, to 5, to 7 /
        and 100% to 0 and 200% to 6 and 300% /
        to 3 and 600% to 1 thru 2, to 8 thru 9 /
        for friction
OK, "FRICTION" CREATED BY RENUMBERING "COVERMAP"
        ASSIGNING    100 TO 0    = NON-YALE LAND
        ASSIGNING    600 TO 1    = WET LAND
                     THRU 2      = OPEN FIELD
        ASSIGNING    300 TO 3    = RED PINE
        ASSIGNING      0 TO 4    = HARDWOODS
                     THRU 5      = MIXED GROWTH
        ASSIGNING    200 TO 6    = WHITE PINE
        ASSIGNING      0 TO 7    = HEMLOCK
        ASSIGNING    600 TO 8    = STRUCTURE
                     THRU 9      = ROAD

.
spread development thru friction for proximitymap
OK, "PROXIMITYMAP" CREATED BY SPREADING "DEVELOPMENT"
        THRU "FRICTION"

.
contour proximitymap
OK, "PROXIMITYMAP" DISPLAYED AS A CONTOUR PLOT
```

AN OPERATIONAL GIS FOR FLATHEAD NATIONAL FOREST

Judy A. Hart, David B. Wherry
Washington State University Computing Service Center
Pullman, WA 99164-1220
and
Stan Bain
Flathead National Forest
P.O. Box 147, Kalispell, MT 59901

ABSTRACT

Flathead National Forest is utilizing Geographic Information System techniques to establish an updateable, forest-wide, geographic database. Hungry Horse Ranger District, discussed in this paper, is the first of five ranger districts to be completed. GIS data layers for the Hungry Horse District include: digital terrain data (elevation, slope and aspect); Landsat data classified by spectral class and then stratified by elevation and aspect to depict vegetation associations; timber compartments; timber harvest history; land types; land ownership; administrative areas such as wilderness areas, recreation areas, wild and scenic areas, and experimental forest boundaries; mean annual precipitation; drainage basins; streams, rivers, lakes and islands; roads and trails; as well as data planes depicting distances to water and distances to roads. Techniques used to compile the database are discussed; various data planes are depicted through illustrations; and, applications of the GIS database are described.

INTRODUCTION

Flathead National Forest (FNF) encompasses approximately 2.6 million acres in northwest Montana (Figure 1) with Forest headquarters in Kalispell. Bordered on the north by Canada and on the northeast by Glacier National Park, FNF is confronted with the problem of managing an area with diverse concerns. To balance the needs of timber, wildlife, and recreation in an economically and ecologically sound manner, it is increasingly important that map and tabular information be available to resource managers. FNF has adopted the use of a Geographic Information System (GIS) as one means of supplying information in a timely and economical fashion.

Primary data compilation for FNF's GIS database started in 1982 in the Hungry Horse Ranger District (HHRD). The completed HHRD database has served as a demonstration of GIS capabilities to other ranger districts and regional Forest Service offices. In 1984, Forest managers scheduled the establishment of 17 to 25 primary data planes for each of FNF's four remaining ranger districts over the next three years. Management problems differ from one ranger district to the next. As such, GIS data planes will be compiled to meet the needs of each individual ranger district.

Among Forest Service GIS efforts, this project is unique in two respects: the Forest Service has not acquired any additional computer hardware or software to reach the present operational stage; and,

Forest Service personnel perform all project work rather than having the project contracted to an outside service agency. Personnel at FNF headquarters perform Landsat classification, map digitization, data management, project implementation coordination and applications through data plane manipulation. As the database for each ranger district reaches the operational stage, ranger district personnel receive training in data plane combination and query techniques enabling data manipulation capabilities at the ranger district office. During the planning and startup stages, the overall project design and guidance with implementation techniques required collaboration with data processing and image analysis consultants at Washington State University Computing Service Center (WSUCSC). The need for consulting from WSUCSC staff diminishes steadily as Forest Service personnel gain skill and familiarity with the systems.

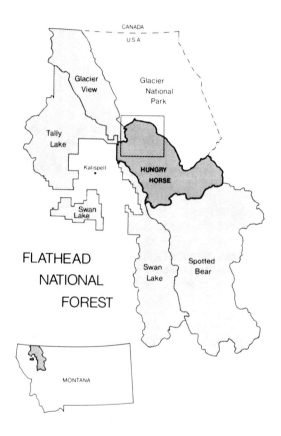

Figure 1: Flathead National Forest. The boxed area at the north portion of the Hungry Horse Ranger District delineates the area shown in the remainder of the figures.

METHODS

Processing and manipulation of the database is accomplished through the facilities in the Digital Image Analysis Laboratory (DIAL) at the Computing Service Center in Pullman, Washington. DIAL systems include Video Image Communication and Retrieval/Image Based Information System (VICAR/IBIS), a batch oriented system, which runs on an Amdahl 470/V8 mainframe; and, an International Imaging Systems-System 511, an interactive system, which is resident on a dedicated PDP 11/34 mini-computer. The FNF's Data General NV8000 is used as a Remote Job

220

Entry/Remote Job Output station via teleprocessing to WSUCSC's mainframe environment.

VICAR/IBIS, the primary processing system used in this project, utilizes raster-based GIS techniques. Raster format data can be conceptualized as a two-dimensional matrix of data elements called pixels. Each pixel stores one data value and the location of the pixel in the data matrix denotes its geographic location. The GIS data planes, a collection of spatially corresponding raster files, are manipulated mathematically using digital image processing algorithms (Jaffray, Hansen & Hart, 1984).

Data Acquired in Digital Form

Defense Mapping Agency (DMA) digital terrain data and Landsat Multispectral Scanner (MSS) data were acquired in computer compatible format on magnetic tapes. Both DMA and Landsat datasets are standard products distributed by U.S. Geological Survey and National Oceanographic and Atmospheric Administration respectively. For compatibility with VICAR/IBIS, the format of each dataset was restructured with specialized VICAR/IBIS "logging" programs (Hart & Wherry, 1984).

Digital Terrain Data Processing. Four 1 x 1 degree DMA digital elevation blocks were geometrically corrected to a Universal Transverse Mercator (UTM) map projection and mosaicked into a 2 x 2 degree, 50 by 50 meter pixel image encompassing the HHRD. Best results were achieved in subsequent processing when DMA 16 bit pixel accuracy was preserved during this process.

Images depicting slope magnitude and slope direction (aspect) were modelled from the DMA elevation mosaic. Slope and aspect models were compressed to 8 bit pixel accuracy for storage efficiency and data generalization purposes. Data were represented as 40 vertical feet per pixel value; 1 degree of slope per pixel value; and 5 degrees of declination per pixel value on the final elevation, slope, and aspect GIS planes.

Landsat 2 MSS Data Processing. Landsat acquisition 22354-17420 (03 July 84, EDIPS format) encompasses all of the HHRD and Glacier National Park. This scene exhibited no cloud cover or data anomolies. Landsat classification and subsequent stratification procedures were performed by FNF personnel in cooperation with Glacier National Park researchers who are developing a GIS database in parallel with FNF (Haraden, 1984, and Key, et al., 1984).

The multispectral classification techniques employed in this project most nearly resemble the multi-cluster blocks technique described by Hoffer (1979). Numerous training areas of known cover type were processed through a clustering algorithm and the consequent spectral statistics were evaluated and compared. From several hundred original spectral classes, 99 were chosen for the final classification statistics set. The resulting classification of the entire Landsat scene did not adequately discriminate between land cover types. The spectral classes contained confusion between water and shadow, inadequate distinction between agricultural classes and natural grasslands, and confusion between certain timber categories and shrubs.

The 99 spectral classes were combined with elevation and aspect data forming 3,600 unique spectral-elevation-aspect categories. Contingency tables were constructed to correlate spectral-elevation-aspect categories with ground data and each spectral-elevation-aspect category was assigned to one of 189 land cover classes. This stratification process alleviated the confusion problems inherent in the multispectral classification and improved the quality of the classified Landsat product.

The southern two-thirds (the United States portion) of the classified, stratified Landsat scene was geometrically corrected to a UTM map projection at a 50 by 50 meter pixel resolution. A set of 38 control points (approximately one control point per 7-1/2 minute quad area) were selected in the DIAL facility by FNF and Glacier National Park personnel. Figure 2 shows a black and white rendition of the 189 land cover categories in the geometrically corrected, stratified, Landsat classification.

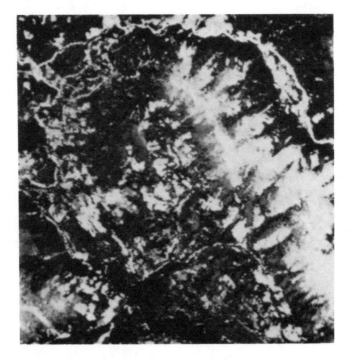

Figure 2: Landsat Classification. Stratified by elevation and aspect, registered to UTM map projection; 189 classes.

Data Digitized from Maps

Maps are digitized at the FNF headquarters offices in Kalispell using a Numonics model 2401 coordinate digitizer which is on-line to the Data General computer. Line segments are recorded in tablet inches, converted to latitude/longitude and subsequently to UTM coordinates, then edges are matched between separate quad sheets. The data files are mailed to WSUCSC on computer tape for incorporation into the GIS database. All but two data planes originated from 1:24,000 USGS quads.

Timber Harvest History. Records of timber harvest in the HHRD date back to 1943. The timber harvest data plane identifies each area by year and type of activity (thinning, selective cut, clearcut, etc.). For purposes of manipulation, summation, or display, data can be grouped as needed by any permutation of year and type of activity. Examples of year/type combinations are: all clearcuts of 1982; all areas which were thinned between 1970 and 1980; or, all types of logging operations for 1954.

Landtypes. Landtypes, outlined in Table 1, identify the opportunities and limitations associated with the physical characteristics of FNF. The landtype classification system is based on geologic processes (as reflected by physiography), soil types, and the factors which determine the behavior of ecosystems (i.e., climate, vegetation, relief, parent materials, and time) (Proposed Forest Plan, 1983). The Landtype data was derived from 1:63,360 scale maps.

```
| Table 1:  Landtype Classification System
|
| Group  1 -- Flood Plains
| Group  2 -- Wetlands
| Group  3 -- Avalanche Fans, Talus
| Group  4 -- High Alpine Basins
| Group  5 -- Mass Failure Lands
| Group  6 -- Alluvial Fans, Outwash Plains, and Reworked Tills
| Group  7 -- Silty Glacial Tills
| Group  8 -- Sandy Glacial Tills
| Group  9 -- Clayey Glacial Tills
| Group 10 -- Steep Glacial Tills
| Group 11 -- Residual Soils, 20 to 40% slopes
| Group 12 -- Residual Soils, 40 to 60 % slopes
| Group 13 -- Alpine Ridges and Rocklands
| Group 14 -- Glacial Trough Walls
| Group 15 -- Fluvial Breaklands
| Group 16 -- Structural Breaklands
```

Land Ownership. The Land Ownership data plane delineates private land holdings within the FNF boundaries. In the HHRD there are 16,600 acres of private land.

Administrative Areas. The Administrative Areas data plane delineates areas such as the Coram Experimental Forest (8,020 acres), Wild and Scenic River System (13,838 acres), and wilderness and primitive areas. The HHRD shares administrative responsibilities with other ranger districts for the Great Bear Wilderness (286,700 acres) and the Jewel Basin Hiking Area (15,368 acres).

Precipitation. Mean annual precipitation data derived from snow surveys and Soil Conservation Service historical records were drafted on a 1:63,360 scale map. The precipitation isolines were then digitized and entered into a GIS data plane. Precipitation in the FNF ranges from 20 inches to 120 inches per year.

Timber Compartments. Timber Compartments (Figure 3) are management units of timber stands used for timber inventory purposes. Boundaries of these units were determined by ridges and streams. The Timber Compartments data layer correlates this database to the Forest Service Timber Stand Database at Fort Collins, Colorado.

Drainage Basins.   The Drainage Basin data plane shows the watershed boundaries in the HHRD (Figure 4).   Information from the GIS database is often summarized by watershed area using a polygon overlay process.

Roads and Trails.   Figure 5 illustrates the road network in the HHRD.   Line segments denoting segments of road are classified by seven different schemes as shown in Table 2.   Trails occupy a separate data plane as do each of the seven road classification schemes.

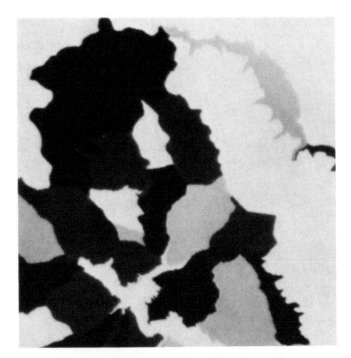

Figure 3: Timber Compartments. Management units for timber inventory.

Figure 4: Drainage Basins. Watershed boundaries.

Figure 5:
Road System.
All existing roads.

Table 2:   Road Classification Schemes

Road Function Class		Road Service Level	
1	Other	1	paved, double lane
2	Arterial	2	gravel, double lane
3	Collector	3	gravel, single lane
4	Local	4	dirt

Road Management Reason		Road Management Jurisdiction	
1	public safety	1	County
2	resource protection	2	State
3	soil and watershed	3	Forest Service
4	management direction	4	State forest
5	wildlife	5	private
6	administrative	6	Class D permit

Road Management Device		Road Management Date	
1	Barrier	1	closed year long
2	Controlled gate	2	closed 4/1 - 7/1
3	Service barrier	3	closed 10/15 - 11/30
4	Gate	4	closed 12/1 - 5/15
5	Signed	5	closed 10/15 - 5/15
6	Controlled by sign	6	closed 10/15 - 7/1
7	Controlled by barrier	7	closed 9/1 - 7/1

Road Maintenance Level	
1	custodial care only
2	limited passage
3	low usage/care
4	moderate usage/care
5	maximum usage/care

Hydrology. Streams are recorded as first through fifth order, perennial or ephemeral. Line segments denoting the South and Middle Forks of the Flathead River are given different values; and, islands are differentiated from lakes. Figure 6 illustrates the hydrology data plane.

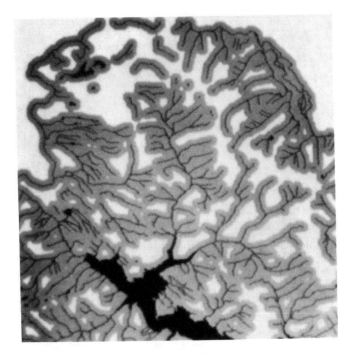

Figure 6:
Hydrology. Streams, rivers, lakes and islands; 600 meter distance to water is shaded.

Distance Function Data Planes. Graduated distance corridors were modelled from the hydrology and roads data planes. Pixel values in the distance corridor image indicate how far any pixel is from a road or hydrologic feature. The distance modelling is accomplished with iterative applications of a convolution filter to the data plane of interest. Corridors identify distances between 0 and 1,200 meters.

APPLICATIONS OF THE GIS

In one application utilizing the HHRD database, a data plane depicting distances of zero to 600 meters from water was digitally overlaid with another data plane delineating timber harvest areas. The resulting image and acreage summary quantifies areas which present erosion hazards in that particular watershed.

A road viewshed plan presently being constructed will be used by the FNF landscape architect to design scenic and recreational qualities along a heavily traveled reach of highway. In this application, a distance corridor around arterial roads is superimposed with information from the data planes delineating hydrology, timber harvest, and generalized Landsat classification. Information from these images will be used to plan management of timber harvest methods along the road to enhance the recreational and visual experience and provide a pleasing spatial arrangement to the traveler.

Generation of a map showing areas critical for elk calving habitat is a good example of the utility of the FNF GIS. Modelled after Langley

(1983), Table 3 outlines the selection criteria used in manipulating the HHRD data layers. An acreage summary of the three suitability categories revealed 253 acres of Optimum, 1,798 acres of Acceptable, and 94 acres of Marginal elk calving areas in the HHRD. The data plane depicting elk calving areas will be used in conjunction with the distance to road data plane to evaluate road closure in areas of critical habitat.

Table 3: Elk Calving Criteria

Optimum
  elevation: 4,000 - 4,800 feet
  aspect: flat or south facing
  proximity to water: less than 300 meters
  vegetation: non-forest; less than 50 acres in size

Acceptable
  elevation: 3,500 - 5,000 feet
  aspect: southwest or southeast facing
  proximity to water: 300 - 750 meters
  vegetation: non-forest; greater than 50 acres in size

Marginal
  elevation: above 5,000 feet
  aspect: west or east facing
  proximity to water: 750 - 1,200 meters
  vegetation: timber

CONCLUSIONS

The demand for information from the FNF GIS database grows steadily with each successful application of manipulation and query capabilities. The implementation of this GIS has passed the all important test of user acceptance and is becoming an integral part of strategies for accessing resource management alternatives.

The resolution and accuracy of the GIS database meets or exceeds the quality of FNF databases constructed by traditional methods. In addition, the computer assisted GIS approach affords FNF personnel a quicker and more economical means of updating the database. Manual cartographic updates are replaced by data file editing and rerunning a series of computer jobs.

FNF personnel estimate that the GIS will save a significant amount of money and time and provide the ability to explore a larger variety of management alternatives. The GIS will allow planners and managers to analyze alternatives on critical lands where vegetative manipulation and other project proposals can achieve the desired objectives of land management economies and efficiencies.

REFERENCES

Haraden, R.C. 1984, "Development of Non-Renewable Resources" in Proceedings of Conference on the Management of Biosphere Reserves, Great Smoky Mountains National Park, Gatlinburg, TN

Hart, J.A. and D.B. Wherry ed., 1984, VICAR/IBIS User Reference Manual, Volumes 1-4, Washington State University Computing Service Center, Pullman, WA 99164-1220, 876 pages

Hoffer, R.M. 1979, "Computer-aided Analysis of Remote Sensor Data -- Magic, Mystery, or Myth?" in Proceedings of Remote Sensing for National Resources, University of Idaho, Moscow, ID, pp. 156-179

Jaffray, M.J., R.S. Hansen, and J.A. Hart 1984, "A Geographic Information System Application for Spokane County" in Proceedings 22nd Annual Conference of the Urban and Regional Information Systems Association, Seattle, WA, pp. 154-163

Key, C.H., S. Bain, D.R. Killerude, and D.B. Wherry 1984, "Regional Databases through Interagency Cooperation" poster presentation at Conference on the Management of Biosphere Reserves, sponsored by UNESCO MAB Secretariat and National Park Service, Great Smoky Mountains National Park, Gatlinburg, TN.

Langley, P.G. 1983, Multiresource Inventory Methods Pilot Test, Final Report, U.S. Department of Agriculture, Forest Service, Nationwide Forestry Applications Program, Houston, TX, Report No. NFAP 309

Proposed Forest Plan 1983, U.S. Department of Agriculture, Forest Service, Flathead National Forest, Kalispell, MT, p. II-28

BIOGRAPHICAL SKETCH

Judy A. Hart is an Image Processing Analyst, Graphics and Image Analysis Group, Washington State University, Computing Service Center, Pullman, Washington.

David B. Wherry is the Project Manager of the Graphics and Image Analysis Group at Washington State University, Computing Service Center, Pullman, Washington.

Stan Bain is Supervisory Cartographer Technician, Flathead National Forest, USDA Forest Service, Kalispell, Montana.

PART 7
GLOBAL STUDIES

CONCEPTS FOR A GLOBAL RESOURCES INFORMATION SYSTEM

F.C. Billingsley and J.L. Ureña

Jet Propulsion Laboratory, California Institute of Technology
Pasadena, California

ABSTRACT

The objective of the Global Resources Information System (GRIS) is to establish an effective and efficient information management system to meet the data access requirements of NASA and NASA-related scientists conducting large-scale, multidisciplinary, multi-mission scientific investigations. Using standard interfaces and operating guidelines, diverse data systems can be integrated to provide the capabilities to access and process multiple geographically dispersed data sets and to develop the necessary procedures and algorithms to derive global resource information.

INTRODUCTION

Future National Aeronautics and Space Administration (NASA) flight missions and, in particular, the planned Space Station will provide an unprecedented potential for advancement in the Earth Sciences in the next decade. However, the new requirements that will be levied by the scientific community will also create a challenge for the implementation of information systems that will improve scientific productivity to a level that allows for the realization of this potential.

In man's attempt to understand the Earth and its environment, enormous amounts of data are being gathered from many different sources. The extreme dispersion and sheer quantity of this data historically have made it extremely difficult for space and terrestrial applications scientists to access data of particular relevance to their work or even to learn of its existence. Problem areas of greatest concern have been the poor quality, poor usability, and inaccessibility of relevant data.

The demands upon NASA's information interface to the user community are evolving along with the research objectives. The complexity of the environmental and resource processes that affect the habitability of the Earth requires a multidisciplinary approach to the understanding of the natural phenomena and their dynamics. Increasingly, the requirements involve comprehensive, multi-mission, multidisciplinary investigations that require data from many sources, both from space sensors and from conventional sources. In addition, it has become apparent that certain disciplines interact, using parameters generated within the purview of other disciplines. There are commonalities of needed source data and parameters among the uses and users in many Earth Science discipline areas. One study[1] shows that, considering all disciplines and parameters, on the average over 3 1/2 disciplines are served by each parameter derived, and each parameter could be measured by over ten sensors.

The task of locating and assembling data from several heterogeneous sources, and then integrating the data into usable correlated data sets, has become a technically formidable undertaking. This work frequently consumes a sizable fraction of the total resources available for a research investigation.

The earth-watching satellites of NASA and other national and international agencies provide a wide spectrum of detailed data about the earth's resources. Heretofore, the state of the art in information science was such that, to perform broad-scale scientific research, drawing upon multiple data sets from different sensors and satellites meant long, laborious, expensive periods of data gathering and manipulation. Today, with advances in the technologies of data storage, networking, and data management, it is possible to develop a system within practical costs to allow efficient remote access to multiple distributed data bases and to enable productive global-scale research.

In recognition of this growing need, the House Committee on Science and Technology of the United States Congress directed NASA to prepare a "preliminary program plan for a Global Resources Information System utilizing an interactive network of relevant data bases including program scope, technology, needs, and resource requirements."

In 1978, recognizing the need and the technological opportunity, NASA initiated a coordinated, multidisciplinary activity aimed at making satellite data accessible on-line to the NASA applications research community and providing within the system the processing capabilities necessary to enable global-scale, multidisciplinary research based on multi-sensor data sets. This activity involved the systematic development of improved techniques for data-base management, networking, data standards, and advanced processing systems. The scope of the task is large and complex; therefore, it was structured around a series of distinct science-discipline-oriented "pilot systems" and supportive technology development tasks.

Today, pilot systems have begun to serve the needs of the atmospheres/climate and oceanographic communities; the Pilot Climate Data System and the Pilot Ocean Data System are now operative and supporting scientific research. This activity continues with the development of two new pilot systems oriented toward the land-related science (Pilot Land Data System) and planetary science (Pilot Planetary Data System) research communities, respectively, together with on-going work with applications disciplines to develop and demonstrate improved tools and techniques for researchers in earth resources, oceanography, atmospheric physics, climate research, and planetary science.

These pilot information systems, together with the supporting technology developments necessary to accomplish the stated goals, can provide the foundation for a Global Resources Information System (GRIS), a broad-based, multidisciplinary information system to facilitate integrated research involving global data sets potentially available from satellite remote sensing and other conventional sources to address global processes involving the land, air, and water.

In its initial stages, the Global Resources Information System will serve NASA and NASA-supported and cooperating scientists involved in applied science research within selected universities and other U.S. institutions and federal agencies. However, the global nature of the GRIS concept and of future flight missions (like the Space Station) that it will be supporting dictates that the scope will have to be broadened increasingly to include scientists and institutions with data bases and research interests in key foreign countries as well.

The range of science disciplines and investgators involved, the number of agencies that will need to cooperate and/or participate, and the international scope of the Global Resources Information System pose a significant challenge. However, it is a challenge that must be accepted if we expect to be prepared to use the full potential that space systems of the next decade represent for the advancement of the sciences of the Earth and the understanding of our planet.

SCIENCE RATIONALE

The science rationale for a Global Resources Information System has at its foundation the concept that the Earth is one large, complex life-support system made up of discrete, interacting parts. The functioning and interrelations of these parts (lithosphere, hydrosphere, biosphere, and atmosphere) is not clearly understood. In addition, humanity is subtly and surely introducing changes that impact the balance between and within these parts. As mankind's technological potential has expanded, so too has man's inherent responsibility to employ technology wisely. To many scientists involved in research on all aspects of the Earth's global resource base, it has become clear that in this century, the capability to significantly alter this planet's environment on a global scale has finally been achieved. What may be less apparent to these same environmental scientists is that research has also developed new tools and techniques that offer the potential to observe, analyze, and perhaps control the end results of action affecting the environment. To date, humanity's actions regarding the environment have been haphazard: Man has acted first and only later recognized and attempted to assess the consequences of those actions. Organic carbon deposited by natural processes over millions of years has been extracted from the earth in dozens of years. Until recent years man has used energy stored in fossil reserves, depositing vast quantities of carbon in the atmosphere (estimated to be about four billion tons per year) without regard to the consequences on climate.

The effects of man's actions may be subtle. Agricultural practice, favoring cultivation of legumes, enhances the rate at which nitrogen is fixed by natural processes. In combination with nitrogen fixed inadvertently by combustion or deliberately during the manufacture of fertilizers, this nitrogen may put this planet's global life-support system in new domains, with as yet unpredictable consequences for the air, sea, soil, and biota. Humanity's actions may also be direct. The greatest feat of global engineering, conversion of an estimated 10% of total planetary land area from natural vegetation to agriculture, was carried out without regard to the large-scale consequences. Early farmers did not file environmental impact statements. Even if they had done so, their documents probably would have provided little insight into the long-range consequences of agricultural land conversion. The scientific and technical basis for such a global assessment was lacking then, and to a real extent, is still lacking.

This has been in part due to a lack of adequate data of the type, scale, and quality to address scientific issues that are global in nature. In addition, the complex nature of these problems requires that a multidisciplinary approach be taken. In a society in which specialization has been the modus operandi, interdisciplinary research of the type needed to address global problems is rare indeed. It is, then, a basic premise supporting the need for the Global Resources Information System, that these large-scale (global) problems require multidisciplinary research. It also seems that little truly multidisciplinary, large-scale research is occurring.

Reasons for this lack of research are many, varied, and complex. They range from basic technical problems associated with the collection, archiving, networking, and processing of large data volumes to fundamental factors involved in having scientists with diverse backgrounds meaningfully interact to achieve research goals. Yet, it is just this type of research, in which satellite and aircraft remote sensing is supported by surface sampling and backed by effective information systems, that can provide significant benefits today. Some examples of global-scale problems that could benefit from such a multidisciplinary, remote-sensing, total-information science approach are described below. The fundamental characteristics of these problems are that they are global in scale and require a concerted, long-range research effort to achieve significantly improved understanding.

In each of these examples of global problems, global models are required. Currently, global models for the most part are not available to support many lines of scientific inquiry, and, in many cases, those that are available are inadequate. Global models inherently require global data for their solution. Use of global data requires that they be planned for, acquired, then located and assembled. In this global modeling problem, even if the models were formulated, the quality data required to fully support them are not currently available.

Global Carbon Dioxide Cycle

Several large models are required: Global atmospheric circulation, interaction with the ocean, effect of terrestrial vegetation, release from fossil fuels, other sources of natural carbon dioxide, biological productivity, and relation to other global chemical (P,N,S) cycles, to name several. For this discussion, the thrust is whether remote sensing can provide the necessary data to solve these models, and what is necessary for an adequate information system. This will require such varied information as circulation patterns; latitudinal temperature variations; forest distributions by types and rates of change; evaluation of cultural practices as they affect forest product disposition of carbon in various forest regimes; carbon transfer at the air-sea interface as pH, temperature, and sea state vary; ocean precipitation of carbonates; and fossil-fuel practices. The multidisciplinary nature of the problem and wide gamut of data required must be addressed.

Acid Rain and Pollution Dispersion.

Information concerning circulation patterns is required. In addition, knowledge will be required of a number of individual parameters: fuel components and their dispersion, depending upon energy practices; man-made sources such as automotive combustion; possible natural sources; effects on soil pH and water pH; efficacy of natural buffering; the identification of specific sources by trace components; and the relative importance of S and N. Can remote sensing of

trace materials be of assistance? Given the problems of repeat coverage due to orbit constraints, are better orbits possible? How about a synchronous polar orbit repeatedly going across Kansas with trace pollutant sensors? Could laser-based, absorption-line detectors be of use? If one addresses this problem, what would one do specifically; and what remote sensing, supported by what type of information system, would be needed?

Biogeochemical Cycle

The biochemical cycle is the very foundation of our existence. To paraphrase recent studies: "In the coming years, we will be able to predict how much food the Earth will produce while most of that potential food is still 'on the hoof' or in the ground; this will help us to feed the hungry. We will know in advance how much fresh water, timber, oil, and metallic minerals we can safely consume--this will help us to harmonize growth...."[2] Again, a multi-disciplinary approach is required as well as the use of global models and a wide variety of data. If we strip-mine Montana for coal, for example, how soon can the wheat and cattle culture again become practical? The concern is with both aquatic and terrestrial cycles and their vitality with changes in their environment: e.g., upwellings; water temperature and other coastal processes; water circulation; El Nino effects; any interactions with the global CO_2 problem; effects of soil pH; rainfall (acid or not); hydrology and water-table problems; primary productivity in the wetlands; leaf chemistry; and the nutrient value in crops and development of crops to suit environmental conditions. Figure 1, from a recent

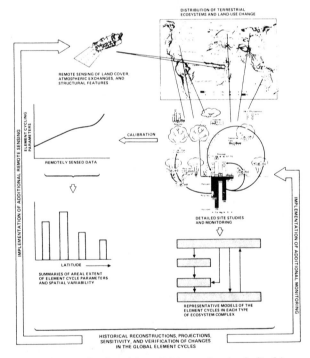

Figure 1. Relationship of Biogeochemical Cycling Research Areas

NASA study on Global Habitability[3], illustrates the interdisciplinary nature of the problems and suggests the use of the remote sensing component. To aid in the understanding of the biogeochemical cycle and the renewable resources problems, how can a coordinated (across disciplines) remote sensing/information systems approach aid in solving the present models and allow new models, now untenable, to be developed?

Deforestation, Desertification, and Habitabilty

Remote sensing has aided in the understanding of the problems of deforestation, desertification, and habitability. Again, the problem is multidisciplinary. Some of the factors involved include soil conditions; general rainfall and temperature conditions; population pressure and cultivation practices; forest stripping for fuel, wood products, or for other uses of the land; land suitability for other uses; and changes in rainfall due to changes in vegetative cover. This problem interacts with and uses many of the same data as the biogeochemical-cycle problem: the global CO_2 and weather and climate problems. Worldwide vegetative maps can be assembled, showing changes over time. Worldwide data bases of temperature, rainfall, and other parameters can be assembled. Many of the problems are largely social. In this context, a question is: How can the social sciences benefit from remote sensing with an overall information science context?

Weather and Climate Monitoring and Prediction

Weather and climate monitoring and prediction have been a major contribution of remote sensing. There is no need for an apologia for this use. The emphasis for this discussion is the interaction of studies conducted in this discipline with the others, for example, the use of global circulation and precipitation models with the global CO_2 and habitability problems. Research has shown the potential of correcting Landsat images by flagging areas of high cirrus clouds, which distort normal spectral responses. Can the data distribution techniques developed for this discipline be used to retain more of the old weather data for historical studies?

Other major problems could be discussed, such as: global navigation, global sea conditions and the effects of shipping, ice processes, global topographic mapping, physical oceanography, basic geodesy. All have been, and will be, aided by remote sensing and advanced information systems.

The need for global research is vital. The potential provided by remote sensing and information systems is high. Indeed, when thoughtfully analyzed, the two technologies are inextricably linked. Remote sensing data must reach a user in an appropriate form and time frame. It often achieves its maximum effectiveness when combined with other data in spatially referenced (geographic) information systems. Such information systems achieve their maximum effectiveness when the data they contain are accurate and current. The linkage of these two technologies has not occurred because of fundamental incompatibility. This linkage is a matter of necessity if these global issues are to be realistically addressed.

PRELIMINARY SYSTEM CONCEPT

The prime motivation for the GRIS is the establishment of an effective data-management system to meet the data-access requirements of the scientific community in its multidisciplinary and multi-mission investigations in the Earth Sciences. Such investigations, commonly known as correlative research, use data from many sources for a wide range of applications in a variety of disciplines.

In addition, every discipline can use data common to other disciplines. For example, the climatologist is interested in hydrology, land cover, the upper atmosphere, solar physics, and indeed, in the behavior of other planetary atmospheres as well as the behavior of the Earth's atmosphere and oceans. Each discipline generally must make some of its observations on its own, yet each should not have to make all its observations on its own.

Vast amounts of data on the Earth and its environment are being collected in centers throughout the nation and around the world. However, investigators are, for a variety of reasons, often often unable to use more than a small fraction of the existing relevant data. The potential data consumer (investigator) must determine:

(1) What data collections exist that have data relevant to the investigation?

(2) Which specific data sets within the relevant collection are pertinent?

(3) How can the specific data sets be obtained?

(4) What must be done to convert the data into a useful format?

(5) What calibrations and corrections have been applied? What are the known problems with the data?

(6) What support services and background information exist that would facilitate multi-user, multidisciplinary use of the data?

Because each of the several data sources used in an investigation may have its own access protocols and product forms, the overall process of finding out what data are available, obtaining the data, and integrating the data into a compatible set is currently a technically formidable task. Frequently, this task results in the development of extensive application-unique data access and conversion capabilities.

The GRIS concept is a strategy for lowering the information exchange barriers that exist between participating producers of data and information, and the existing and potential users of that data and information. Using standard interfaces and operating guidelines, diverse data systems can be integrated to provide the capabilities to access and process multiple geographically dispersed data sets and to develop the necessary procedures and algorithms to derive global resource information.

To accomplish these objectives, a completed GRIS, as currently envisioned, must be capable of performing the following functions:

(1) Interface to a geographically dispersed, heterogeneous user community via a common-carrier, electronic communications network.

(2) Provide access to data bases that contain both space and terrestrial data in varying formats and that are managed by dissimilar data management systems.

(3) Provide substantial ease and uniformity of user access to the data bases.

(4) Provide extensive assistance in the location of data.

(5) Enable data to be transferred in a timely and efficient manner.

(6) Allow concurrent access to the service by multiple remote and local interactive users.

(7) Allow full control by participating organizations over access to and utilization of their computing and data resources.

(8) Minimize time, cost, and operational impact of adding (or removing) users or data centers to the GRIS.

(9) Develop and maintain data and communication standards.

(10) Support on-line, cataloged processes to perform special functions, allowing controlled computing resource sharing when appropriate.

(11) Perform services such as data integration or active archiving of data, when such services are desired.

(12) Perform the administrative and management functions necessary for the coordination of the information network.

The intent of the GRIS is to supplement and coordinate the functions performed by the various member data systems to which it is related, not to duplicate those functions. The capabilities as listed above will come about as the GRIS encourages shared, mutual development amongst the Pilots and equivalent non-NASA data systems, and supplements these as required to provide capabilities beyond those provided by the member systems. A further intent is to recognize and, to as great an extent as possible, use the results of the extensive on-going work by other organizations such as the Consultative Committee for Space Data Systems (CCSDS), International Standards Organization (ISO), the data base community, and the communications community.

An important feature is that the data sets accessed via GRIS will remain under the management and control of the organizations and data centers in which they reside. No data will be managed directly by the GRIS, with the exception of catalogs and directories of information about available data bases and participating organizations and centers and their resources, unless a specific archive is placed under GRIS management for special purposes.

Discipline-oriented data will reside in discipline-oriented systems. These discipline-oriented systems are based on the concept of a Scientific Data Management Unit (SDMU)[4]. An SDMU is an organizational entity responsible for developing and maintaining a "live" data archive in support of a specific scientific objective. A live archive contains data that are in active use, together with a knowledgeable support staff. The staff makes decisions about the structure, organization, and content of the data base required to support the discipline investigations. The SDMU is typically associated with a central processing and analysis facility that is a principal user of the data. Other remote users are interconnected through the network systems.

SDMUs and associated computing facilities are created wherever there exists a critical organizational mass with deliverable research objectives.

Examples of SDMUs include NASA's Pilot Ocean Data System (PODS), Pilot Climate Data System (PCDS), the proposed Pilot Land Data System and Pilot Planetary Data System, presently under development, and other research institutions with specialized data bases.

Figure 2 is an overview of the GRIS concept. It includes the following elements:

(1) Archives to retain usable data sets and make them readily available (i.e. existing Pilot System's archives).

(2) Discipline-oriented information extraction and processing systems (i.e. existing Pilot Systems) to

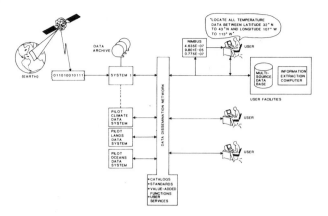

Figure 2. Global Resources Information System (GRIS) Overview

prepare data sets, operate R & D models, develop user-oriented products and displays, and perform scientific analysis.

(3) A common data cataloging and dissemination network service to allow users to locate, order, access, exchange, and integrate data quickly and at low cost. It will access the existing archives' cataloging systems and provide directory information to allow direct user access to these catalogs. This would supplement rather than replace the current mailing of tapes and disks.

SYSTEM OVERVIEW

Global Resources Information System Drivers

A set of generic data handling and analysis requirements results from consideration of the large-scale science problems that will be characteristic of future global remote sensing such as will occur from, or in association with, the Space Station:

(1) Ability to locate required data sets in the various catalogs located in the United States and other countries.

(2) Ability to move data sets rapidly from the archives to locations where the research is being conducted.

(3) Ability to register, calibrate, and modify data sets to standards rapidly with minimal manual intervention.

(4) Ability to perform processes on the group of assembled data sets in near real time, both locally and remotely.

(5) Ability to communicate research data and technical information between scientists locally and remotely in near real time.

GRIS General System Architecture

Integrated software packages known as Geographic Information Systems (GIS) have been developed that can perform all phases of storage, maintenance, retrieval, and analysis of spatial data. The function that sets them apart from data-base management systems is their ability to perform various degrees of analysis, at the users' direction. Accordingly, this analysis will be incorporated within the definition of the GRIS. The general functions of a GIS, of whatever size, are described below.

A GRIS may be thought of as a super GIS, serving many users and accessing many data bases in a systematic, standardized manner. A potential data query and location scenario and the corresponding flow/block diagram will serve as a basis for discussion of factors that relate to the use of geocoded data in such a system (Figure 3). Note first that the GRIS will not take over the the established archives. Rather, it will interface with these and provide the coordinating and standardizing so sorely needed. It will have no control over these archives but must understand their structure and cataloging to provide the necessary interfaces. Its structure will be much like that of the telephone service: A centralized organization must be present to organize, provide standards, implement the network and interfaces, perform research to provide future improvements, and provide specialized services. But, just as a telephone customer directly telephones his party without visibly invoking all of the company structure, the GRIS user will be provided with facilitating services as well as specialized ones. Thus, a distributed heterogeneous set of data bases is turned into an information system. At the same time, however, the need for timely responses is evident; this may require that the GRIS itself contain an archive available for rapid turnaround.

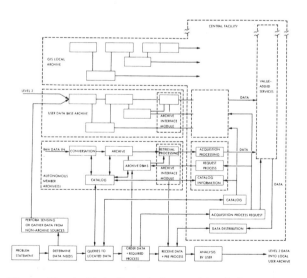

Figure 3. A Distributed Geographic Information System

A walk-through of the diagram will be useful. A user will determine what data are needed to answer a given research problem (lower left). He may query the central GRIS catalog, which in turn has access to archive catalogs such as those located with the Pilots. He may also query directly the catalogs of any known archives. The located data are then ordered via either the GRIS or the member archive operations data base management system. Accompanying the order would be requests for the desired retrieval processing to be done in the archive interface module at the archive. The data would be sent either to the requester directly, or, if he has requested further value-added processing, to the GRIS for this processing. On receipt of the data, he would proceed with his own processing. The GRIS may serve as a temporary or permanent archive of selected data, as might the user's own facility. Either would maintain archival data in a form compatible with that of the professional member archives to provide maximum commonality and minimum retrieval conversions.

In this scenario, the established archives would not change their data structures and would only have to establish catalog interfaces suitable for electronic query by the GRIS or the user.

The archive interface modules of the GIS would perform the necessary standardizing. Their location would be with each archive, to provide a consistent interface to the network.

Data Archive Principles

To make such a system practical, certain principles should be followed in designing any archive and its data structure:

(1)　To avoid aggregation/disaggregation problems, the data should be stored at the resolution commensurate with its data content, usually at the resolution at which it was obtained.

(2)　If practical, remove the intra-image distortions so that only affine low-order corrections will later be required.

(3)　Store with the data the precision information required to register the data to a well understood reference during retrieval, such as latitude/ longitude (i.e., georeference the data). This also provides the required reference information for the data location queries by geographic coordinates.

(4)　Store with the data all relevant ancillary information, such as calibrations, sensor data, processing history.

(5)　During data retrieval, geocode the data. That is, reproject and rescale the data to the grid requested by the analyst, and provide the ancillary data.

(6)　Supply the data in a standard format independent of the incoming data type or sensor, with suitable annotation to allow the analyst to use the various types interchangeably.

The GRIS will endeavor to encourage all member archives to adhere to these principles. This will allow the GRIS or the individual user to easily combine data from two or more SDMU data bases.

SYSTEM MAJOR COMPONENTS

The GRIS consists of the following major components, many of which will be provided by the individual SDMUs:

Archive Nodes

For this discussion, an archive is a data repository that potentially can be connected to the system for electronic catalog conversations and potential data transfer. The initial supposition is that the data will be required from a number of currently operating archives, that future archives will be established for future NASA missions, that some archives such as the National Space Science Data Center (NSSDC) contain somewhat specialized data, and that other archives such as Eros Data Center (EDC) contain data of more general interest. As these archives are selected to be included, individual decisions must be made concerning the interface modes. The general requirements for an archive node will be to service the catalog searches and requests, to locate and prepare for distribution, and to distribute the requested data. The data reformatting is proposed at the archive nodes to allow all transmissions to be in common style, perhaps using the Standard Format Data Units (SFDUs)[5] being developed. Eventually it will be desirable to maintain an on-line data browse for selected data.

The Network

The network would consist of all of the data transport mechanisms as appropriate: mail, leased line, dedicated line, arrangements for Domsat, etc. It could be built on networks such as the NASA Program Support Communications (PSC) facilities, and could contain local area networks to connect users to the long-line network. It would interface to each node with specialized hardware, and service both text communications (such as the catalog conversations) and data transfer. As the capabilities can be increased, data transfer progression might be: tabular and low volume point or polygon data, then browse images, then larger images. Interactive capability would initially only be the catalog conversations, then expand to catalog browse, selected data browse, and (much later) to interactive data analysis.

Support Services Administration

This component would consist of the directory/ catalog/dictionary services required to access the various data bases required. It would set the standards for this service and provide catalog query translations, access to the various catalogs, and a supercatalog function. It would also provide interface specifications for the alphanumeric terminals to the member nodes and a search function to locate and connect to new data bases as new data are requested. In short, it is the nerve center of the system: It will provide the administrative services for the system, including accounting, billing, and standards maintenance. It will include general user support functions such as the administrative user interfaces to the system and the general operations functions required to keep the system running. During implementation, this component would also make the necessary interagency and international arrangements.

System Central Services

The System Central Services are those data preprocessing services implemented by the system, as requested by the Steering Committee. This function will provide value-added services of selected kinds for system and non-system data upon request from the users. It will also build the capability to perform value-added services of an algorithmic kind. This service provides the registration, rectification, mosaicking, media conversions (such as map digitizing), data format conversions, and related operations. It will be the data outlet for the system for those data sets not transmitted directly from the archives to the users and it will maintain an archive of selected data that it has processed to avoid the necessity of reprocessing. It will build these capabilities on the existing capabilities at the various centers and will provide the interface requirements and specifications for the data terminals.

User Interface Nodes (SDMUs)

User interface nodes contain the interfaces between the network and the local equipment, the local GIS, analysis hardware and software, and the staff expertise. This is necessary to fulfill the desire for a distributed system, not just a networked data base. Network-interfacing equipment would be under the control of the network function, but all other functions would be under local control to encourage local developments.

Compatible, modular hardware at node locations will allow interchange of algorithms and of hardware designs, thus allowing developments from one node to be more easily transported to another. This hardware must interface with the terminal hardware and should support a local GIS. The data analysis developments being performed at the various NASA centers and at participating universities will need to be coordinated and arrangements made to modularize, package, and document for distribution those algorithms and procedures that are deemed to be generally useful. All centers are developing or have in place hardware and software systems; these developments must continue and the available capabilities used where practical.

CONCLUSIONS

The underlying motivation of GRIS is to establish an effective and efficient information management system to meet the data access requirements of NASA and NASA-related scientists conducting large-scale, multidisciplinary, multi-mission scientific investigations. The objective of GRIS is to develop an interactive data dissemination network that will provide research access to data on a global scale to serve the needs of the atmosphere/climate, oceanographic, and land-sciences communities.

To test the concept of GRIS, a limited number of multidisciplinary research projects, linking ocean, land, and atmospheric scientists within the United States, should be initiated. These projects would be used to test the feasibility and viability of the GRIS concept. Projects selected should have specific scientifically significant goals and objectives.

The need for systems to facilitate the use of multisource data by multidisciplinary research teams to address global-scale scientific issues is great. Systems must be developed to meet the needs of multidisciplinary research into global problems. We feel the approach proposed herein addresses this need.

ACKNOWLEDGMENT

The study described in this paper was carried out by the Jet Propulsion Laboratory, California Institute of Technology, under contract with the National Aeronautics and Space Administration.

The authors wish to acknowledge the significant contributions to this paper from Dr. John E. Estes and Dr. Jeffrey L. Star from the University of California at Santa Barbara.

(1) OSTA Commonality Analysis, Final Report, NASA-CR-69805, Vol. 1, OAO Corp., Beltsville, Maryland, June 1, 1981.

(2) "Space, A Resource for Earth," American Institute of Aeronautics and Astronautics (AIAA), 1290 Avenue of the Americas, New York 10019, April 1977.

(3) "Land-Related Global Habitability Science Issues," NASA Office of Space Science and Applications, NASA Technical Memorandum 85841, June 1983.

(4) Data Management and Computation, Volume 1: Issues and Recommendations, Committee on Data Management and Computation, Space Science Board, National Research Council, National Academy Press, Washington D.C., 1982.

(5) Recommendation for Space Data System Standards: Standard Format Data Units-Concept and Primary Label, Consultative Committee for Space Data Systems, Panel 2: Standard Data Interchange Structures, January 1984.

238

John E. Estes
Jeffrey L. Star
Geography Department
University of California
Santa Barbara, CA 93106
Philip J. Cressy
NASA Goddard Space Flight Center
Greenbelt, MD 20771
Michael Devirian
NASA Headquarters
Washington, DC 20546

Pilot Land Data System

PLDS will be a limited-scale distributed information system to explore scientific, technical, and management approaches to satisfy land science research needs.

Introduction

S ATELLITE REMOTE SENSING can provide information of tremendous value for the Earth Sciences. However, realization of this potential requires information systems not currently available. Technological advances now make it possible to design a data system to meet the land scientists' most critical information systems needs, and prepare the community for the Space Station era.

proaches to satisfy the needs of the land science research community. Accomplishing this goal will require land and information scientists working closely together to understand the needs of the users of scientific data. The system must support the full spectrum of functions needed to conduct land science investigations, including data location, acquisition, processing, and transfer. Properly developed, PLDS will provide a sound technical basis for a future, fully operational Land Data System.

ABSTRACT: *Beginning in 1983 and continuing until the present, the National Aeronautics and Space Administration (NASA) coordinated a series of meetings to develop initial plans for a Pilot Land Data System (PLDS). PLDS is intended to improve the ability of NASA and NASA-sponsored researchers to conduct research on land processes. The meetings have coordinated planning, concept development, and implementation activities, and examined research and information science requirements, and strategies for system evaluation. PLDS will be a limited-scale distributed information system to explore scientific, technical, and management approaches to satisfy land science research needs. Implementation is beginning in FY85. PLDS can pave the way for a Land Data System, and possibly the Earth Observing Information System of the space station era, by improving access to data and analysis capabilities, fostering an environment in which information synthesis can occur at scales not previously possible.*

Under the sponsorship of the NASA Information Systems Office, the Universities Space Research Association (USRA) assembled a working group to examine the need for a Pilot Land Data System (PLDS). The working group included discipline scientists, information scientists, and management personnel from universities, private industry, and the federal government.

Participants at meetings determined that the goal of the pilot program should be to establish a limited-scale, distributed information system to explore ap-

Such a system could serve as a key component of an Earth Observing Information System (EOIS), which will support science in the space station era of the 1990's.

Study of environmental processes on the Earth's surface requires a multidisciplinary approach. This has been recognized in the definition of several major programs, such as the International Geosphere Biosphere Program (IGBP), Global Habitability and Global Biology, and the International Satellite Land Surface Climatology Program

PHOTOGRAMMETRIC ENGINEERING AND REMOTE SENSING,
Vol. 51, No. 6, June 1985, pp. 703-709.

0099-1112/85/5106-0703$02.25/0

(ISLSCP; see Waldrop (1984), NASA (1983a), NASA (1983b), NASA (1984b)). This recognition has given rise to the concept of multidisciplinary information systems. Integration of the PLDS (as the precursor to a Land Data System) with other discipline data systems can provide a foundation for such integrated systems as EOIS and the Global Resources Information System (Billingsley *et al.*, 1984).

It was recognized from the outset that developing a PLDS is a complex task. PLDS must be a distributed system, because both the data and the users are geographically distributed. This complicates network and communications designs dramatically, compared with other NASA pilot data systems (for example, the Pilot Ocean Data System at the Jet Propulsion Laboratory, or the Pilot Climate Database System at Goddard Space Flight Center). Also, due to the multi-disciplinary and inter-institutional nature of land sciences research, this proposed system must be based on cooperation among NASA Centers and other institutions.

A number of principles were adopted for PLDS planning and design:

- Data bases tend to remain most viable when maintained by active researchers with a long term commitment to the use and sharing of the data (CODMAC, 1982);
- PLDS will serve the data and information systems needs of NASA and NASA-related scientists working on land science projects;
- PLDS represents a research and "proof-of-concept" tool;
- Long-term goals must be defined, both in terms of major Earth science issues to be examined and feasible tools for the task;
- PLDS must exploit components in place at participating institutions, and testing of the elements of the PLDS must build upon ongoing research programs; and
- System development should be based on available, well-understood technology. Close coordination with NASA computer science and communications research must provide the mechanism for incorporation of new technology and upgrades of the PLDS/LDS.

The material which follows details the science objectives, whose pursuit (let alone successful accomplishment) requires the development of an advanced information system. An abridged science scenario is included to provide an example of the type of requirements which could be levied on a Land Data System. A conceptual overview of a future Land Data System is presented next. This is followed by a brief description of the proposed development of the Pilot Land Data System. Finally, a brief discussion of the conclusions of the PLDS Working Group is presented and the current status of PLDS is described.

At present, PLDS has been funded for the initial planning phase. Management responsibilities for technological areas within PLDS have been assigned to different organizations within NASA, and a science steering group has been assembled. System implementation is beginning in 1985.

BACKGROUND

The launch of Landsat 1 stimulated major advances in the science and technology of remote sensing. These, as well as comparable advances in information sciences, are changing the nature of land science. Traditionally, field studies in land sciences have been limited in focus to a few variables in a small geographic area. This was due in large part to the problems of obtaining and working with large volumes of data.

Through the Landsat program, the land science community now has observational tools at scales appropriate to examine the critical processes that define a "real world" system. However, full realization of the scientific potential of satellite remote sensing has been handicapped by inadequate information systems. The ability to access and exchange both data and software is hindered by lack of both communications and standards. Appropriate computational resources are often lacking. Scientists are now required to devote a significant portion of their efforts to data acquisition and preparation. Better integration of remote sensing and information technologies can overcome these barriers.

In recognition of the need for improved understanding of large scale Earth processes, there is a movement in scientific research in general (and in NASA Earth science programs) to ask research questions which are both multidisciplinary and global in scale (Gwynne, 1982; NASA, 1983a; NASA, 1983b; NASA, 1984b). The resolution of such large-scale science issues requires interdisciplinary research teams and sophisticated technologies. It was intended from the beginning that these kinds of science issues drive the evolution of the PLDS, and that new advances in information science will, in turn, create a new perspective for looking at critical problems in the land sciences.

There are many environmental problems with important economic, human health, and environmental impacts. An ultimate objective of the scientific community is to understand correctly the factors involved in land processes and to provide a sound predictive modeling capability. Some of the important goals identified (NASA, 1983a) in the land science area are to

- establish methods by which a global carbon budget model may be developed and monitored;
- detect the presence and amounts of pollutants;
- establish the relationship between the energy balance and biophysical conditions on land, and their interrelationships with climate;
- improve the accuracy of models used for prediction of the availability and quality of water, snow, and ice;

- identify the early indicators of change in global element cycles, climate, and hydrology;
- advance the understanding of global and regional geologic and geomorphic structure and process; and
- develop improved methods for assessing and monitoring geologic hazards.

In order to achieve goals of this scale, an efficient processing and information management system is essential. While it is relatively easy to conceive of the general operation of a Land Data System, there is no existing prototype for it. The technology for each element of a Land Data System is understood, but experience in the integration of technologies must be developed before NASA can proceed toward implementing a global scale system (NASA, 1984a). A well-defined pilot data system, serving a relatively small group of users, is a necessary first step.

SCIENCE SCENARIOS

As a pilot program, PLDS cannot meet all possible needs. A limited number of science scenarios were selected from both existing and proposed research projects within the NASA land science community. Through these scenarios, the working group derived the generic information system functions and requirements while keeping an appropriately narrow focus.

The science projects used as examples were

- Vegetation Biomass and Productivity, and Large Area Inventory;
- Biogeochemical Cycling in Forests;
- Land Surface Climatology;
- Hydrologic Modeling and Soil Erosion/Productivity Modeling;
- Multispectral Analysis of Sedimentary Basins; and
- Monitoring Environmental Change

The land surface climatology and sedimentary basins projects were later selected by NASA Office of Space Science and Applications Code E personnel for initial incorporation in PLDS. These projects, in particular, represent a wide mix of requirements from both science and information systems perspectives.

The following is an abridged version of the Land Surface Climatology scenario, developed primarily by researchers at NASA Goddard Space Flight Center. This scenario (as well as the others used to drive PLDS design) was not intended to be a complete research description, but to highlight data access and processing requirements.

The objective of the Land Surface Climatology Project is to develop a better understanding of the interactions among the Earth's biospheric, edaphic, hydrologic, and atmospheric systems, and to determine their roles in influencing climate. An improved understanding of these processes and interactions can best be achieved through the development and validation of terrestrial and climatological

process models which require many diverse types of data.

Investigations in land-surfaces climatology are being supported through a new international program, the international Satellite Land Surface Climatology Project (ISLSCP). ISLSCP is conducted under the auspices of COSPAR and the International Association of Meteorology and Atmospheric Physics. Goddard Space Flight Center and a number of other institutions (including universities, other federal agencies, and international organizations) will be participating in this project.

This research program requires facilities to move information between the data bases and computational systems at physically separate facilities. The development of preprocessing capabilities and interchange standards will greatly facilitate this research. The definition of generic data formats, projections, and file structures could lead towards greater compatability among institutions. From an evaluation of this and other science scenarios, a number of information systems' support needs were developed.

REQUIRED FUNCTIONS

Figure 1 shows a general model for an information system to support land sciences research. This diagram generalizes the steps in this type of research, and identifies the functions which could be supported by a PLDS. Table 1 summarizes the results of analyzing each project in this manner. Priorities were assigned by discipline scientists as follows:

1—enable the scientists to do the research.
2—enhance the scientists' ability to do the research.

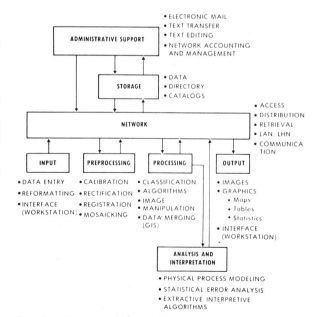

FIG. 1. Functions of an Information System to Support Land Science Research.

TABLE 1.

Processing Functions	Scenario						Total
	1	2	3	4	5	6	
Input							
Data Encoding	1	2	2	1	2	2	10
Data Reformatting	1	2	1	1	2	2	9
Preprocessing							
Data Calibration	2	3	1	1	2	2	11
Image Registration	1	1	1	1	1	2	7
Image Mosaicking	1	1	3	2	1	2	10
Processing							
Multi-source geocoded data overlay	1	1	1	1	1	1	6
Image and statistical processing (software sharing)	1	1	2	1	3	1	9
Analysis							
Statistical analysis	2	2	4	2	4	4	18
Modeling	1	2	4	2	4	4	17
Output							
Image	1	2	3	1	3	4	14
Statistical (tabular)	2	2	3	2	4	4	17
Tables and figures (graphic)	2	2	3	2	4	4	17
Storage media—CCT, disc	2	3	3	2	4	2	16
Network Storage							
Directory	1	1	3	1	2	2	10
Catalog	2	1	2	1	2	2	10
Data	2	2	1	2	3	2	12
Network Distribution							
Access to archive data	1	1	1	1	1	2	7
Networking of processing	1	1	1	1	3	2	9
Shared peripherals for output	2	2	2	1	4	2	13
Network Administrative Support							
Electronic mail	2	1	3	3	2	4	15
Text transfer	2	1	3	3	2	4	15

Legend: 1—Enable Scientific Research
2—Enhance scientist ability to do research
3—Research could be accomplished now but Support Service would be useful
4—PLDS support not required

3—research can already be accomplished but support would be useful.

4—PLDS support is not required.

Table 2 shows the functions in order of priority. The science scenarios require PLDS for data storage, input, preprocessing, and distribution on a high priority basis. Lower priority is assigned to support for analysis and output as well as network administration (although some of these items may be implicitly required to support the functions that were assigned higher priority).

CHARACTERISTICS OF A LAND DATA SYSTEM

Consideration of the science scenarios led the working group to a description of an information system for the future, a Land Data System (LDS). The overall goal of LDS would be to provide a powerful and responsive system to support land science research, facilitate understanding of the land resource complex, and provide general access to relevant data sets and processing capabilities.

Important characteristics of LDS include

- User-friendly interfaces for novice and expert users;

- Systematic archiving, maintenance, and access to relevant data;
- Data management and manipulation tools;
- Simple acccess to existing bibliographic information systems; and
- Maintenance of the history of data sets (origin, calibration information, and so on).

A system exhibiting these characteristics could change the character of land science research. Such a system would enable multi-disciplinary, multi-institutional research which is not now practical, and could allow experiments to be conducted in near-real time when required.

As currently envisioned, the LDS would consist of five major subsystems: Data Management, Communications and Networking, Intense Computational Processing, User Interface, and Input/Output Interface.

The Data Management Subsystem would provide both data and information about the data to the scientists. Ultimately, users would communicate with the subsystem using natural language. The subsystem would also store and update large amounts of data and support many users concurrently.

The Communications and Networking Subsystem

242

TABLE 2. ORDERED RANKING OF INFORMATION SYSTEM FUNCTIONS

	Total
Multi-source geocoded overlay	6
Access to archived data	7
Image registration	7
Data reformatting	9
Software sharing	9
Networking of processing	9
Mosaicking of images	9
Directory of information	10
Calibration of data	10
Data encoding	10
Data storage	12
Shared peripherals for output	13
Image output production	14
Electronic mail	15
Text transfer (compatible text editing)	15
Output storage media	16
Tabular output production	17
Graphic output production	17
Modelling	17
Statistical analysis	18

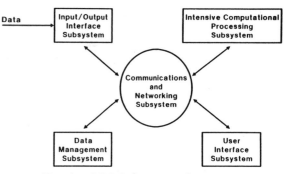

FIG. 2. LDS Subsystems Overview.

would support a near-real-time interface between the other subsystems. Such communications would be supported by several technologies, including packet switching network communications, local area networking, and satellite communications.

In an LDS, data manipulation and analysis would occur on a number of the subsystems. The Intensive Computational Processing Subsystem would provide the power of large-scale computers to users around the network for technically demanding tasks such as image interpretation and pattern recognition.

The User Interface Subsystem would consist of a range of microprocessor and minicomputer workstation types with differing capabilities. Workstations will be connected to the other subsystems by means of the Communication and Networking Subsystem and will either interface to local processing facilities or serve as free-standing processing stations.

The Input/Output Interface Subsystem would connect the overall LDS with outside computer systems and data sources. This subsystem will perform reformatting, modification, and data manipulation, and allow the overall LDS to communicate efficiently.

A functional overview of the LDS concept is seen in Figure 2, and possible structure of a node is shown in Figure 3. The detailed node in the latter could be a NASA Center, but nodes are also expected to be located at universities and other institutions or agencies.

PILOT LAND DATA SYSTEM DEVELOPMENT

PLDS development is based on three fundamental principles. First, the system will build on existing capabilities where possible, minimizing costs and permitting rapid concept testing. Second, a structured system engineering effort at the onset of the project helps to keep long-range goals in view. Finally, new technologies should be regularly reviewed for integration where appropriate. It is important to note that researchers will continue their ongoing efforts to improve data management and communications; the PLDS provides a formalism and a focus for further progress.

Functional requirements for PLDS can be summarized as goals by the end of fiscal year 1987:

- Establish communications capabilities;
- Build directories and catalogs of data sets at NASA Centers, participating universities, and other agencies;
- Develop an efficient data management system;
- Demonstrate remote access and use of data;
- Demonstrate that remote requests for value-added services (calibration and rectification, for example) can be answered in a timely way; and
- Demonstrate the expandability of the system.

In any pilot study, there must be periodic benchmarks to measure progress. Technical measures (e.g., data transmission volumes) can be used to evaluate some aspects of the system. While scientific achievements may not lend themselves to similar quantification, periodic evaluations and peer review, as well as publications in reviewed journals, can serve to evaluate program progress.

When fully implemented, PLDS will be capable of supporting a subset of the NASA-sponsored land science community. Systems concepts for the phases of PLDS, and the transition to an operational LDS (which could begin in 1989), are illustrated in Figure 4. Because technologies used in the PLDS are in a state of rapid development, development must proceed with a full awareness of the volatility of these technologies in order to prevent built-in obsolescence.

Implementation of PLDS has begun. A science working group has been convened under Dr. Robert Price of NASA Goddard Space Flight Center (GSFC). Dr. Paul Smith (also of GSFC) heads the technology working group. The sedimentary basin study at Jet Propulsion Laboratory and the land surface clima-

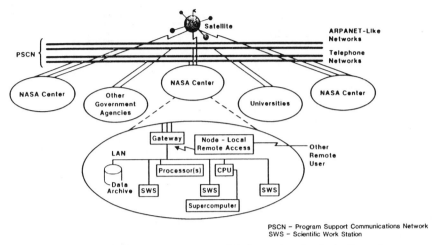

PSCN – Program Support Communications Network
SWS – Scientific Work Station

FIG. 3. LDS Communications and Information Analysis Overview.

tology project at GSFC are being used as initial science scenarios to drive pilot development. PLDS implementation is under the coordinated management of the Earth Science and Applications Branch and the Information Systems Office of NASA Headquarters Code E.

In addition, a science steering group under Dr. Ray Avidson (Washington University, St. Louis) has been convened. Specific development is occurring in the areas of systems engineering, communications, work stations, image processing, and intensive computation. Early efforts are directed towards a demonstration of the ability to link scientists at a number of institutions, and facilitate their research through improved communications.

CONCLUSION AND SUMMARY

Satellite remote sensing is a unique tool, providing data of a type and on a scale previously unobtainable. Yet, particularly with Space Station and the Earth Observing System (EOS) on the horizon, the applications of satellite remote sensing are handicapped by inadequate information systems. Future systems must not stop at the ground receiving station, but must fully integrate the flow of information to meet the user's ultimate needs.

There is a need to improve the ability of NASA and NASA-sponsored scientists to use remotely sensed and other land resource science data. Unless the ability to handle these and other land science data is established now, effective use of data from future systems (e.g., Moderate Resolution Imaging Spectrometer, High Resolution Imaging Spectrometer, High Resolution Multifrequency Microwave Radiometer, and Synthetic Aperture Radar) will be severely impacted. PLDS can permit researchers to better address important, multi-disciplinary science questions, and can lead to improved understanding of many land processes. PLDS can be a means of increasing scientific productivity through better use of information science technology.

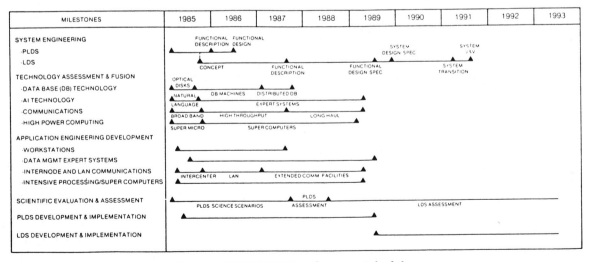

FIG. 4. PLDS/LDS Development Schedule.

244

Based upon these conclusions and the recommendations of the PLDS Working Group, NASA has begun a detailed planning phase for PLDS. The initial system will be a limited scale, distributed information system, directed towards supporting NASA and NASA-sponsored land science researchers, to permit them to function more effectively as scientists rather than librarians and communications experts. Work has begun on PLDS . The effort involves the cooperation of both discipline- and technology-oriented scientists. The degree to which it can improve the quality of research will be its ultimate measure of success.

The working group established that such a system is needed and will facilitate research. We encourage all land science researchers to monitor the progress of this system, and we encourage your comments.

ACKNOWLEDGMENTS

We gratefully acknowledge the sponsorship and encouragement of the Information Systems Office, NASA Headquarters and Dr. Caldwell McCoy; the Office of Space Science and Applications, Mr. Alexander Tuyahov; Ms. Janet Franklin and Universities Space Research Association; and especially the members of the Pilot Land Data System Working Group, whose contributions and synthesis we have summarized here.

REFERENCES

Billingsley, F. C., J. L. Urena, J. E., Estes, and J. L. Star, 1984. *Global Resources Information System: A Concept Paper*. NASA Jet Propulsion Laboratory Publication D-1524.

CODMAC (Committee on Data Management and Computation), 1982. *Space Science Board, National Academy of Sciences, Data Management and Computation, Vol 1: Issues and Recommendations*, National Academy Press.

Gwynne, M. D., 1982. The Global Environment Monitoring System (GEMS) of UNDP. *Environmental Conservation* 9(1):35-41.

NASA (National Aeronautics and Space Administration), 1983a. *Land-Related Global Habitability Science Issues*. Land-Related Global Habitability Sciences Working Group. NASA Technical Memorandum 85841. 112 p.

NASA (National Aeronautics and Space Administration), 1983b. *Global Biology Research Program; Program Plan*. NASA Technical Memorandum 85629. 112 p.

NASA (National Aeronautics and Space Administration), 1984a. *The Pilot Land Data System: Report of the Program Planning Workshops*. Pilot Land Data Systems Working Group. NASA Technical Memorandum 86250. 170 p.

NASA (National Aeronautics and Space Administration), 1984b. *Earth Observing System: Science and Mission Requirements Working Group Report*. NASA Technical Memorandum 86129. 107 p.

Waldrop, M. M., 1984. An inquiry into the State of the Earth. *Science* 226:33-35.

(Received 13 October 1984; revised and accepted 11 February 1985)

MAP OF DESERTIFICATION HAZARDS

EXPLANATORY NOTE

May, 1984

Prepared for: UNITED NATIONS ENVIRONMENT PROGRAMME

PREFACE

Activities on the preparation of a World Map of Desertification Hazards have been initiated in accordance with a recommendation adopted by the United Nations Conference on Desertification (UNCOD), and approved by the Thirty-Second Session of the General Assembly, that the Food and Agricultural Organization of the United Nations (FAO), in cooperation with the United Nations Environment Programme (UNEP), the United Nations Education,Scientific and Cultural Organization (UNESCO), and the World Meteorological Organization (WMO), prepare, publish and distribute a desertification hazards map of the world at a scale of 1:5 million (1).

On the basis of this recommendation, activities were initiated to identify a set of critical indicators of desertification through the aegis of a UNEP/FAO Desertification Assessment and Mapping Project launched in 1979. Under Phase I of this Project, a provisional methodology for assessment and mapping of desertification was compiled and subsequently tested in nine countries; a World Map of Desertification Hazards has been initiated under Phase II of the Project. Work in this phase resulted in the completion of a global assessment of soil elements related to desertification assessment. This inventory is based on the analysis of existing soils data contained in the FAO/UNESCO 1:5 million Soil Map of the World (2). A more comprehensive and detailed set of information has been assembled and analyzed for the African continent. The following maps result from these analyses.

Desertification Hazards - Africa. Scale 1:5 million. Two map sheets and one legend sheet.

Soil Elements used in assessing Desertification and Degradation - World. Scale 1:10 million. Five map sheets (North America, South America, Europe/Asia, Africa and SE Asia/Australia), and one legend sheet; derived from FAO/UNESCO Soil Map of the World, scale 1:5 million.

These maps have been prepared through the creation of a Geographical Information System (GIS) resource data base, analysis and digital cartography by the Environmental Systems Research Institute, Redlands, California. The methodology for the assessment of desertification hazard was prepared and the maps edited by Dr. Todor G. Boyadgiev of the Soil Resources, Management and Conservation Service, Land and Water Development Divison, FAO, Rome, who may be contacted for further technical information.

CONTENTS

DATA BASE DESIGN AND IMPLEMENTATION

DATA BASE DESIGN

An automated Geographic Information System (GIS) was designed and implemented to support the current and ongoing investigation of desertification. As part of this, a spatial data base with an intelligent data structure was developed, and a systematic overlay and analysis of relevant mapped information was conducted. The data base was developed at varying intensities. For Africa, it was designed to provide for a full assessment of desertification hazards and to account to a wide range of land analysis, planning, and management interests. It is comprised of some thirty geographic elements. These were photo-verified, rectified, and composited prior to automation. All related elements were cross-compared in order to eliminate conspicuous boundary discrepancies. The data base for the world contains soils information.

Data were mapped and automated at a scale of 1:5 million using the world basemap created by the American Geographical Society (AGS) in 1942. The automation procedures were designed to preserve the spatial integrity of the mapped data. Map automation was conducted by means of manual digitizing and automated laser scanning. All file processing was conducted in a vector format. Editing was conducted throughout the mapping and automation processes. The map data were merged by continent and iterative modeling was conducted. A seven-stage analysis of desertification hazard was conducted for Africa. An analysis of soil elements used in assessing desertification and degradation was conducted for the world. The results of these analyses were converted to raster format for display by means of maps produced by a laser plotter. Areal statistics were produced which indicated the distribution of ratings by select land areas.

DATA BASE IMPLEMENTATION

Data base development was implemented in five steps. These are outlined in Figure 1, and are described below:

Step 1: Data Assembly

Data were prepared by the Soil Resources, Management and Conservation Service, Land Water Development Division, FAO, Rome according to the methodology for the compilation of a desertification hazard map of Africa and a map of soil elements used in assessing desertification and degradation for the world. These included film copies of the printing plates for the Soil Map of the World, film copies of the plates for the Potential Population Supporting Capacity Maps, and maps of climatic data, vegetation, and animal densities. Some data for Africa, including topographic maps, remotely-sensed imagery, and supplementary data were collected from international and national sources.

Step 2: Data Standardization

All data were standardized prior to automation. For Africa, the process involved six tasks:

Reformatting. Cartographic features were clearly expressed as points, lines, or closed polygons. Attribute information was expressed in alpha/numeric code.

Rescaling. Where necessary source data were photographically enlarged or reduced to a scale of 1:5 million.

Verification. Topical source data were compared to the imagery, basemaps, and topographic maps prior to final delineation. The water and stream delineations on the basemaps were updated.

Rectification. Mapped data were rectified to the AGS basemaps and topographic maps by comparing known features such as rivers, ridges, and roads. Key latitude-longitude registration points were marked on each map.

Figure 1
Data Base Implementation Flow Chart

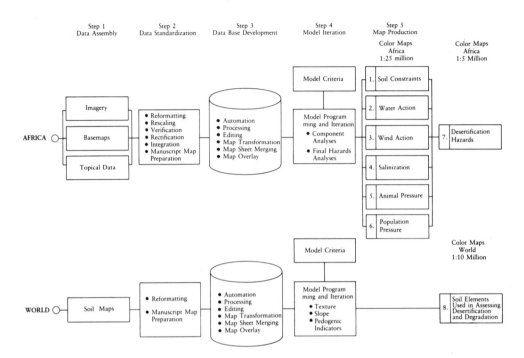

Integration. All related data were manually cross-compared in order to resolve boundary discrepancies and to eliminate small mismatches between the elements which were composited on each of the overlay sheets.

Manuscript Map Preparation. The data for Africa were drafted on twelve stable-base manuscript maps for each of the three map sheets at 1:5 million. Attribute code sheets were prepared for each of these. For the world outside of Africa, stable-base manuscript maps were prepared for the water and soil films. Alpha/numeric attribute code sheets were keyed to the map unit polygons.

Step 3: Data Base Development
This process involved the following tasks:

Automation. The cartographic features on the manuscript maps were automated by high resolution manual digitizing and laser scanning. The coordinates of features which were represented on more than one map were entered once, thereafter being joined automatically to other pertinent overlays. All coordinate information was topologically structured in a vector format to facilitate cartographic manipulation using ARC software. Attribute codes were keypunched and put into a tabular format to facilitate relational manipulations using INFO software.

Processing. The automated cartographic files were processed to ensure true point match, line connectivity, and polygon closure. The automated code files were associated with their related cartographic features.

Editing. The files were edited by both automatic and manual means. The cartographic files were checked to ensure an exact match to the maps which were automated. The code files were automatically checked to identify undocumented and invalid codes.

Map Transformation. The cartographic files, referenced by means of latitude-longitude coordinates, were molded to a true Miller oblated stereographic projection in the Eastern Hemisphere and a true bipolar oblique conic conformal projection in the Western Hemisphere using computerized algorithms.

Map Sheet Merging. Cartographic data on adjacent map sheets were merged. True coordinate match of features crossing sheet boundaries was effected. The original eighteen map sheets of the FAO Soil Map of the World were aggregated to the five continental areas illustrated in Figure 2. This formed the framework for the presentation of the world analysis at 1:10 million. Africa was subsequently divided into two parts for presentation at 1:5 million.

Figure 2
Map Sheet Merging

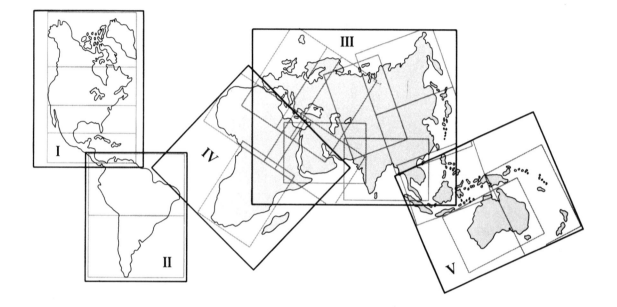

Map Overlay. The polygon map coverages for Africa were automatically overlain. The resulting files contained all of the climatic, soils, vegetation, land use, and human and animal population data required for the assessment of desertification in Africa.

Step 4: Model Iteration

Based on criteria in the methodology developed by FAO, conceptual models were programmed, and were tested using the map sheet for northeast Africa. Most models went through three or four iterations, each iteration involving a review of maps and statistical outputs from the computer analysis. The models for Africa drew upon the components from the overlaid file for the continent. The soil-related components of the models were first run in a master soil file for the continent and then associated with other factors such as climate, vegetation, land use, and population. Six component analyses served as the basis for the final assessment of desertification hazard. The models for world soil texture, slope, and pedogenic processes and regimes were applied to a unified file containing the attributes for all of the unique soils in the world. This file was then related to each of the cartographic soil units on the map sheets of the world and the final map files were produced.

Step 5: Map Production

The results of the analyses were generalized for the final map presentations in order to ensure cartographic legibility. For the 1:5 million maps of Africa, polygons falling below the following minimum sizes were eliminated automatically: 12.5 sq. km. for islands, glaciers and lakes; 125 sq. km. for hazard rating classes and factors. For the 1:10 million maps of the world, the following spatial resolutions were effected: 50 sq. km. for islands, glaciers, and lakes; 500 sq. km. for slope/texture groups and pedogenic processes and regimes. The final files were converted from a vector to a raster format for the final presentation. A computer-driven laser film recorder was used to produce color separates for the final maps. These separates produced the following set of proofs:

Map of Desertification Hazards. Africa. Component Analyses 1:25 Million. Final Hazards Analyses 1:5 Million.

Soil Elements used in Assessing Desertification and Degradation. World. 1:10 Million.

Windowed areas from both the 1:5 million scale final analyses for Africa and the 1:10 million scale analysis for the world are included with this text. All of the component analyses for Africa are also included.

DATA ANALYSIS AND RESULTS

DESERTIFICATION HAZARDS ANALYSIS - AFRICA

CONCEPT AND DEFINITIONS

The concepts and definitions concerning desertification are detailed in the 'Provisional Methodology for Assessment and Mapping of Desertification' (3).

Desertification hazard refers to the natural susceptibility of the land to desertification and man-made factors.

In the context of the presently described activities, desertification is considered as a comprehensive expression of economic and social processes, as well as those natural or induced processes which destroy the equilibrium of soil, vegetation, air and water in the areas subject to edaphic and/or climatic aridity. Continued deterioration leads to a decrease in/or destruction of the biological potential of the land, deterioration of living conditions and increase of desert landscapes (3).

In areas not subject to edaphic and/or climatic aridity, any hazard of loss of biological productivity is considered, reported and inventorized as degradation hazard.

An important difference between soil degradation and desertification is that soil degradation is not necessarily continuous; it takes place over relatively short periods, and can be reversed. Also desertification or the danger of it, is confined to the arid, semi-arid and sub-humid areas, whereas soil degradation can occur in all climates. Continued degradation of soil properties will eventually lead to desertification. Furthermore, certain processes important to the concept of soil degradation are not considered in desertification, i.e., waterlogging, depletion of plant nutrients, and acidification (3).

METHODOLOGY

Desertification hazards have been assessed as a function of:

- the soil status
- the vulnerability of land to desertification processes, and
- animal and population pressures on the land.

The following information has been used for inventorizing and geolocating these factors, for analysis and for the compilation of a desertification hazards map of Africa:

- FAO/UNESCO Soil Map of the World (1:5 million)

- USDA-SCS World Soil Map (1:1 million) (partly used)

- FAO/UNESCO Vegetation Map of the Mediterranean zone and of Africa south of the Tropic of Cancer (1:5 million and 1:10 million)

- FAO Climatic Maps on mean annual rainfall (isohyet zones at 50,100 and 200 mm intervals), mean wind velocity, number of wet days and P/PET ratio (precipitation/potential evapotranspiration) (1:5 million)

- FAO data and maps on present animal density (1:10 million)

- FAO data and maps on present population density and potential population supporting

253

capacities by length of growing period zones (LGP) and major climates (1:10 million).

- World Atlas of Agriculture (1.5 million).

The contributions and quantification of each of the main factors used in the analysis, i.e., soil status, land vulnerability and animal and population pressure, are described below.

Soil Status

Soil status is assumed an indicator of biological productivity. In the present study, this indicator has been quantified through compilation of a soil constraints index.

Soil Constraints Index

Taking into consideration the soil requirement of crops (7), the soil units of the FAO/ UNESCO Soil Map of the World have been evaluated with regard to suitability for specific rainfed crop production and a combined soil constraint index of such production has been attributed to each soil unit. These indices àre taken as a reflection of the status of land degradation and are considered as indicators of desertification.

It is assumed that the effect of climatic aggressivities and animal and human population pressures will be greater on degraded land than on land with high suitability for crop production.

Class limits established to quantify soil status are as follows:

Range in Values of Computed Soil Constraints Index	Soil Constraints
≤4.0	Slight
4.1 - 12.0	Moderate
12.1 - 20.0	Severe
>20.0	Very Severe

The results of the soil status assessment for Africa are presented in Table 1 and Map 1.

Vulnerability of Land to Desertification

Vulnerability of land to desertification processes is quantified with regard to water action, wind action and salinization.

Water Action

Water action has been evaluated through consideration of the following characteristics:

- Mean annual precipitation and amount of sediment yield (water erosion) with natural vegetation cover

- Monthly and annual precipitation and coefficients of variation

- Components and properties of the soil, soil erodibility, soil units and pedogenic conditions.

Indices for sediment yield (water erosion), resulting from the interaction of these factors have been computed for each rainfall isoyhets zone and a water action index subsequently evaluated for every soil mapping unit, taking into consideration the topography.

Class limits established to quantify the water action are as follows:

Range in Values of Computed Water Action Index	Water Action Hazards
≤ 1.900	Slight
1.901 - 5.700	Moderate
5.701 - 9.500	Severe
>9.500	Very Severe

The results of the water action assessment for Africa are presented in Table 1 and Map 2.

Wind Action

Variables taken into consideration to assess wind action comprise: wind speed; soil moisture regime; soil texture and surface roughness; vegetation and general land use.

The relationships between the following have been used to evaluate the effect of wind speed on wind erosion:

- annual average wind speed (velocity) and distribution of classes of wind (from nonactive to very active) as a percentage of the total number of wind observations

- classes of wind and sand movement

These relations provided an equation between average value of wind speed and sand movement.

The quantified amount of sand movement has been subsequently modified by soil moisture regime as reflected by number of wet days.

A wind erodibility index has been computed for every soil mapping unit and finally modified according to type of vegetation and type of land use.

Class limits established to quantify the effects of wind action are as follows:

Range in Values of Computed Wind Action Index	Wind Action Hazards
≤3.00	Slight
3.001-9.000	Moderate
9.001-15.000	Severe
>15.000	Very Severe

The results of the wind action assessment for Africa are presented in Table 1 and Map 3.

Salinization

Salinization hazard has been evaluated for each soil mapping unit by each major climate and aggregated growing period zones using the following variables:

- The maximum quantity of soluble salts in surface horizons of Solonchaks as indicators of salt accumulative processes

- The P/PET ratios of major climates

- The maximum capillary rise in relation to groundwater depth, soil texture and topography

- The existing salinization in the soil as indicators of specific condition for salt accumulation (depth and mineralization of groundwater, mineralogical composition of the surface rocks, fan piedmont, deltaic deposits, closed depressions).

Class limits established to quantify the effects of salinization are as follows:

Range in Values of Computed Salinization Index	Salinization Hazards
≤ 3.250	Slight
3.251-13.000	Moderate
13.001-39.000	Severe
>39.000	Very Severe

The results of the salinization assessment for Africa are presented in Table 1 and Map 4.

Animal and Population Pressure on the Land

Man-made factors influencing desertification have been quantified by estimation of present animal and human population densities and comparison with the potential livestock and human carrying capacities. These following data and analyses have been used to evaluate animal and population pressures on the land.

Animal Pressure

Animal pressure has been quantified on the basis of estimates of present animal density compared with assessments of potential livestock carrying capacity.

Present Animal Density:

Information on the geographical distribution of cattle, goats and sheet populations in Africa (5) and FAO statistics for 1975 (6) provided the basic data used to establish the map inventory of present animal density. Conversion of different types of livestock into a standard livestock unit was realized through application of the following conversion (multiplication) factors to the various animal populations: 0.1 for goats and sheep; 0.3 for pigs; 0.8 for cattle and asses; 1.0 for buffaloes, horses and mules; and 1.1 for camels. Annual maintenance dietary requirement for one livestock unit was fixed at 1350 feed units.

Potential Livestock Carrying Capacity:

Calculation of potential livestock carrying capacity by major climate and lengths of growing period zones was based on the estimation of climatic potential for fodder production, using the linear equation of le Houerou and Hoste (8) for Mediterranean (winter rainfall) and Sahelio-Sudanian (summer rainfall) areas. These equations were modified for the cool subtropical and tropical climates.

To arrive at land potentials, soil conditions of the soil mapping units were taken into account by considering how well the constituent soils matched or did not match with soil requirement for grassland production (4). The climatic potential was accordingly increased or downgraded (plus 25% to minus 50%) to arrive at the land potential. From the potential fodder production assessment, the consumable fodder in feed units was derived using information on the mapped vegetation and general land use types. For conversion of the estimates of feed units into livestock carrying capacity, annual maintenance requirements were applied.

Animal Pressure:

Comparison of present animal density and potential livestock carrying capacity provided the estimates of animal pressure on the environment used in the study.

The following class limits were established:

Ratio of Present Livestock Density/ Potential Livestock Carrying Capacity	Animal Pressure
≤ 0.40	None to Slight
0.41-1.60	Moderate
1.61-4.80	Severe
> 4.80	Very Severe

Data on assessed animal pressures on land in Africa are presented in Table 1 and Map 5.

Population Pressure

Population pressure on the environment was evaluated by using data from FAO/UNFPA/IIASA Project INT/75/P13 'Potential Population Supporting Capacities of Lands in the Developing World' (9).

Present Population Density:

Information was obtained by reaggregating administrative area statistical data (standardized to 1975) to establish the map inventory of present population density.

Potential Population Supporting Capacity:

The methodology developed to assess potential population supporting capacities computes potential crop (and calorie equivalent) production through length of growing period zone and major climate delineation and analysis, matching identified soil and climatic requirements of crops with the soil and climatic conditions of the zones.

The results are totalled to arrive at calorie (protein) potentials. Once the maximum potential calorie (protein) production combination is ascertained, including the present and projected contribution from irrigated areas, application of FAO/WHO country specific per capita calorie (protein) requirements allows computation of the potential population supporting capacities.

Population Pressure:

Comparison of the potential population supporting capacities with 1975 population data, provides the assessment of population pressure on the environment used in the study.

Class limits established for evaluation of the population pressure on the environment in Africa are as follows:

Ratio of Present Population Density/ Potential Population Supporting Capacity	Population Pressure
≤ 1.000	None to Slight
1.001-4.000	Moderate
4.001-12.000	Severe
>12.000	Very Severe

Data on assessed population pressures on land in Africa are presented in Table 1 and Map 6.

Desertification Hazards Summary

The results of the assessment are presented in a 1:5 million scale Desertification Hazards Map of Africa showing the locations and extents of lands variously subject to the main desertification processes.

Desertification hazard has been assessed by combining the previously described indices and ratios on:

- soil constraints
- water action
- wind action
- salinization
- population pressure
- animal pressure

257

into an overall index of desertification hazard, grouped into the following four classes:

Range in Values of Computed Overall Desertification Index	Desertification Hazards
≤ 42.5	None to Slight
42.51 - 85.0	Moderate
85.1 - 127.5	Severe
>127.5	Very Severe

According to the concept accepted in the 'Provisional Methodology' (3), desertification occurs in arid, semi-arid and sub-humid areas, or on that part of the earth characterized by a P/PET ratio inferior to 0.75 and an aridity index in general superior to 1.5.

These parameters correspond with the 180 days length of growing period limit, while a limit of hyper-arid areas is associated with a 0-day growing period boundary. Accordingly, the final hazard analysis on the map is shown separately for the following climatic groupings:

a) zones without a growing period - for the most part completely desertified but with localized areas subject to desertification;

b) zones with 1 to 180 days of length of growing period - prone to desertification;

c) zones with more than 180 days in length of growing period and areas with high altitude and low temperatures - subject to land degradation are excluded from the desertification hazards analysis.

The two processes contributing most to the evaluation of overall hazard are shown on the map.

Results on the desertification hazards by country in Africa are presented in Table 2 and on the 1:5 million map. Map 7 is a window illustrating desertification hazards in the same scale as presented in the wall map and includes the map legend.

ANALYSIS OF SOIL ELEMENTS USED IN ASSESSING DESERTIFICATION AND DEGRADATION

This analysis is a prerequisite to compilation of World Map of Desertification.

The FAO/UNESCO 1:5 million scale Soil Map of the World provides necessary information on the location and extent of soil units, miscellaneous land units, soil phases and dominant slope and texture classes. This information is the base used for compilation of the Desertification Hazards Map of Africa.

Equally important data required for computation of salinization, wind erosion, water erosion, animal pressure and population pressure on the land and evaluation of desertification hazards, is not presently available for the rest of the world. For this reason, the soil elements used for assessment of desertification in Africa are presented separately for the remainder of the world in illustration of the base data presently available.

The legends and methodology used in this presentation are as follows:

Soil Texture:
The soil texture of each mapping unit is assessed and presented through quantification of the soil unit composition of each mapping unit and the texture of the component soils. The results are classified into seven texture groups as presented in the legend accompanying

Map 8, namely:

- Textural Group 1: Mapping units where more than 75% of the soils are coarse-textured.

- Textural Group 2: Mapping units where more than 75% of the soils are medium-textured.

- Textural Group 3: Mapping units where more than 75% of the soils are fine-textured.

- Textural Group 4, 5 and 6: Mapping units where 50 to 75% of the soils are respectively coarse, medium and fine-textured.

- Textural Group 7: Mapping units with a balanced (equal) distribution of coarse, medium and fine-textured soils.

Slope

A similar principle to that applied for texture assessment, has been applied for assessment of slope conditions. The following class limits are recognized:

Slope Group a: Mapping units where more than 60% of the soils occupy level to gently undulating land (between 0 and 8% slope) and with less than 20% of the soils occuring on steeply dissected to mountainous slopes (more than 30% slope).

Slope Group b: Mapping units dominated by soils on rolling to hilly land (between 8 and 30% slope) and with less than 80% of the soils occuring on slopes between 0 and 8% slope and less than 50% of the soils on slopes more than 30%.

Slope Group c: Mapping units where more than 50% of the soils occur on steeply dissected to mountainous topography (more than 30% slope).

Pedogenic Indicators

Pedogenic indicators which are of importance to desertification have been classified at two levels according to the degree of severity of the processes or regimes. Those of the first level (the more severe manifestation of the processes) are presented on the map by black letters, while those at the second level are indicated by gray letters.

Pedogenic processes and regimes used as indicators for assessing desertification hazard, and shown on the map of soil elements used in assessing desertification, are as follows:

Salinization Indicators

Salt flats and Solonchaks (first level); saline phase, sodic phase, Solonetz, Thionic Fluvisols, Takyric Yermosols and Solodic Planosols (second level);

Compaction and Cementation Indicators

For this category, the nature of cementation is taken in consideration as follows:

- gypsic cementation: petrogypsic phase (first level); gypsic subgroup of Yermosols and Xerosols (second level);

- calcareous cementation: petrocalcic phase (first level); calcic subgroup of Yermosols, Xerosols, Kastanozems, Chernozems, Cambisols, Luvisols and Gleysols (second level);

- silica cementation: duripan phase (first level); fragipan phase (second level);

- ferric cementation: petroferric phase (first level); petric phase, ferralic, ferric and plinthic subgroup of Arenosols, Cambisols, Luvisols, Gleysols such as all Acrisols, Nitosols and Ferralsols (second level).

Wind Ablation and Accumulation Indicators
Dunes or shifting sand (first level); eutric Regosols, cambic, luvic and albic Arenosols and Yermosols with coarse texture and stony and/or lithic phase (second level).

Water Erosion Indicators
Rock debris or desert detritus and Lithosols (first level); the stony and lithic phase not associated with Yermosols with coarse texture (second level).

Aridic Conditions Indicators
Calcaric subgroup of Fluvisols, Regosols and Phaeozems, Rendzinas, haplic Kastanozems and haplic Chernozems (first level); luvic and haplic Yermosols and haplic Xerosols (second level).

Low Temperature Regime Indicators
Gelic subgroup of Gleysols, Regosols, Cambisols, Planosols and Histosols or areas with permafrost (first level); areas with intermittent permafrost (second level).

The extent (km2) of these indicators of desertification (i.e., slope, texture and pedogenic processes) by continent is presented in Table 3. Map 8 is a window showing the soil elements used in assessing desertification and includes the map legend.

Because of the complexity of reproduction, Maps 7 and 8 could not be included in this reprint.

Conclusions (Tentative)

Following the World Map of Desertification at 1:25 million established for the UNCOD in 1977 (10), the present Desertification Hazards Map of Africa is a new step in the field of collection and interpretation of information for desertification-prone regions and compilation of maps in larger scales. The basic document incorporated into the data base was the FAO/UNESCO Soil Map of the World. This inventory has been interpreted separately to establish a World Map on 'Soil Elements used in Assessing Desertification and Degradation' to serve in preparation of desertification maps for the rest of the world.

However, more work to combine these elements with data concerning geo- physical, bioclimatic and socio-economic conditions is necessary. Much of the data used for assessing desertification in Africa is not presently available for other continents. It will be necessary to collect, homogenize and geolocate such information in order to apply the methodology to the rest of the world.

The Desertification Hazards Map of Africa, which reflects the multi- factoral hazard of degradation on the environment, permits location of homogeneous areas with the same degree of desertification hazards and identification of the main causes of desertification. The critical zones, with moderate, severe and very severe desertification hazard, shown on the map will guide implementation of necessary measures to improve these areas, depending on the determinative desertification processes involved.

The Desertification Hazards Map of Africa can be used as a framework for more detailed surveys in selected areas. In applying the existing knowledge, the information on the map can not only be used to guide measures to stop the physical processes of desertification, but to also enhance awareness of the harm done to the fragile ecosystems of dry lands by existing economic and social activities.

Recent technological advances now permit geographic data to be compiled, stored, and displayed accurately and automatically. The resulting data bases may be distributed easily in the form of magnetic tapes. New information can be added to the system, and entire systems can be combined to form larger data bases.

After being automated, stored geographic information can be used to test hypotheses. Single theme maps can be overlaid automatically to produce multivariable maps. Models can combine existing information and re-display it in novel ways. Data can be used synergistically to graphically portray scientists' insights about a geographic area. One-time geographic studies are being replaced by process-oriented efforts wherein data are systematized for long-term iterative analysis.

LIST OF REFERENCES

1. UNCOD
   1978

   Round-up, Plan of Action and Resolutions, 29 August-9 September. New York.

2. FAO/UNESCO
   1971-81

   Soil Map of The World - 1/5 000 000, vol. 1-X, Unesco. Paris.

3. FAO/UNEP
   1983

   Provisional Methodology for Assessment and Mapping of Desertification. Rome.

4. FAO/UNFPA
   1980

   Land Resources for Populations of the Future. Report on the Second FAO/UNFPA Expert Consultation. Rome.

5. OAU STRC
   1980

   Africa Distribution of Cattle, Sheep and Goats - three maps at 1/10 000 000. Nairobi.

6. FAO
   1978

   FAO Production Yearbook, vol. 31. Rome.

7. FAO
   1978

   Report on the Agro-ecological Zones Project, vol. 1, Methodology and Results for Africa. Rome.

8. Le Houerou, N.H. and Hoste F.H.
   1977

   Rangeland Production and Annual Rainfall Relations in the Mediterranean Basin and in the African Sahels - Sudanian Zone.
   Journal of Range Management, vol. 30, No. 3. Addis Ababa.

9. FAO/UNFPA/ IIASA
   1982

   Potential Population Supporting Capacities of Lands in the Developing World. Rome.

10. UNCOD
    1977

    World Map of Desertification, explanatory notes, 29 August - 9 Sept.

Table 1
Average Data for Africa by Process.

Component Analyses	None to Slight Sq Km*	None to Slight %	Moderate Sq Km	Moderate %	Severe Sq Km	Severe %	Very Severe Sq Km	Very Severe %
Soil Constraints	1,258	5	11,261	40	11,962	43	3,446	12
Water Action	23,573	84	3,514	13	550	2	290	1
Wind Action	19,043	68	6,361	23	1,057	4	1,466	5
Salinization	20,932	75	2,167	8	1,802	6	3,026	11
Animal Pressure	10,952	39	11,865	43	2,871	10	2,239	8
Population Pressure	15,335	55	8,924	32	2,952	10	716	3

*1,000's

Table 2
Degree of Desertification Hazards by Country in Africa

Country	None to Slight* % Area	Moderate % Area	Severe % Area	Very Severe % Area
Algeria	4.1	28.3	38.9	28.7
Angola	85.8	11.4	2.6	.2
Benin	79.0	21.0	-0-	-0-
Botswana	39.3	60.7	-0-	-0-
Burundi	100.0	-0-	-0-	-0-
Cameroon	94.9	5.1	-0-	-0-
Canary Islands	-0-	28.2	54.1	17.7
Cape Verde	-0-	-0-	-0-	100.0
Central African Republic	96.6	3.4	-0-	-0-
Chad	21.8	28.9	39.7	9.6
Comoros Islands	100.0	-0-	-0-	-0-
Congo	100.0	-0-	-0-	-0-
Djibouti	-0-	90.4	6.9	2.7
Egypt	<.1	23.4	36.2	40.3
Equatorial Guinea	100.0	-0-	-0-	-0-
Ethiopia	44.4	36.2	15.0	4.4
Gabon	100.0	-0-	-0-	-0-
Gambia	44.4	56.0	-0-	-0-
Ghana	95.6	4.4	-0-	-0-
Guinea	97.5	2.5	-0-	-0-
Guinea Bissau	98.8	1.2	-0-	-0-
Ivory Coast	100.0	-0-	-0-	-0-
Kenya	13.0	64.3	21.0	1.7
Lesotho	26.9	57.2	-0-	15.9
Liberia	100.0	-0-	-0-	-0-
Libya	.5	28.3	48.4	22.8
Madagascar	91.4	6.1	2.4	<.1
Madeira	-0-	-0-	100.0	-0-

Table 2 (Continued)
Degree of Desertification Hazards by Country in Africa

	Desertification Hazards Rating			
	None to Slight*	Moderate	Severe	Very Severe
Country	% Area	% Area	% Area	% Area
Malawi	94.5	5.5	-0-	-0-
Mali	12.8	45.1	6.0	36.1
Mauritania	5.7	17.0	23.0	54.4
Mauritius	100.0	-0-	-0-	-0-
Morrocco	34.0	27.1	35.5	3.4
Mozambique	79.9	20.1	.1	-0-
Namibia	25.5	50.2	24.3	.1
Niger	.1	17.9	52.9	29.1
Nigeria	62.8	31.4	5.8	-0-
Reunion	100.0	-0-	-0-	-0-
Rwanda	100.0	-0-	-0-	-0-
Sao Tome and Principe	100.0	-0-	-0-	-0-
Senegal	26.7	72.0	1.3	-0-
Sierra Leone	100.0	-0-	-0-	-0-
Somalia	7.9	56.7	34.2	1.2
South Africa	11.4	17.5	33.3	37.8
Sudan	34.1	33.8	7.7	24.4
Swaziland	69.6	30.4	-0-	-0-
Tanzania	65.4	33.4	1.2	-0-
Togo	100.0	-0-	-0-	-0-
Tunisia	13.7	25.9	42.6	17.8
Uganda	80.2	19.2	.6	-0-
Upper Volta	41.7	58.3	-0-	-0-
Western Sahara	-0-	11.7	69.8	18.5
Zaire	100.0	-0-	-0-	-0-
Zambia	97.1	2.9	-0-	-0-
Zimbabwe	39.2	55.0	5.8	-0-

*Includes areas not rated for desertification hazards. Land degradation hazards may occur.

Table 3
Texture, Slope and Pedogenic Indicators by Continent

	Africa		North** America		South** America		Europe/ Asia**		SE Asia/ Australia	
	Sq Km*	%	Sq Km	%	Sq Km	%	Sq Km	%	Sq Km	%
Textural Group										
1	3,603	13	703	3	774	4	2,939	5	749	7
2	11,139	40	13,531	57	5,916	33	31,479	56	1,893	18
3	516	2	459	2	885	5	2,164	4	140	1
4	4,205	15	4,590	19	1,279	7	3,323	6	1,330	13
5	3,962	14	3,031	12	2,771	16	6.035	11	3,295	32
6	3,935	14	1,412	6	6,006	34	9,802	18	2,589	25
7	567	2	172	1	107	1	32	<1	342	3
Slope Group										
a	16,057	57	8,544	36	9,433	53	24,100	43	3,466	34
b	9,171	33	11,960	50	6,200	35	20,873	38	5,738	55
c	2,699	10	3,394	14	2,105	12	10,801	19	1,134	11
Pedogenic Processes and Regimes										
High Severity Class										
Z	177	1	12	<1	249	1	1,692	3	85	1
G	173	1	-0-	-0-	-0-	-0-	293	1	-0-	-0-
C	740	3	108	<1	195	1	195	<1	139	1
D	9	<1	95	<1	-0-	-0-	-0-	-0-	1,949	19
F	752	3	-0-	-0-	-0-	-0-	53	<1	92	1
E	2,326	8	53	<1	423	2	1,642	3	<1	<1
W	2,307	8	1,400	6	1,149	7	8,921	16	548	5
A	176	1	889	4	80	<1	1,895	3	63	1
P	-0-	-0-	1,695	7	-0-	-0-	6,916	13	-0-	-0-
Low Severity Class										
Z	998	4	182	1	741	4	3,295	6	1,196	12
G	172	1	-0-	-0-	-0-	-0-	69	<1	-0-	-0-
C	695	2	288	1	146	1	1,927	4	280	3
D	-0-	-0-	184	1	-0-	-0-	12	<1	27	<1
F	6,495	23	980	4	6,642	37	1,610	3	637	6
E	3,563	13	380	2	217	1	1,180	2	185	2
W	3,327	12	9,187	38	1,544	9	9,874	18	715	7
A	568	2	497	2	760	4	774	1	16	<1
P	-0-	-0-	970	4	-0-	-0-	-0-	-0-	-0-	-0-

*1,000's.
**Preliminary

Note: Data in Tables 1, 2, and 3 calculated using basemaps which are not equal area projections.

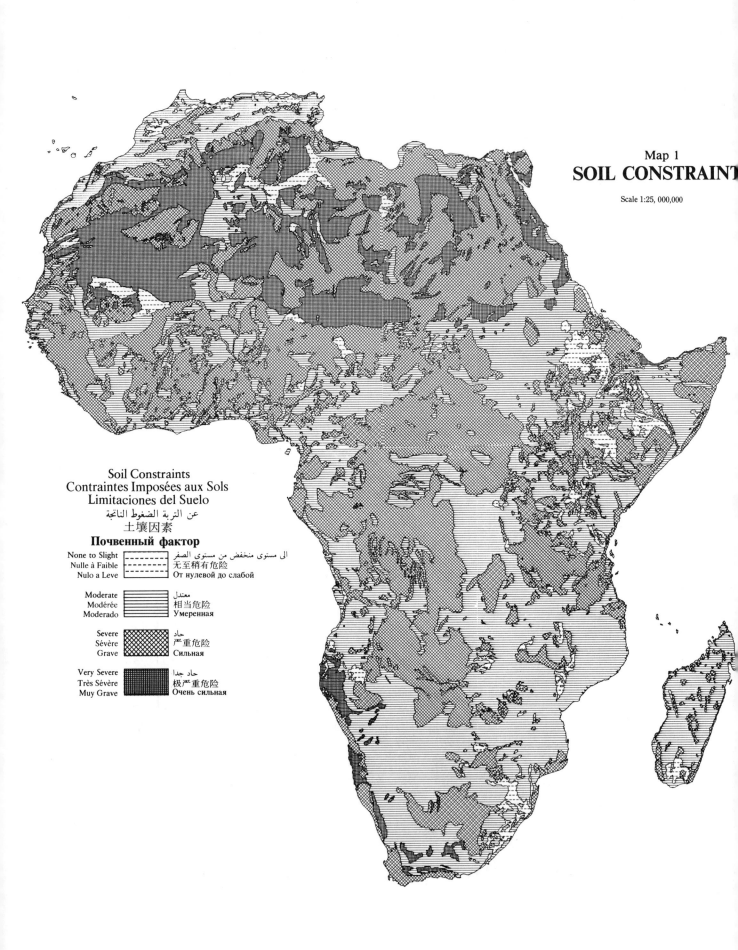

Map 1
SOIL CONSTRAINT

Scale 1:25, 000,000

Soil Constraints
Contraintes Imposées aux Sols
Limitaciones del Suelo
عن التربة الضغوط الناتجة
土壤因素
Почвенный фактор

None to Slight	الى مستوى منخفض من مستوى الصفر	
Nulle à Faible	无至稍有危险	
Nulo a Leve	От нулевой до слабой	

Moderate	معتدل	
Modérée	相当危险	
Moderado	Умеренная	

Severe	حاد	
Sévère	严重危险	
Grave	Сильная	

Very Severe	حاد جدا	
Très Sévère	极严重危险	
Muy Grave	Очень сильная	

266

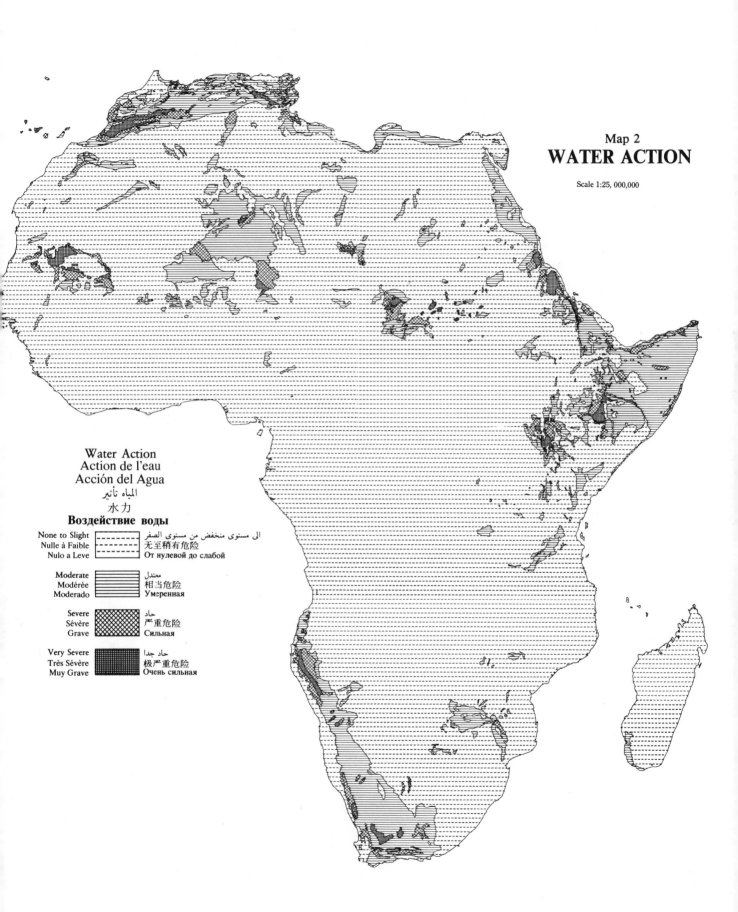

Map 2
WATER ACTION

Scale 1:25, 000,000

Water Action
Action de l'eau
Acción del Agua
المياه تأثير
水力
Воздействие воды

None to Slight		الى مستوى منخفض من مستوى الصفر
Nulle à Faible		无至稍有危险
Nulo a Leve		От нулевой до слабой
Moderate		معتدل
Modérée		相当危险
Moderado		Умеренная
Severe		حاد
Sévère		严重危险
Grave		Сильная
Very Severe		حاد جدا
Très Sévère		极严重危险
Muy Grave		Очень сильная

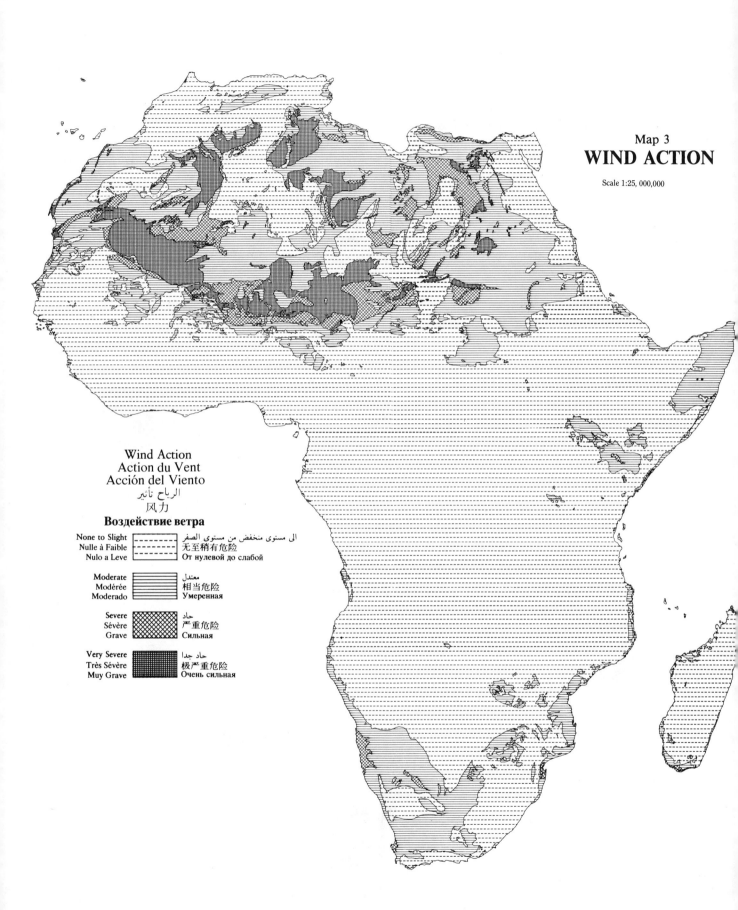

Map 3
WIND ACTION

Scale 1:25, 000,000

Wind Action
Action du Vent
Acción del Viento
الرياح تأثير
风力
Воздействие ветра

None to Slight		الى مستوى منخفض من مستوى الصفر
Nulle à Faible		无至稍有危险
Nulo a Leve		От нулевой до слабой

Moderate		معتدل
Modérée		相当危险
Moderado		Умеренная

Severe		حاد
Sévère		严重危险
Grave		Сильная

Very Severe		حاد جدا
Très Sévère		极严重危险
Muy Grave		Очень сильная

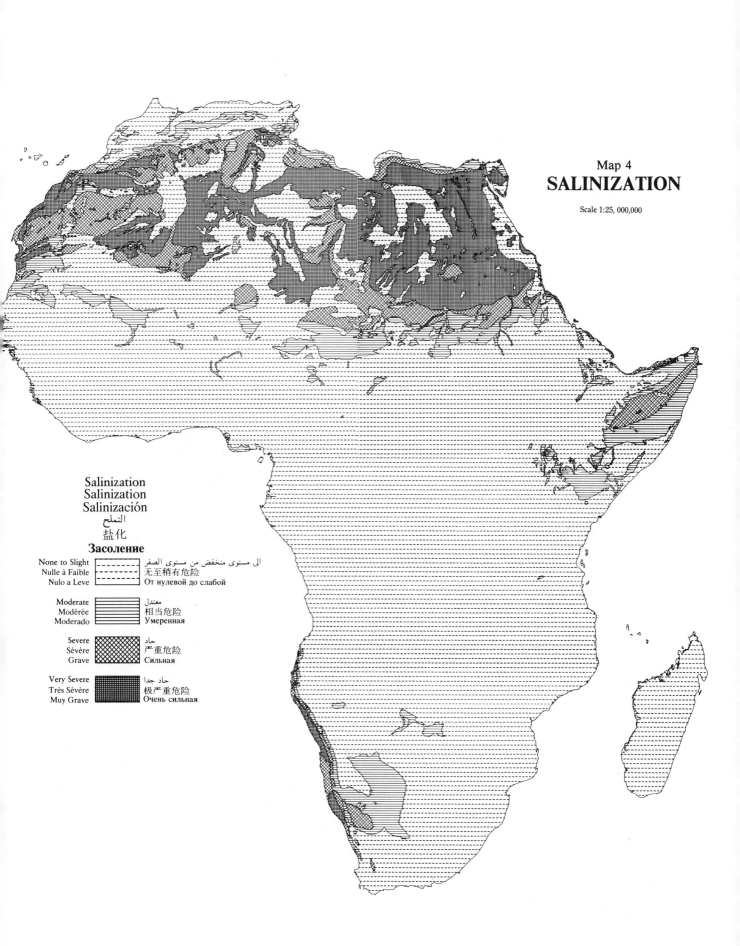

Map 4
SALINIZATION

Scale 1:25, 000,000

Salinization
Salinization
Salinización
التملح
盐化
Засоление

None to Slight		الى مستوى منخفض من مستوى الصفر
Nulle à Faible		无至稍有危险
Nulo a Leve		От нулевой до слабой

Moderate		معتدل
Modérée		相当危险
Moderado		Умеренная

Severe		حاد
Sévère		严重危险
Grave		Сильная

Very Severe		حاد جدا
Très Sévère		极严重危险
Muy Grave		Очень сильная

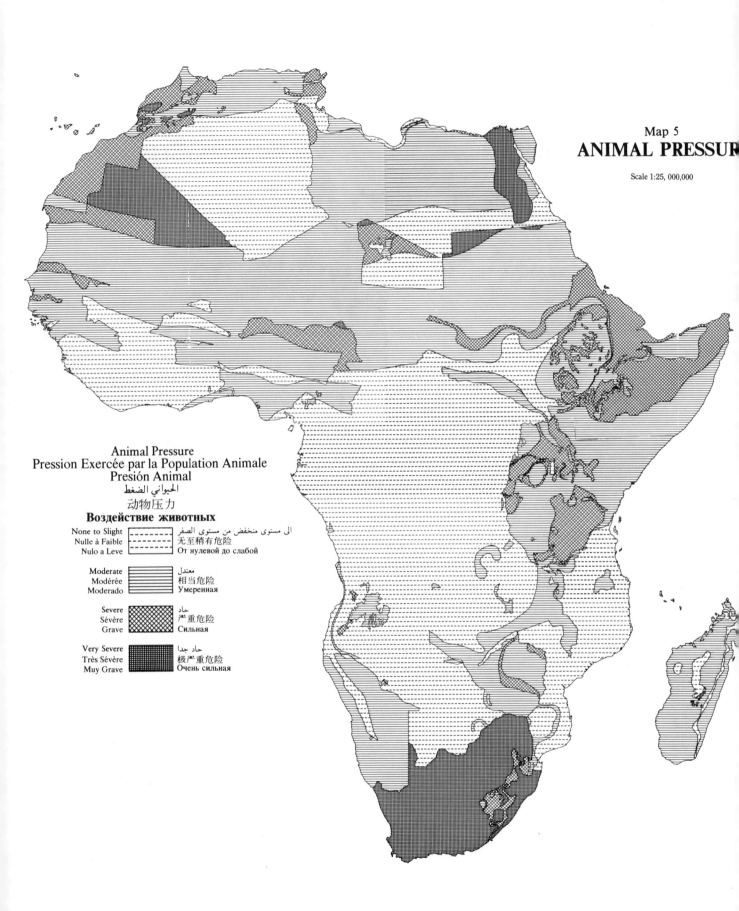

Map 5
ANIMAL PRESSUR

Scale 1:25, 000,000

Animal Pressure
Pression Exercée par la Population Animale
Presión Animal
الحيواني الضغط
动物压力
Воздействие животных

None to Slight Nulle à Faible Nulo a Leve		الى مستوى منخفض من مستوى الصفر 无至稍有危险 От нулевой до слабой
Moderate Modérée Moderado		معتدل 相当危险 Умеренная
Severe Sévère Grave		حاد 严重危险 Сильная
Very Severe Très Sévère Muy Grave		حاد جدا 极严重危险 Очень сильная

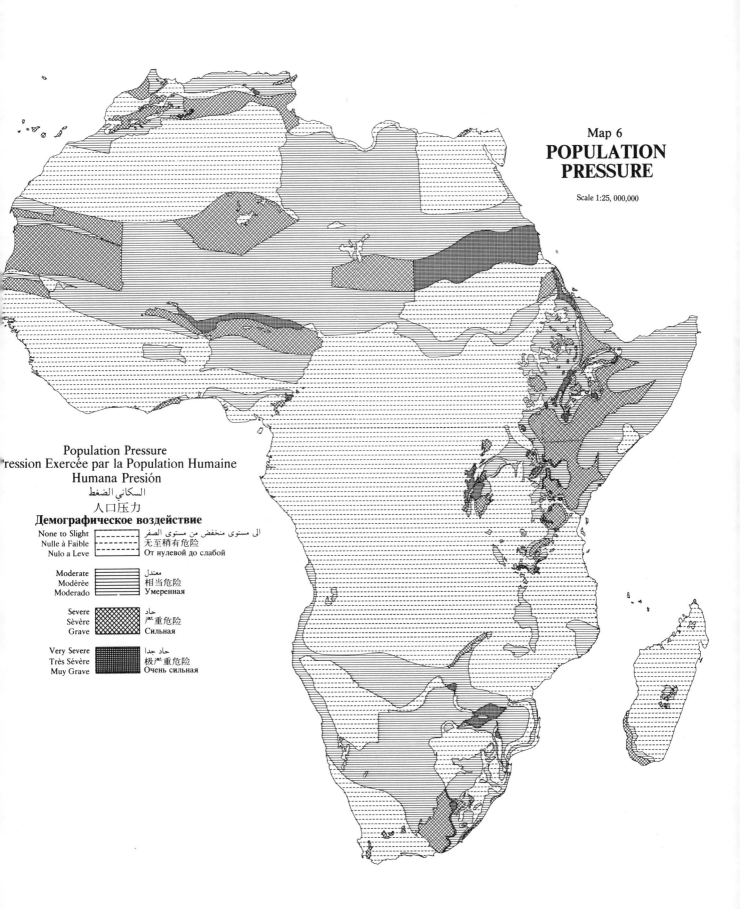

Map 6
POPULATION PRESSURE

Scale 1:25, 000,000

Population Pressure
Pression Exercée par la Population Humaine
Humana Presión
السكاني الضغط
人口压力
Демографическое воздействие

None to Slight	الى مستوى منخفض من مستوى الصفر
Nulle à Faible	无至稍有危险
Nulo a Leve	От нулевой до слабой

Moderate	معتدل
Modérée	相当危险
Moderado	Умеренная

Severe	حاد
Sévère	严重危险
Grave	Сильная

Very Severe	حاد جدا
Très Sévère	极严重危险
Muy Grave	Очень сильная

BIBLIOGRAPHY

A Selected Bibliography
on the
Applications of Geographic Information Systems
for
Resource Management

By
William J. Ripple and Joseph A. Bernert
Environmental Remote Sensing Applications Laboratory
Oregon State University
Corvallis, Oregon 97331

May 1986

This bibliography represents a selection of references on the applications of Geographic Information Systems for resource management. The bibliograpy contains 144 citations and is not intended to be all-inclusive, but rather a list of selected significant works published from January 1980 through December 1985. Books, major journals and symposium proceedings were examined. For works prior to 1980, the reader is directed to review "Literature for Geographic Information Systems" by Thomas K. Peucker (1980).

This bibliography was produced as a project of the publications subcommitte of the Geographic Information Management Systems (GIMS) committee with membership from the American Society for Photogrammetry and Remote Sensing and the American Congress on Surveying and Mapping.

Adams, J.R., T.H. Cahill, T.J. Logan, D.R. Urban, and S.M. Yaksich, 1982. A Land Resource Information System (LRIS) for water quality management in the Lake Erie Basin. _Journal of Soil and Water Conservation_, 37:45-50.

Adams, J.R., Logan, T.J., Urban, D.R. and S.M. Yaksich, 1982. Erosion control potential with conservation tillage in the Lake Erie Basin: Estimates using the universal soil loss equation and the Land Resource Information System (LRIS). _Journal of Soil and Water Conservation_, 37:50-57.

Adams, J.R. and C.J. Merry, 1980. Application of a land resource information systen (LRIS) and the CAPDET computer assisted procedure for the design and evaluation of wastewater treatment systems model to facilities planning and land treatment of municipal wastewater. _Proceedings of the Fourteenth International Symposium on Remote Sensing the Environment_, San Jose, Costa Rica. Environmental Research Institute of Michigan, Ann Arbor, Michigan, pp. 663-673.

Adeniyi, P.O., 1980. Land-use change analysis using sequential aerial photography and computer techniques. _Photogrametric Engineering and Remote Sensing_, 46(11):1447-1464.

Ader, R.R. and J.B. Johnson, 1982. Analysis of wetland changes on an active delta with a geographic information system. _Proceedings of the Fall Meeting of the American Congress on Surveying and Mapping and the American Society for Photogrammetry and Remote Sensing_, Hollywood, Florida, pp. 16-27.

Ader, R.R. and F. Stayner, 1982. The role of the USFWS Geographic Information System in coastal decision making. _Proceedings of the Fifth Auto-Carto Meeting_, Crystal City, Virginia, pp. 1-12.

Anderson, D.R. and M.J. Sorensen, 1984. Development of a geographic data base for natural resource management on state owned lands. _Proceedings of the Annual Conference of the Urban and Regional Information Systems Association_, Seattle, Washington, pp. 446-457.

Bain, S., D.B. Wherry, and J.A. Hart, 1985. An operational GIS for Flathead National Forest. _Proceedings of the Seventh Auto-Carto Meeting_, Washington, D.C., pp. 244-253.

Barker, C.P., 1982. FRIS: St Regis' forest management approach. _Remote Sensing for Resource Management_. Soil Conservation Society of America, Ankeny, Iowa, pp. 454-470.

Beard, M.K., N.R. Chrisman, and T.D. Patterson, 1984. Integrating data for local resource planning: A case study of sand and gravel resources. _Proceedings of the Annual Conference of the Urban and Regional Information Systems Association_, Seattle, Washington, pp. 513-524.

Berry, J.K., 1981. Spatial information systems: Instant maps for analyzing resource data. Special Libraries Journal, July, pp. 261-269.

Best, R.G. and F.C. Westin, 1984. GIS for soils and rangeland management. Proceedings of the Ninth Annual William T. Pecora Memorial Symposium on Remote Sensing, Sioux Falls, South Dakota, pp. 70-74.

Betts, R., 1985. Geographic information system: Permanent resource for urban planners. Plan and Print, October.

Billingsley, F.C. and J.L. Urena, 1984. Concepts for a global resource information system. Proceedings of the Ninth Annual William T. Pecora Memorial Symposium on Remote Sensing, Sioux Falls, South Dakota, pp. 123-131.

Blanchard, W.A. and S.J. Walsh, 1982. Development of the Oklahoma geographic information retrieval system for resource management and environmental assessment. Oklahoma Geological Notes, 42(6):224-229.

Brannon, D.P. and G.J. Irish, 1985. Satellite remote sensing and geographic information systems applied to biogeochemical flux estimates. Proceedings of the Eighth Biennial International Estuarine Research Conference.

Brooks, W.T., 1985. Implementation of an integrated resource information system and its application in the management of fish and wildlife resources in Alaska. Proceedings of the Seventh Auto-Carto Meeting, Washington, D.C., pp. 47-52.

Campbell, R.W. Jr. and S.A. Sader, 1980. Design of a nationwide natural resource inventory and information system for Costa Rica. Proceedings of the Fourteenth International Symposium on Remote Sensing the Environment, San Jose, Costa Rica, Environmental Research Institute of Michigan, Ann Arbor, Michigan, pp.325-340.

Chrisman, N.R., 1984. The role of quality information in the long-term functioning of a geographic information system. Cartographica, 21(2&3):79-88.

Civco, D.L. and M.W. Lefor, 1981. Computerized natural resources information for the management of university owned lands in Connecticut. Proceedings of the In-Place Resource Inventories: Principles and Practices, Orono, Maine, pp. 1018-1028.

Costello, M., 1984. Geographic Information System Development: The Boston Experience. Proceedings of the Fall Meeting of the American Congress on Surveying and Mapping and the American Society for Photogrammetry and Remote Sensing, San Antonio, Texas, pp. 294-300.

Crain, I.K. and C.L. MacDonald, 1984. From land inventory to land management. Cartographica, 21(2&3):40-46.

Crossfield, J.K. and D.D. Moyer, 1983. Developing a county land records management system. Proceedings of the Fall Meeting of the American Congress on Surveying and Mapping and the American Society for Photogrammetry and Remote Sensing, Salt Lake City, Utah, pp. 216-225.

Curran, R.P., 1981. Resource inventory and geographic information system design to facilitate the Adirondack Parks Agency's planning capabilities of natural resources. Proceedings of the In-Place Resource Inventories: Principles and Practices, Orono, Maine, pp. 435-448.

Currie, A.L., 1984. Spatial information systems and their geological applications. Geological Association of Canada, Mineralogical Association of Canada joint annual meeting, London, Ontario, Canada, May 14-16, 1984.

Dangermond, J., 1984. Selecting new town sites in the United States using regional data bases. Computer Graphics and Environmental Planning, E. Teicholz and B.J.L. Berry, eds., Prentice Hall, Englewood Cliffs, New Jersey, pp. 119-140.

Dangermond, J. and C. Freedman, 1984. Findings regarding a conceptual model of a municipal data base and implementation for software design. Proceedings of the International Symposium on Spatial Data Handling, Zurich, Switzerland, pp. 479-497.

Dangermond, J. and L.K. Smith, 1981. Concepts and issues in large area natural resource information systems. Proceedings of the In-Place Resource Inventories: Principles and Practices, Orono, Maine, pp. 435-448.

DeVries, M., F. Westin, F. Redondo, C. Gargantini, N. Marlenko, G. Vassallo and M. Wehde, 1982. The use of a geographic information system to combine land use information derived from Landsat with soils data to stratify an area in Argentina for crop forecasting. Proceedings of the Sixteenth International Symposium on Remote Sensing of Environment, Environmental Research Institute of Michigan, Ann Arbor, Michigan, pp. 381-391.

Dolan, C.W., 1985. The need for a modern land data system by the public administrator. Surveying and Mapping, 43:141-152.

Donahue, J.C., 1983. The role of the land surveyor in the design and control of land information systems. Proceedings of the Fall Meeting of the American Congress on Surveying and Mapping and the American Society for Photogrammetry and Remote Sensing, Salt Lake City, Utah, pp. 216-225.

Drinnan, C.H., 1982. Implementation of land records information systems using Informap system. Proceedings of the Annual Conference of the Urban and Regional Information Systems Association.

Drinnan, C.H., 1985. Mapping information management system requirements for master plan applications. Proceedings of the Annual Meeting of the American Congress on Surveying and Mapping and the American Society for Photogrammetry and Remote Sensing, Washington, D.C., pp. 298-310.

Ehler, C.N., D.J. Basta and T.F. LaPointe, 1982. An automated data system for strategic assessment of living marine resource in the Gulf of Mexico. Proceedings of the Fifth Auto-Carto Meeting, Crystal City, Virginia, pp. 83-91.

Eidenshink, J.C. and M.E. Wehde, 1981. Use of remote sensing inputs in geographic information systems for watershed management. Proceedings of the Seventh Annual William T. Pecora Memorial Symposium on Remote Sensing, Sioux Falls, South Dakota, pp. 482-493.

Eidenshink, J.C. and D.C. Sjaastad, 1985. Updating range surveys using a geographic information system. Proceedings of the Tenth Annual William T. Pecora Memorial Symposium on Remote Sensing, Fort Collins, Colorado, pp. 135-142.

Eli, R.N., 1983. Geographic information systems in coal mine applications. Proceedings of the first conference on use of computers in the coal industry, Society of Mineral Engineers, Sanford, pp. 263-274.

Estes, J.E., J.L. Star, P.J. Cressy and M. Derivan, 1985. Pilot land data system. Photogrametric Engineering and Remote Sensing, 51(6):703-709.

Fay, T.H., J. Mazemetz, D.M. Sandford, V. Taneja and P. Walsh, 1982. An intelligent earth sensing information system. Photogrammetric Engineering and Remote Sensing, 48(2):281-286.

Faust, N.C., S.R. Wheeler, W.M. Finlay and W.H. Clerk, 1983. Development of REMIDAS. A multi-source integrated Analysis System for forestry applications. Proceedings of the Fall Meeting of the American Congress on Surveying and Mapping and the American Society for Photogrammetry and Remote Sensing, Salt Lake City, Utah, pp. 636-645.

Feineman, D.R., 1980. A case study in computer mapping in petroleum exploration. Computer Mapping of Natural Resources and the Environment, Volume 10, Harvard Library of Computer Graphics, 1981 Mapping Collection, Harvard University, Cambridge, Massachusetts, pp. 41-51.

Fellows, J.D., 1983. A geographic information system for regional hydrologic modelling. Unpublished Doctoral Dissertation, University of Maryland, College Park, Maryland.

Frazier, J.W. and N.F. Henry, 1982. A geographic information system at the tax plot (parcel) level. Proceedings of the Fall Meeting of the American Congress on Surveying and Mapping and the American Society for Photogrammetry and Remote Sensing, Hollywood, Florida, pp. 148-158.

George, L.R. and L. Brietenwischer, 1984. Abilene area information management system. Proceedings of the Fall Meeting of the American Congress on Surveying and Mapping and the American Society for Photogrammetry and Remote Sensing, San Antonio, Texas, pp. 218-227.

Giles, R.H. Jr. and J.E. De Steigueur, 1981. Introduction to computerized land-information systems. Journal of Forestry, 79:734-737.

Gold, C.M., 1980. Drill-hole data validation for subsurface stratigraphic modeling. Computer Mapping of Natural Resources and the Environment, Volume 10, Harvard Library of Computer Graphics, 1981 Mapping Collection, Harvard University, Cambridge, Massachusetts, pp. 52-58.

Gose, J.B. and J.B. Munson, 1981. A resource data base with APL. Proceedings of the In-Place Resource Inventories: Principles and Practices, Orono, Maine, pp. 991-993.

Graetz, R.D. and J.F. O'Callaghan, 1982. The development of a land image-based resource information system (LIBRIS) and its application to the assessment and monitoring of Australian arid rangelands. Proceedings of the Sixteenth International Symposium on Remote Sensing the Environment, Environmental Research Institute of Michigan, Ann Arbor, Michigan, pp. 257-275.

Griffith, C., 1980. Geographic information systems and environmental impact assessment. Environmental Management. New York, Springer, 4(1):21-25.

Hanigan, F.L., 1983. Houston's metropolitan common digital base: Four years later. Surveying and Mapping, 43:141-152.

Hanigan, F.L., 1983. Metrocom: Houston's Geographic Information municipal management system. Proceedings of the Fall Meeting of the American Congress on Surveying and Mapping and the American Society for Photogrammetry and Remote Sensing, Salt Lake City, Utah, pp. 205-215.

Hanigan, F.L., 1984. TACIMS: A topographic and cadastral
    information management system for Jeddah. Proceedings of the
    Fall Meeting of the American Congress on Surveying and Mapping
    and the American Society for Photogrammetry and Remote Sensing,
    San Antonio, Texas, pp. 305-312.

Harrison, J., 1983. Maintaining a database on the world's
    protected areas land and biological conservation. Parks,
    7(4):3-5.

Hart, J.A., D.B. Wherry and S. Bain. 1985. An operational GIS for
    Flathead National Forest. Proceedings of the Seventh
    Auto-Carto Meeting, Washington, D.C., pp. 244-253.

Harvey, R.B., 1985. The use of orthophotography and GIS
    techonology to conduct a storage drainage utility impervious
    surface analysis: A case study. Proceedings of the Annual
    Meeting of the American Congress on Surveying and Mapping and
    the American Society for Photogrammetry and Remote Sensing,
    Washington, D.C., pp. 271-278.

Henderson, J., 1985. The utility of a GIS in evaluating the
    accuracy of classified Landsat land cover type maps on the
    Kenai Peninsula. Proceedings of the Seventh Auto-Carto
    Meeting, Washington, D.C., pp. 257-264.

Henderson, J., 1984. Applications of geographic information system
    and remotely sensed digital data in the national wildlife
    refuge planning process in Alaska. Proceedings of the Annual
    Conference of the Urban and Regional Information Systems
    Association, Seattle, Washington, pp. 493-501.

Henry, B.B., 1984. Automated design in Minneapolis and Hennepin
    County Minnesota. Photogrametric Engineering and Remote
    Sensing, 50(12):1747-1751.

Hertz-Brown, E. and J.R. Williams, 1981. Application of a
    geographic and resource information system to the spruce
    budworm problem in Maine. Proceedings of the In-Place
    Resource Inventories: Principles and Practices, Orono, Maine,
    pp. 1041-1045.

Hodler, T.W., 1983. Cartographic modeling of hydrothermal resource
    potential. Cartographica, 20(3):31-44.

Hill, J.M., C.A. Harlow and P. Zimmerman, 1983. Geographic
    information systems as applied to the manipulation of
    environmental data. Environmentalist, 3:33.

Hopkins, C.D., 1980. Land suitability analysis: Methods and
    interpretation. Landscape Research, 5:8-9.

Hotrabhavananda, T., 1980. Development of automated techniques for creating a natural resources information system. Unpublished Doctoral Dissertation, University of Missouri, Columbia, Missouri.

Imhoff, M.L., G.W. Petersen, S.G. Sykes and J.R. Irons, 1982. Digital overlay of cartographic information on Landsat MSS data for soil surveys. Photogrammetric Engineering and Remote Sensing, 48(8):1337-1342.

Jackson, M.J., S.B. Bell and B.M. Diaz, 1983. Database development in the experimental cartographic unit. Cartographica, 20(3):55-64.

Jett, S.C., A.D. Weeks and W.M. Grayman, 1980. Geographic information systems in hydrologic modeling. Proceedings of the Hydrologic Transport Modelling Symposium, American Society of Agricultural Engineers, pp. 127-137.

Johannsen, C., J. Pan, T. Barney and G. Koelin, 1984. A data base management system for Missouri's natural resource inventory. Proceedings of the Ninth Annual William T. Pecora Memorial Symposium on Remote Sensing, Sioux Falls, South Dakota, pp. 25-28.

Key, G.S., 1981. Generalized natural resources data base. Proceedings of the In-Place Resource Inventories: Principles and Practices, Orono, Maine, pp.1074-1080.

Kleckner, R.L., 1984. Federal mineral land information systems. Proceedings of the Annual Meeting of the American Congress on Surveying and Mapping and the American Society for Photogrammetry and Remote Sensing, Washington, D.C., pp. 20-27.

Klock, G.O., P. Gum and L.E. Jordan, III, 1985. Developing a resource management data base for the Okanogan National Forest from multispectral imagery and the use of GIS. Proceedings of the Tenth Annual William T. Pecora Memorial Symposium on Remote Sensing, Fort Collins, Colorado, pp. 144-152.

Koel, G.T., and E.A. Cook, 1984. Applications of geographic information systems for analyis of radio-telemetry data on wildlife. Proceedings of the Ninth Annual William T. Pecora Memorial Symposium on Remote Sensing, Sioux Falls, South Dakota, pp. 154-158.

Koller, G.R., 1981. Interpretation and display of the NURE data base using computer graphics. Computer Mapping of Natural Resources and the Environment, Volume 15, Harvard Library of Computer Graphics, 1981 Mapping Collection, Harvard University, Cambridge, Massachusetts, pp. 41-49.

Logan, T.J., D.R. Urban, J.R. Adams and S.M. Yaksich, 1982. Erosion control potential with conservation tillage in the Lake Erie Basin; estimates using the universal soil loss equation and the land resource information system (LRIS). Journal of Soil and Water Conservation, 37:50-55.

Loveland, T.R. and G.E. Johnson, 1983. The role of remote sensing and other spatial data for predictive modeling: The Umatilla, Oregon example. Photogrammetric Engineering and Remote Sensing, 49(8):1183-1192.

Maggio, R.C., R.D. Baker and M.K. Harris, 1983. A geographic data base for Texas pecan. Photogrammetric Engineering and Remote Sensing, 49(1):47-52.

Marble, D.F., 1984. Geographic information systems: An overview. Proceedings of the Ninth Annual William T. Pecora Memorial Symposium on Remote Sensing, Sioux Falls, South Dakota, pp. 18-24.

Marble, D.F., H.W. Calkins, and D.J. Peuquet, ed., 1984. Basic Readings in Geographic Information Systems. SPAD Systems Ltd.

Marble, D.F. and D.J. Peuquet, 1983. Geographic information systems and remote sensing. Manual of Remote Sensing, R.N. Colwell, editor-in-chief, second edition, American Society for Photogrammetry and Remote Sensing, Falls Church, Virginia, pp. 923-958.

Mark, D.M. and J.P. Lauzon, 1985. Approaches for quadtree- based Geographic Information Systems of continental or global scales. Proceedings of the Seventh Auto-Carto Meeting, Washington, D.C., pp. 355-364.

Mathews, E., 1983. Global vegetation and land use: New high resolution data bases for climate studies. Journal of Climate and Applied Meteorology, 22(3):474-487.

Maw, K.D. and J.A. Brass, 1981. Forest management applications of Landsat data in a geographic information system, Proceedings of the Seventh Annual William T. Pecora Memorial Symposium on Remote Sensing, Sioux Falls, South Dakota, pp. 330-340.

McCulluch, S., 1984. Development of the Texas Natural Resource Inventory and Monitoring System (TNRIMS). Proceedings of the Fall Meeting of the American Congress on Surveying and Mapping and the American Society for Photogrammetry and Remote Sensing, San Antonio, Texas, pp. 751-761.

McFarland, W.D., 1981. Geographic data bases for natural resources. Remote Sensing for Resource Management, Soil Conservation Society of America, Ankeny, Iowa. pp.41-50.

McLaughlin, J. and M. Salem, 1982. Multipurpose land information systems: A Canadian perspective. Journal of Surveying and Mapping DICV-ASCE, 108:1.

Mead, D.A., 1981. Statewide natural resource information systems. A Status Report. Journal of Forestry, 79:369-372.

Mead, D.A. 1982. Assessing data quality in geographic natural resource information systems. Remote Sensing for Resource Management, Soil Conservation Society of America, Ankeny, Iowa, pp. 51-59.

Merchant, J.W., 1983. A prototype Landsat based rangeland resource information system in Cimarron National Grassland. Proceedings of The Fall Meeting of the American Congress on Surveying and Mapping and the American Society for Photogrammetry and Remote Sensing, Salt Lake City, Utah, pp. 110-119.

Merchant, J.W. and E.A. Roth, 1981. Inventory and evaluation of rangeland in the Cimarron National Grassland, Kansas. Proceedings of the Seventh Annual William T. Pecora Memorial Symposium on Remote Sensing, Sioux Falls, South Dakota, pp. 104-113.

Meyers, W.L. and J.J. Kolenik, 1984. Applications of a geographic information system to management of the Penn State University Experimental Forest. Proceedings of the Ninth Annual William T. Pecora Memorial Symposium on Remote Sensing, Sioux Falls, South Dakota, pp. 174-182.

Miller, B.M. and M.A. Domaratz, 1984. The applications of spatial information technology to petroleum resource assessment analysis. Proceedings of the Ninth Annual William T. Pecora Memorial Symposium on Remote Sensing, Sioux Falls, South Dakota, pp. 303-310.

Miller, S.W., 1982. A raster encoded polygon data structure for land resource analysis applications. Proceedings of the Fifth Auto-Carto Meeting, Crystal City, Virginia, pp. 529-536.

Miller, S.W., 1984. A spatial data structure for hydrologic applications. Proceedings of the International Symposium on Spatial Data Handling, Zurich, Switzerland, pp. 267-288.

Miller, S.W. and J.L. Goldberg, 1984. Integrating digital map data from the national topographic map series with other data sets for land and water resource assessment. Proceedings of the Fall Meeting of the American Congress on Surveying and Mapping and the American Society for Photogrammetry and Remote Sensing, San Antonio, Texas, pp. 489-499.

Miller, W.A., M.B. Shasby, W.G. Rohde and G.R. Johnson, 1982. Developing in-place data bases by incorporating digital terrain data into the Landsat classification process. Proceedings of the In-Place Resource Inventories: Principles and Practices, Orono, Maine, pp. 511-518.

Mitchell, R.J. and P.C. Johnson, 1985. Towards a common land information system – a full multipurpose cadastre. Proceedings of the Annual Meeting of the American Congress on Surveying and Mapping and the American Society for Photogrammetry and Remote Sensing, Washington, D.C., pp. 106-111.

Monmonier, M.S. 1983. Raster-Mode area generalization for land use and land cover maps. Cartographica, 20(4):65-91.

Moreland, R.M., and E.A. Fosnight, 1981. Remote sensing data integration into a geographic information system for the creation of a biogenic hydrocarbon inventory of the San Francisco Bay area. Proceedings of the Seventh Annual William T. Pecora Memorial Symposium on Remote Sensing, Fort Collins, Colorado, pp. 278-289.

Murray, R., 1984. Segmentation of computer classified Landsat multispectral scanner data into spatially-connected regions of elk habitat components. Proceedings of the Ninth Annual William T. Pecora Memorial Symposium on Remote Sensing, Sioux Falls, South Dakota, pp. 208-212.

O'Callaghan, J.F. and K.N. O'Sullivan, 1982. Image-based geographic information systems for mineral exploration: The Broken Hill Project. Proceedings of the Sixteenth International Symposium on Remote Sensing of Environment, Volume I, Environmental Research Institute of Michigan, Ann Arbor, Michigan, pp. 133-136.

Olson, R.J. and R.L. Burgess, 1981. Analyses of ecoregions utilizing the geoecology data base. Proceedings of the In-Place Resource Inventories: Principles and Practices, Orono, Maine, pp. 149-156.

Palmer, D. and J. McLaughlin, 1984. Land-related information networks: Accessing user requirements. Proceedings of the Annual Meeting of the American Congress on Surveying and Mapping and the American Society for Photogrammetry and Remote Sensing, Washington, D.C., pp. 101-109.

Pascucci, R. and A. Smith, 1982. Application of an automated geographic information system to remote sensor resources exploration, Proceedings of the Sixteenth International Symposium on Remote Sensing of Environment, Volume II, Environmental Research Institute of Michigan, Ann Arbor, Michigan, pp. 775-784.

Peucker, T.K., 1980. Literature for geographic information systems. Urban, Regional and State Government Applications of Computer Mapping, Volume 11, Harvard Library of Computer Graphics, 1980 Mapping Collection, Harvard University, Cambridge, Massachusetts, pp. 175-179.

Pierce, R.R. and B.Q. Rado, 1981, Georgia Resource Assessment Project: Institutionalizing Landsat and Geographic Data Base Techiques. Computer Mapping of Natural Resources and the Environment, Volume 15, Harvard Library of Computer Graphics, 1981 Mapping Collection, Harvard University, Cambridge, Massachusetts, pp. 139-154.

Reed, C.N., 1981. Micro processor based geoprocessing systems. Proceedings of the In-Place Resource Inventories: Principles and Practices, Orono, Maine, pp. 1068-1073.

Reed, C. and W. Brooks, 1980. The wetland analytical mapping system (WAMS) and map overlay and statistical system (MOSS) production environment. Proceedings of the Annual Meeting of the American Congress on Surveying and Mapping and the American Society for Photogrammetry and Remote Sensing, St. Louis, Missouri, pp. 308-314.

Relp, M.G., Treinish, L.A. and P.H. Smith, 1984. The pilot climate data system, Proceedings of the Ninth Annual William T. Pecora Memorial Symposium on Remote Sensing, Sioux Falls, South Dakota, pp. 132-139.

Rhyason, D.B. and T. Salmer, 1985. An evaluation of the city of Edmonton's geographic base information system (G.B.I.S) after 7 years. Proceedings of the Annual Conference of the Urban and Regional Information Systems Association, Ottawa, Ontario, pp. 28-40.

Ripple, W.J. and S.B. Miller, 1982. Remote sensing and computer modeling for water quality planning in South Dakota. Remote Sensing for Resource Management, Soil Conservation Society of America, Ankeny, Iowa, pp. 309-328.

Rogoff, M.J., 1982. Computer display of soil survey interpretations using a geographic information system. Soil Survey and Land Evaluation, 2(2):37-41.

Ross, J., 1985. Detecting land use changes on Omaha's urban fringe using a geographic information system. Proceedings of the Seventh Auto-Carto Meeting, Washington, D.C., pp. 463-471.

Rowland, E.B., C.W. Smart and R.L. Jolly, 1982. TVA's geographic information system: An integrated resource data base to aid environmental assessment and resource management. Proceedings of the Fifth Auto-Carto Meeting, Crystal City, Virginia, pp. 429-436.

Salmen, L., J. Gropper, J. Hamill and C. Reed, 1977. Natural Resource Geographic Data Bases for Montana and Wyoming. U.S. Fish and Wildlife Service, Fort Collins, Colorado, Report no. FWS/OBS-77155, pp. 1-71.

Schneider, J.B., 1983. Mapping congestion patterns on urban highway networks. Computer Graphics and Environmental Planning, E. Teicholz and B.J.L. Berry, eds., Prentice Hall, Englewood Cliffs, New Jersey, pp. 202-223.

Shasby, M.B., R.R. Burgan and G.R. Johnson, 1981. Broad area forest fuels and topographic mapping using digital Landsat and terrain data. Machine Processing of Remotely Sensed Data, pp. 529-538.

Shasby, M.B., C. Markon, M.D. Fleming, D.L. Murphy and J.E. York, 1983. Land cover and terrain mapping for development of digital data bases for wildlife habitat assessment in the Yukon Flats National Wildlife Refuge, Alaska. Geological Survey Circular 0911, pp. 41-43.

Shultz, F.P. and J. Lyle, 1983. Computerized land use suitability mapping. The Cartographic Journal, 20:39-49.

Smith, A.Y. and R.J. Blackwell, 1980. Development of an information base for watershed monitoring. Photogrametric Engineering and Remote Sensing, 46(8):1027-1038.

Solomon, D.S., 1981. Spruce budworm impact on the spruce-fir forest of Maine: An integration of spatial information systems. Proceedings of the In-Place Resource Inventories: Principles and Practices, Orono, Maine, pp. 1046-1050.

Somers, R.M., 1984. Calgary's computer mapping and geoprocessing projects: System development and evaluation. Proceedings of the Annual Conference of the Urban and Regional Information Systems Association, Seattle, Washington, pp. 164-175.

Spanner, M.A., A.H. Strahler and J.E. Estes, 1982. Soil loss prediction in a geographic information system format. Proceedings of the Sixteenth International Symposium on Remote Sensing of Environment, Environmental Research Institute of Michigan, Ann Arbor, Michigan, pp. 89-102.

Star, L.E. and K.E. Anderson, 1981. Some thoughts on cartographic and geographic information systems for the 1980's. Proceedings of the Seventh Annual William T. Pecora Memorial Symposium on Remote Sensing, Sioux Falls, South Dakota, pp. 41-55.

Star, L.E., M.J. Cosentino and T.W. Foresman, 1984. Geographic information systems: Questions to ask before it's too late. Proceedings of the Ninth Annual William T. Pecora Memorial Symposium on Remote Sensing. Sioux Falls, South Dakota, pp. 194-197.

Sturdevant, J.A., 1981. The development and application of a county-level geographic data base. Proceedings of the Seventh Annual William T. Pecora Memorial Symposium on Remote Sensing, Sioux Falls, South Dakota, pp. 383-392.

Sturdevant, J.A. and R.L. Kleckner, 1984. Spatial analysis requirements for a federal mineral land information system. Proceedings of the Ninth Annual William T. Pecora Memorial Symposium on Remote Sensing, Sioux Falls, South Dakota, pp. 168-173.

Swain, P.H. 1984. Advanced computer interpretation techniques for earth data information systems. Proceedings of the Ninth Annual William T. Pecora Memorial Symposium on Remote Sensing, Sioux Falls, South Dakota, pp. 184-190.

Teicholz, E. and B.J.L. Berry, eds., 1983. Computer Graphics and Environmental Planning. Prentice Hall, Englewood Cliffs, New Jersey, p. 150.

Tessar, P.A., E.C. Palmer and W.J. Ripple, 1982. Remote sensing as a tool for resource management by state governments. Remote Sensing for Resource Management, Soil Conservation Society of America, Ankeny, Iowa, pp. 519-531.

Tilmann, S.E. and D.L. Mokma, 1980. Description of a user-oriented geographic information system. Proceedings of the Sixth Annual Symposium on Machine Processing of Remotely Sensed Data, West Lafayette, Indiana, pp. 248-257.

Tomlin, C.D., 1983. Digital cartographic modeling techniques in environmental planning. Unpublished Doctoral Dissertation, Yale University, New Haven, Connecticut, p. 290.

Tomlin, C.D., S.H. Berwick and S.M. Tomlin, 1983. Cartographic analysis of deer habitat utilization. Computer Graphics and Environmental Planning, E. Teicholz and B.J.L. Berry, eds., Prentice Hall, Englewood Cliffs, New Jersey, pp. 141-150.

Tomlin, S.M. and C.D. Tomlin, 1982. Computer assisted spatial allocation of timber harvesting activity. Proceedings of the Fifth Auto-Carto Meeting, pp. 677-686.

Tomlinson, R.F. and A.R. Boyle, 1982. The state of development of systems for handling natural resources inventory data. Cartographica, 18(4):65-95.

Tomlinson, R., A.R. Boyle and A. Aldred, 1980. A Study of Forest Inventory Data Handling Systems for the Province of Saskatchewan. Associates Consulting Geographers, Ottawa, Canada.

Van Berkel, P., 1981. Computer mapping aspects of the New Zealand land resource inventory. New Zealand Cartographic Journal, 11:43-50.

van der Meulen, 1984. Spatial data handling for urban and regional planning. Proceedings of the International Symposium on Spatial Data Handling, Zurich, Switzerland, pp. 436-448.

Vang, A.H., R.E. Rouse and M.W. Perry, 1984. Integration of digital reference files for improved resource information system management. Proceedings of the International Symposium on Spatial Data Handling, Zurich, Switzerland, pp. 299-313.

Walsh, S.J., 1985. Geographic information systems for natural resource management. Journal of Soil and Water Conservation, 40:202-295.

Walsh, S.J., S.J. Stadler and M.S. Gregory, 1984. Modeling of regional evapotranspiration through a geographic information system approach. Proceedings of the International Symposium on Spatial Data Handling, Zurich, Switzerland, pp. 252-266.

Wehde, M.E., 1982. Grid cell size in relation to errors in maps and inventories processed by computer mapping processing. Photogrametric Engineering and Remote Sensing, 48(8):1289-1298.

Wehde, M.E., K.D. Dalsted and E.K. Worcester, 1980. Resource applications of computerized data processing: The AREAS example. Journal of Soil and Water Conservation, 35:36-40.

Westin, F.C., T.M. Brandner, and M.E. Wehde, 1981. Updating land use data aquired from Landsat with soil data using an information system. Proceedings of the Seventh Annual William T. Pecora Memorial Symposium on Remote Sensing, Sioux Falls, South Dakota, pp. 455-464.

Wilson, P.N., 1984. Multiple purpose information system. Proceedings of the Annual Conference of the Urban and Regional Information Systems Association, Seattle, Washington, pp. 137-143.

Young, J.M., and F. Gossette, 1984. An automated approach to estimate time-specific population densities for metropolitan areas. Proceedings of the Ninth Annual William T. Pecora Memorial Symposium on Remote Sensing, Sioux Falls, South Dakota, pp. 159-166.